Lecture Notes in Bioinformatics 3695

Subseries of Lecture Notes in Computer Science

T0225706

Series Editors

Sorin Istrail, Celera Genomics, Applied Biosystems, Rockville, MD, USA
Pavel Pevzner, University of California, San Diego, CA, USA
Michael Waterman, University of Southern California, Los Angeles, CA, USA

Volume Editors

Michael R. Berthold
University of Konstanz
Department of Computer and Information Science
Box M712, 78457 Konstanz, Germany
E-mail: berthold@ieee.org

Robert Glen
Clare College
Trinity Lane, Cambridge CB2 1TL, UK
E-mail: rcg28@cam.ac.uk

Kay Diederichs
University of Konstanz, Department of Biology
Box M647, 78457 Konstanz, Germany
E-mail: Kay.Diederichs@uni-konstanz.de

Oliver Kohlbacher
University of Tübingen
Wilhelm Schickard Institute for Computer Science
Department for Simulation of Biological Systems
Room C318 Sand 14, 72076 Tübingen, Germany
E-mail: oliver.kohlbacher@uni-tuebingen.de

Ingrid Fischer
University of Erlangen–Nürnberg, Informatik 2
Martensstr. 3, 91058 Erlangen, Germany
E-mail: idfische@informatik.uni-erlangen.de

Library of Congress Control Number: 2005932805

CR Subject Classification (1998): H.2, H.3, H.4, J.3

ISSN   0302-9743
ISBN-10   3-540-29104-0 Springer Berlin Heidelberg New York
ISBN-13   978-3-540-29104-6 Springer Berlin Heidelberg New York

Springer is a part of Springer Science+Business Media

springeronline.com

© Springer-Verlag Berlin Heidelberg 2005
Printed in Germany

Typesetting: Camera-ready by author, data conversion by Scientific Publishing Services, Chennai, India
Printed on acid-free paper   SPIN: 11560500   06/3142   5 4 3 2 1 0

Michael R. Berthold   Robert Glen
Kay Diederichs   Oliver Kohlbacher
Ingrid Fischer (Eds.)

Computational
Life Sciences

First International Symposium, CompLife 2005
Konstanz, Germany, September 25-27, 2005
Proceedings

 Springer

Preface

The integration of knowledge in the life sciences is continuing apace with ever-increasing importance being placed on computer-based methods of data capture, analysis, and knowledge representation. Today, our many different sciences are providing us with a sea of information: it is the handling of this influx that is becoming a key discovery and regulatory question. The solutions to these problems will result in advancements to all of the involved sciences and will be highly influential both in the selection of the areas scientists seek to investigate and also on their success. For this to happen, it is crucial to establish an open and lively exchange between computer scientists, biologists, and chemists. To encourage precisely this type of exchange, crossing the borders of the sciences, we organized the 1st Symposium on Computational Life Science in Konstanz, Germany (September 25–27, 2005). The main objective of the symposium was to form bridges, bringing together scientists from a variety of disciplines to exchange ideas and research efforts and to talk about the problems in areas of research that up until now have not been visible at an interdisciplinary level.

Our conference program shows that the scientific mix worked out very well. From 49 submissions, 21 were selected for presentation at the symposium, covering areas ranging from high-level system biology to data analysis related to mass spec traces.

As a supplement to the regular conference program, we dedicated one section to papers presented in the framework of a workshop on Distributed Data Mining in the Life Sciences (LifeDDM), organized by Giuseppe Di Fatta. The workshop focused on the merging field of high-performance, distributed, parallel and grid-based data mining methods and the applications in the life sciences. This added yet another facet of computational life science research to the program.

Selecting the papers included in this volume would not have been possible without the help of an international Program Committee that put in countless hours to create a minimum of three detailed reviews for each paper! And, of course, a successful conference relies on many individuals working hard behind the scenes. We would like to thank first and foremost Heather Fyson for local organization and keeping everybody on track. Peter Burger worked tirelessly on the Web pages promoting the conference and Thorsten Meinl was the man behind the electronic review system. Last, but certainly not least, thanks go to Ingrid Fischer for putting together this volume!

July 2005

Michael R. Berthold
Kay Diederichs
Robert Glen
Oliver Kohlbacher

Organization

General Chair
Michael R. Berthold
University of Konstanz, Germany
Michael.Berthold@uni-konstanz.de

Program Chairs
Robert Glen
Unilever Center
Cambridge, UK
rcg28@cam.ac.uk

Kay Diederichs
University of Konstanz, Germany
kay.diederichs@uni-konstanz.de

Oliver Kohlbacher
University of Tübingen, Germany
oliver.kohlbacher@uni-tuebingen.de

Publication Chair
Ingrid Fischer
University of Erlangen-Nuremberg, Germany
Ingrid.Fischer@informatik.uni-erlangen.de

Local Chair
Heather Fyson
University of Konstanz, Germany
Heather.Fyson@uni-konstanz.de

Workshop Chair
Giuseppe di Fatta
University of Konstanz, Germany
Guiseppe.de.Fatta@uni-konstanz.de

Publicity Chair
Allan Tucker
Brunel University, UK
allan.tucker@brunel.ac.uk

Submission Chair
Thorsten Meinl
University of Erlangen-Nuremberg, Germany
Thorsten.Meinl@informatik.uni-erlangen.de

Webmaster
Peter Burger
University of Konstanz, Germany
Peter.Burger@uni-konstanz.de

Program Committee

Herman Berendsen, Univ. Groningen, Netherlands
Alexander Bockmayr, FU Berlin, Germany
Tim Clark, Univ. of Erlangen-Nuremberg, Germany
Thomas Exner, Univ. Konstanz, Germany
Lawrence O. Hall, Univ. South Florida, USA
Hans-Christian Hege, Zuse Institute Berlin, Germany
Kim Henrick, EMBL, UK
Joel Janin, CNRS, France
William L. Jorgensen, Yale Univ., USA
Michael Kaufmann, Univ. Tübingen, Germany
Joost Kok, Leiden Univ., Netherlands
Hans-Peter Lenhof, Univ. Saarbrücken, Germany
Ulf Leser, HU Berlin, Germany
Xiaohui Liu, Brunel Univ., UK
Vladimir Y. Lunin, Russian Academy of Sciences, Russia
Yves Muller, Univ. of Erlangen-Nuremberg, Germany
Peter Murray-Rust, Cambridge Univ., UK
David E. Patterson, Tripos, Inc., USA
Knut Reinert, FU Berlin, Germany
Arno Siebes, Univ. Utrecht, Netherlands
Jens Stoye, Univ. Bielefeld, Germany
Alexandre Urzhumtsev, Univ. Henri Poincaré, France
Peter Willett, Sheffield Univ., UK
Ralf Zimmer, LMU München, Germany

Additional Reviewers

Daniel Baum, Timm Baumeister, Frank Cordes, Patrick May, Alexander Steidinger

Sponsoring Institutions

We thank our sponsors for their support in making the 1st International Symposium on Computational Life Science a successful event: Universitätsgesellschaft Konstanz e.V., ALTANA Pharma AG, Boehringer Ingelheim, Tripos, BioLAGO.

Table of Contents

Systems Biology

Data Analysis and Integration

Structural Biology

Genomics

Computational Proteomics

Molecular Informatics

Molecular Structure Determination and Simulation

Distributed Data Mining

Structural Protein Interactions Predict Kinase-Inhibitor Interactions in Upregulated Pancreas Tumour Genes Expression Data

Gihan Dawelbait[1], Christian Pilarsky[2], Yanju Zhang[1], Robert Grützmann[2], and Michael Schroeder[1]

[1] Bioinformatics Group, Biotechnological Centre of TU Dresden, Germany
[2] Department of Viscersal-Thoracic- and Vascular Surgery,
University Hospital Dresden, Germany
gd@biotec.tu-dresden.de, ms@biotec.tu-dresden.de

Abstract. Micro-arrays can identify co-expressed genes at large scale. The gene expression analysis does however not show functional relationships between co-expressed genes. To address this problem, we link gene expression data to protein interaction data. For the gene products of co-expressed genes, we identify structural domains by sequence alignment and threading. Next, we use the protein structure interaction PSIMAP to find structurally interacting domains. Finally, we generate structural and sequence alignments of the original gene products and the identified structures and check conservation of the relevant interaction interfaces. From this analysis, we derive potentially relevant protein interactions for the gene expression data.

We applied this method to co-expressed genes in pancreatic ductal carcinoma. Our method reveals among others a number of functional clusters related to the proteasome, signalling, ubiquitinisation, serine proteases, immunoglobulin and kinases. We investigate the kinase cluster in detail and reveal an interaction between the cell division control protein CDC2 and the cyclin-dependent kinase inhibitor CDKN3, which is also confirmed by literature. Furthermore, our method reveals new interactions between CDKN3 and the cell division protein kinase CDK7 and between CDKN3 and the serine/threonine-protein kinase CDC2L1.

1 Introduction

Pancreatic ductal adenocarcinoma has an extremely poor prognosis. To improve the prognosis, novel molecular markers and targets for earlier diagnosis and adjuvant and/or neoadjuvant treatment need to be identified. One of the key techniques that has been developed to achieve this goal is DNA microarray profiling, which is used to identify the mechanisms of deregulated molecular functions in pancreatic carcinoma cells. Despite the progress made in recent years in the treatment of various types of cancer, the dismal prognosis of pancreatic ductal adenocarcinoma (PDAC) remains unchanged. In the United States PDAC still

M.R. Berthold et al. (Eds.): CompLife 2005, LNBI 3695, pp. 1–11, 2005.
© Springer-Verlag Berlin Heidelberg 2005

ranks fifth among the leading causes of cancer death, accounting for approximately 30,000 deaths annually. Apart from surgery there is no curative therapy, and even resected patients usually die within 1 year of the operation. In this situation there is an urgent need to understand more about the causes and the pathogenesis of PDAC [6]. Having characterised the gene expression profile of PDAC by various methods, our motivation is to identify the genes and proteins that play an important role in the formation of cancer cells from this data set. Further investigations of this data is hampered by the fact that except for the sequence rather little is known about those genes. Therefore, to interpret the data, there is a need for finding the relation between its elements. Protein interactions provide an important context for the understanding of function. Large-scale protein interaction maps provide a new global perspective with which to analyse protein function. Our method is based on utilising PSIMAP, the Protein Structural Interaction Map [4], a database of all the structurally observed interactions between families and superfamilies of protein domains. It computes domain-domain interactions for all multi-domain and multi-chain proteins in the Protein Data Bank (PDB)[2]. The PSIMAP database consists of 90.000 interactions. In a first attempt to evaluate PSIMAP we used only the genes found as up-regulated in PDAC to identify genes possibly involved in self stimulating signal circuits. To overcome the gap between the known genes sequences and known structures, our method uses GTD [9], a database that applies threading to predict the structure of all the structurally unknown proteins. To allow a closer insight into the behaviour of the genes, our method also uses Gene ontology (GO) functional annotation [3] to annotate the genes. The final visualisation of the protein-protein interaction network along with the genes annotations contributes to a better understanding of the relevant metabolic pathways.

Another major challenge that faces gene expression data analysis is that most of the information is hidden in a huge amount of publications. We also include links to web sites such as Harvester, Google scholar, and IHop to consult relevant literature resources to verify the results and further enhancing the interpretability of the interaction network.

2   Material and Methods

2.1   Tumour Microarray Data Set

The tumour microarray data set consisted of 1627 genes, it was obtained by integrating our various analyses of the gene expression profile of PDAC from Affymetrix GeneChip experiments and the meta-analysis of PDAC gene expression profiles from public available data of other projects [13]. Here, we only use the genes found to be up-regulated (954 genes) in PDAC to identify genes possibly involved in self stimulating signal circuits.

2.2   Resources

Before we describe our work flow linking expression and interaction data, we briefly summarises the underlying data sources used.

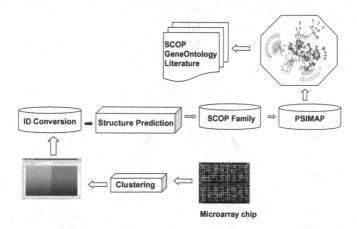

Fig. 1. Method work flow illustrating the different steps performed. Starting with the microarray data provided as a list of genes grouped into up/down regulated clusters, we perform the conversion of gene symbols to their respective ENSEMBL IDs. To obtain the threaded PDB structures and SCOP family assignments of the genes we use the GTD database, the PSIMAP database is queried for obtaining the domain interactions among the genes. The visualisation of the interaction network using PSIEYE and the genes are annotated using GO and links to the literature.

- **PDB.** The Protein Data Bank PDB [2] is a repository for 3D structures. It currently contains some 25.000 structures most of which have been obtained by X-ray crystallography. Around half of the PDB structures are multidomain structures.
- **SCOP.** The structural classification of proteins, SCOP [11], is a hierarchical classification of protein structures at domain level. The hierarchy contains four levels (class, fold, superfamily, family). At the family level domains share a high sequence similarity and hence are structurally are very similar. At superfamily level there is still good structural agreement concerning the overall topology despite possibly low sequence similarity. Domains grouped at family and superfamily level can be considered homologous.
- **GTD.** The Genomic Threading Database (GTD) [9] assigns structural folds to proteins with unknown structure. Annotations of proteomes of all major organisms are available. Annotations are based on threading, which is more sensitive than sequence alignment and can still assign folds correctly despite low similarity.
- **PSIMAP.** PSIMAP, the Protein Structure Interactome Map [12, 4] is a database with all domain-domain interactions in the PDB. Two domains are considered as interacting if there are at least 5 residue pairs within 5 Å.
- **GO.** The Gene Ontology GO [3] is a controlled hierarchical vocabulary for annotation of genes and gene products. GO breaks down into three subontologies for cellular components, biological processes and molecular functions.

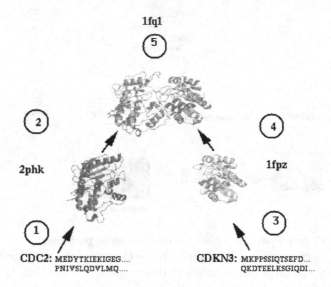

Fig. 2. From co-expressed genes to interactions. CDC2 and CDKN3 are co-expressed in the gene expression data. CDC2 and CDKN3 are assigned by GTD to the PDB structure 2phk and 1fpzA, respectively. These two PDB structures belong the SCOP families d.144.1.7 (Protein Kinase catalytic subunit) and c.45.1.1 (Dual specificity phosphatase-like). PSIMAP, the interaction network, contains an interaction between these two families derived from the PDB structure 1fq1. Structural alignment of 1fq1B/2phk and 1fq1A/1fpzA results in a high alignment score suggesting high structural similarity between the aligned structures and hence the validity of the interaction of CDC2 and CDKN3.

2.3   Work Flow

Figure 1 illustrates the work flow deployed to link the gene expression and protein interaction data.

- **Step 1: From Microarray experiments to set of upregulated genes.** As an initial step, the list of genes obtained from the described tumour microarray data set is used. The genes are clustered into up/down regulated lists. Here, only the upregulated gene cluster is used.
- **Step 2: From genes to structural folds.** Most of the genes from the pancreas data are of unknown structure. For all these genes we apply GTD to assign SCOP structural families to the gene products. Only GTD assignments with confidence rate certain and high are considered.
- **Step 3: From structural folds to domain interactions.** Using the SCOP domain description, we use PSIMAP to identify interacting domains. We consider these interactions at family level.
- **Step 4: From domain interactions to protein interactions.** We consider two proteins as interacting if they contain two domains, whose families

interact structurally according to PSIMAP. As an example, consider Fig. 2. CDC2 is identified as a kinase by GTD and SCOP and CDKN3 as a kinase inhibitor. PSIMAP, the interaction maps, finds the kinase family and inhibitor as interacting based on a corresponding PDB structure.

– **Step 5: Visualisation and GO annotation.** We use an in-house visualisation tool to explore the protein interactions found in the previous step. All proteins are annotated with their corresponding GeneOntology terms. Additionally, we screen out interactions at this stage with one partner annotated as intra- and the other as extra-cellular. Finally, we also include links to relevant web site such as Harvester, Google scholar, and IHop.

3   Results and Discussion

Figure 3 shows the resulting ten different interaction subnetworks. Within the set of upregulated genes, each of the ten subnetworks identifies a group of genes from different functional categories. Out of the ten subnetworks shown in the figure, we will consider the kinase cluster in more detail.

Kinases catalyse the transfer of a phosphate group from a donor, such as ADP or ATP to an acceptor. The kinase cluster is composed of the genes in Figure 4. Cyclins combine with cyclin dependent kinases (CDKs) to form activated kinases that phosphorylate targets leading to cell cycle regulation. A breakdown

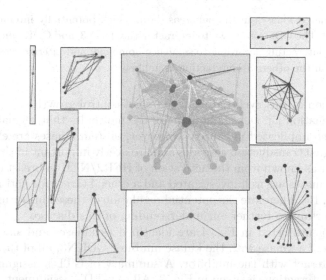

Fig. 3. For the overexpressed genes in the pancreas data set there are ten distinct clusters of interactions. The clusters can be broadly classified as kinase/inhibitor, G proteins, proteasome, ubiquitin, cystatin, serine proteases, canonica RBD, antibody domains, FAD synthatase isoform1, N-acetyl transferase NAT).

Symbol	Gene name
CDKN3	Cyclin-dependent kinase inhibitor 3
ACVR1	Activin receptor type I precursor
BMP2K	BMP-2 inducible protein kinase
BUB1	Mitotic checkpoint serine/threonine-protein kinase BUB1
CDC2	Cell division control protein 2 homolog
CDC2L1	PITSLRE serine/threonine-protein kinase CDC2L2
CDK7	Cell division protein kinase 7
CHEK1	Serine/threonine-protein kinase Chk1
CSNK1A1	Casein kinase I
CSNK1E	Casein kinase I
DYRK2	Dual-specificity tyrosine-phosphorylation regulated kinase 2
EPHA4	Ephrin type-A receptor 4 precursor
FYN	Proto-oncogene tyrosine-protein kinase FYN
MAP4K4	Mitogen-activated protein kinase kinase kinase kinase 4
MELK	Maternal embryonic leucine zipper kinase
MST1R	Macrophage-stimulating protein receptor precursor
MYLK	Myosin light chain kinase
NEK2	Serine/threonine-protein kinase Nek2
PRKACB	cAMP-dependent protein kinase
PRKR	Interferon-induced
PRKWNK1	Serine/threonine-protein kinase WNK1
STK17B	Serine/threonine-protein kinase 17B
STK6	Serine/threonine-protein kinase 6
TRIB2	tribbles homolog 2
TTK	Dual specificity protein kinase TTK

Fig. 4. Upregulated kinases in the pancreas data, which potentially interact with inhibitor CDKN3. The CDC2 is known to interact with CDKN3, and CDK7 and CDC2L1 are likely to be interacting according to sequence homology and interface residues conservation. For all the others an interaction is unlikely.

in the regulation of this cycle can lead to out of control growth and contribute to tumour formation [5]. Defects in many of the molecules that regulate the cell cycle have been implicated in cancer. Moreover, protein kinases are elementary switches in signal transduction cascades and are overly important in the development of cancer as known from the activation of HER2/NEU in breast carcinoma [10]. Protein kinases are well investigated and a crucial target for anti-neoplastic therapy [15, 8, 16]. Therefore the potential regulation of these kinases in pancreas cancer is important to further our understanding of this disease.

Since the genes listed in Fig. 4 are identified as kinases and since there is an interaction of a kinase with the kinase inhibitor CDKN3, all of these kinases potentially interact with the inhibitor. A summary of GTD's assignments and the resulting interactions is given in Fig. 7. All of GTD's assignments are made with high confidence and the assigned structures align structurally well with the kinase in PDB structure 1fq1, chain b (all RMSDs are below 2 Å).

One of these interactions is verified in literature [7], namely CDC2 and CDKN3 and in the interaction databases DIP [17] and BIND [1]. CDKN3 has

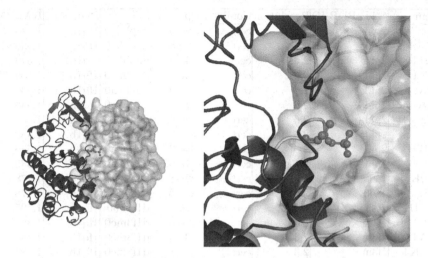

Fig. 5. Left: Cyclin-dependent kinase 2 CDK2 (blue) interacting with the cyclin-dependent kinase inhibitor CDKN3 (yellow/orange). The interfaces are displayed in light blue and yellow, respectively. The phosphorylated threonine of CDK2, which protudes into a pocket of the inhibitor, is shown in red balls-and-sticks mode. PDB ID 1fq1. **Right:** A closeup showing the phosphorylated threonine of CDK2, which protudes into a pocket of the inhibitor.

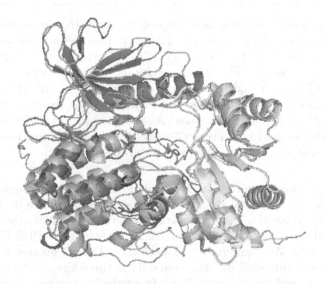

Fig. 6. Structural alignment of the kinases CDK2 (PDB ID 1fq1, chain b) and PDB ID 2phk with alignment RMSD of 1.54929. The inhibitor (PDB ID 1fq1, chain a) is aligned to PDB ID 1fpz, chain a with alignment RMSD of 0.84583. 2phk and 1fpz are the structures assigned by GTD to CDK2 and CDKN3 and 1fq1 is the structure that shows the interaction of the two domains.

Protein	GTD PDB ID	Resolution	Confidence	p_value	SID	PDB ID 2	**RMSD**
CDC2	2phk	2.6	cert	6e-9	d2phka0	1fq1b	1.539
CDC2L1	2phk	2.6	cert	2e-8	d2phka0	1fq1b	1.539
CDk7	2phk	2.6	cert	4e-9	d2phka0	1fq1b	1.539
CHEK1	2phk	2.6	cert	2e-8	d2phka0	1fq1b	1.539
MELK	2phk	2.6	cert	7e-9	d2phka0	1fq1b	1.539
MYLK	2phk	2.6	cert	1e-7	d2phka0	1fq1b	1.539
NEK2	2phk	2.6	cert	5e-9	d2phka0	1fq1b	1.539
PRKR	2phk	2.6	cert	5e-8	d2phka0	1fq1b	1.539
STK17B	2phk	2.6	cert	5e-9	d2phka0	1fq1b	1.539
TTK	2phk	2.6	cert	7e-9	d2phka0	1fq1b	1.539
BMP2K	1a06	2.5	cert	2e-8	d1a0600	1fq1b	1.778
TRIB2	1a06	2.5	cert	6e-8	d1a0600	1fq1b	1.778
CSNK1E	1csn	2.0	cert	1e-8	d1csn00	1fq1b	1.705
ACVR1	1f3m	2.3	cert	2e-8	d1f3mc0	1fq1b	1.884
EPHA4	1f3m	2.3	cert	9e-9	d1f3mc0	1fq1b	1.884
MAP4K4	1f3m	2.3	cert	3e-8	d1f3mc0	1fq1b	1.884
MST1R	1f3m	2.3	cert	2e-8	d1f3mc0	1fq1b	1.884
PRKACB	1f3m	2.3	cert	5e-9	d1f3mc0	1fq1b	1.884
STK6	1f3m	2.3	cert	4e-9	d1f3mc0	1fq1b	1.884
CSNK1A1	1fmk	1.5	cert	4e-8	d1fmk03	1fq1b	1.798
DYRK2	1fmk	1.5	cert	1e-8	d1fmk03	1fq1b	1.798
FYN	1fmk	1.5	cert	2e-9	d1fmk03	1fq1b	1.798
BUB1	1muo	2.9	cert	7e-8	d1muoa0	1fq1b	1.709
PRKWNK1	1muo	2.9	cert	1e-8	d1muoa0	1fq1b	1.709

Fig. 7. Summary of GTD's structural fold assignments for the kinase cluster and their structural alignment with the PDB structure 1fq1 chain b, which contains the interaction between CDK2 and CDKN3. "GTD PDB ID" refers to the structure assigned to the protein, "resolution" indicates the resolution of this structure. All assignments are made by GTD with confidence certain and low P-value. "RMSD" refers to the root mean square deviation of the assigned PDB ID to 1fq1 chain b. The interaction partner CDKN3 (1fq1 chain a) was structurally aligned with the PDB ID 1fpz, chain a, the structure assigned by GTD to CDKN3 (confidence high and p-value of 1e-8). 1fpz, chain a, and 1fq1, cahin a, align with RMSD 0.826.

been shown to interact with, and dephosphorylate the cyclin-dependent kinase CDK2 preventing its activation [14]. This gene was reported to be deleted, mutated, or overexpressed in several kinds of cancers . To validate the interactions we considered the sequence alignments of CDC2's structure (PDB ID 1fq1) with the other kinases. We found only for CDk7 and CDC2L1 greater 40% sequence identity with the threaded structure and only in these kinases the aligned interface residues are well conserved (> 50%). In particular, the key residue threonine 160 (see also Fig. 5) is conserved.

CDK7 is known to be important regulator of cell cycle progression. This protein forms a trimeric complex with cyclin H and MAT1, which functions as a CDK-activating kinase (CAK). This protein is thought to serve as a direct link between the regulation of transcription and the cell cycle. CAK activates

Protein	SID	e-value	Interface SeqId	Thr160 conserved
CDC2	d2phka0	e-108	75.4%	√
CDK7	d2phka0	5e-69	52.6%	√
CDC2L1	d2phka0	6e-70	57.9%	√

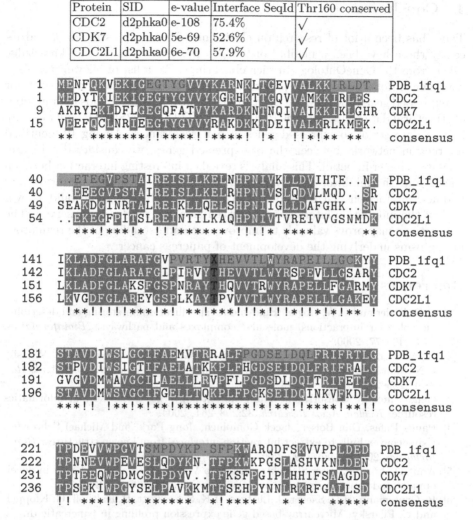

```
  1  MENFQKVEKIGEGTYGVVYKARNKLTGEVVALKKIRLDT.  PDB_1fq1
  1  MEDYTKIEKIGEGTYGVVYKGRHKTTGQVVAMKKIRLES.  CDC2
  9  AKRYEKLDFLGEGQFATVYKARDKNTNQIVAIKKIKLGHR  CDK7
 15  VEEFQCLNRIEEGTYGVVYRAKDKKTDEIVALKRLKMEK.  CDC2L1
     **  ********!!****!!****!  !*  !!*!** **   consensus

 40  ..ETEGVPSTAIREISLLKELNHPNIVKLLDVIHTE..NK  PDB_1fq1
 40  ..EEEGVPSTAIREISLLKELRHPNIVSLQDVLMQD..SR  CDC2
 49  SEAKDGINRTALREIKLLQELSHPNIIGLLDAFGHK..SN  CDK7
 54  ..EKEGFPITSLREINTILKAQHPNIVTVREIVVGSNMDK  CDC2L1
     ***!***!*  !!!*****  !!!!* ****    **    consensus

141  IKLADFGLARAFGVPVRTYXHEVVTLWYRAPEILLGCKYY  PDB_1fq1
142  IKLADFGLARAFGIPIRVYTHEVVTLWYRSPEVLLGSARY  CDC2
151  LKLADFGLAKSFGSPNRAYTHQVVTRWYRAPELLFGARMY  CDK7
156  LKVGDFGLAREYGSPLKAYTPVVVTLWYRAPELLLGAKEY  CDC2L1
     !**!!!!!***!*!  **!*!***!!!*!!!*!!*!*!** !   consensus

181  STAVDIWSLGCIFAEMVTRRALFPGDSEIDQLFRIFRRTLG  PDB_1fq1
182  STPVDIWSIGTIFAELATKKPLFHGDSEIDQLFRIFRALG  CDC2
191  GVGVDMWAVGCILAELLLRVPFLPGDSDLDQLTRIFETLG  CDK7
196  STAVDMWSVGCIFGELLTQKPLFPGKSEIDQINKVFKDLG  CDC2L1
     ***!!  !**!*!*!**!***********!*!**!!****!***!!   consensus

221  TPDEVVWPGVTSMPDYKP.SFPKWARQDFSKVVPPLDED   PDB_1fq1
222  TPNNEVVWPEVESLQDYKN.TFPKWKPGSLASHVKNLDEN  CDC2
231  TPTEEQWPDMCSLPDYV..TFKSFPGIPLHHIFSAAGDD   CDK7
236  TPSEKIWPGYSELPAVKKMTFSEHPYNNLRKRFGALLSD   CDC2L1
     !!  ***!!**  ******  *!****    *  *    *****   consensus
```

Fig. 8. Sequence alignment and interface conservation results of CDC2, CDK7, CDCL1 with 1fq1, chain b. shows the sequence alignment e-value and the percent of conserved residues in the interface (highlighted in green) and in particular, whether the key residue threonine 160 is conserved, which is indeed the case (the coulmn hilighted in Red).

the cyclin-associated kinases cdc2/cdk1, cdk2, cdk4 and cdk6 by threonine phosphorylation. Its expression and activity are constant throughout the cell cycle.

CDC2L1 encodes a member of the p34Cdc2 protein kinase family. p34Cdc2 kinase family members are known to be essential for eukaryotic cell cycle control. The protein kinase encoded by this gene could be cleaved by caspases and was demonstrated to play roles in cell apoptosis.

4 Conclusions

There has been a lot of research on clustering of gene expression data and recently there have been a number of tools to incorporate functional knowledge taken from the GeneOntology in such clusterings. For a list of relevant tools see www.geneontology.org. In this paper we propose to find interaction partners for co-expressed gene products. We illustrate the method, which builds on a number of structural data source such as PDB, SCOP, GTD, and PSIMAP, and evaluate it on a data set of co-expressed genes in pancreas cancer. Among 10 identified interaction networks between the co-expressed genes, we considered a kinase-inhibitor cluster in detail. This analysis reveals a interesting interaction between CDC2 and the inhibitor CDKN3, which is documented in the literature and two new interactions, CDK7 and CDC2L1, which we believe to be valid as over 50% of the interaction interface is conserved as a key residue is conserved. The interactions may prove valuable to improve our understanding of the regulatory mechanisms underlying the development of pancreas cancer.

References

[1] G.D. Bader and C.W. Hogue. Bind–a data specification for storing and describing biomolecular interactions, molecular complexes and pathways. *Bioinformatics*, 16(5):465–77, 2000.

[2] H. M. Berman, J. Westbrook, Z. Feng, G. Gilliland, T. N. Bhat, H. Weissig, I. N. Shindyalov, and P. E. Bourne. The protein data bank. *Nucleic Acids Res*, 28(1):235–242, 2000.

[3] Gene Ontology Consortium. The gene ontology (go) database and informatics resource. *Nucleic Acids Research*, 32:258–261, 2004.

[4] Panos Dafas, Dan Bolser, Jacek Gomoluch, Jong Park, and Michael Schroeder. Using convex hulls to extract interaction interfaces from known structures. *Bioinformatics*, 20(10):1486–1490, 2004.

[5] Amit Deshpande, Peter Sicinski, and Philip W Hinds. Cyclins and cdks in development and cancer: a perspective. *Oncogene*, 24(17):2909–15, 2005.

[6] R. Grutzmann, H.D. Saeger, J. Luttges, H.K. Schackertt, H. Kalthoff, G. Kloppel, and C. Pilarsky. Microarray-based gene expression profiling in pancreatic ductal carcinoma: status quo and perspectives. *International Journal of Colorectal Diseases*, pages 401–413, 2004.

[7] G.J. Hannon, D. Casso, and D. Beach. Kap: A dual specificity phosphatase that interacts with cyclin-dependent kinases. *Proc of the National Academy of Sciences*, 91:1731–1735, 1994.

[8] Rumey Ishizawar and S.J. Parsons. c-src and cooperating partners in human cancer. *Cancer Cell*, 6(3):209–214, Sep 2004.

[9] D. T. Jones. Genthreader: an efficient and reliable protein fold recognition method for genomic sequences. *J Mol Biol*, 287(4):797–815, 1999.

[10] S. Menard, P. Casalini, M. Campiglio, S.M. Pupa, and E .Tagliabue. Role of HER2/neu in tumor progression and therapy. *Cell Mol Life Sci*, 61(23):2965–78, 2004.

[11] Alexey G. Murzin, Steven E. Brenner, Tim Hubbard, and Cyrus Chothia. SCOP: A structural classification of proteins database for the investigation of sequences and structures. *Journal of Molecular Biology*, 247(4):536, 1995.

[12] ' J. Park, M. Lappe, and S.A. Teichmann. Mapping protein family interactions: intramolecular and intermolecular protein family interaction repertoires in the pdb and yeast. *J Mol Biol*, 307(3):929–38, 2001.

[13] C. Pilarsky, M. Wenzig, T. Specht, H. Saeger, and R. Grutzmann. Identification and validation of commonly overexpressed genes in solid tumours by comparison of microarray data. *Neoplasia*, 2004.

[14] RY Poon and T Hunter. Dephosphorylation of cdk2 thr160 by the cyclin-dependent kinase-interacting phosphatase kap in the absence of cyclin. *Science*, 1995.

[15] Tomi K. Sawyer. Novel oncogenic protein kinase inhibitors for cancer therapy. *Curr Med Chem Anti-Canc Agents*, 4(5):449–455, Sep 2004.

[16] Raoul Tibes, Jonathan Trent, and Razelle Kurzrock. Tyrosine kinase inhibitors and the dawn of molecular cancer therapeutics. *Annu Rev Pharmacol Toxicol*, 45:357–384, 2005.

[17] I. Xenarios, L. Salwinski, X.J. Duan, P. Higney, S.M. Kim, and D. Eisenberg. Dip: the database of interacting proteins. *Nucleic Acids Research*, 28(1):289–291, 2000.

Biochemical Pathway Analysis via Signature Mining

Eleftherios Panteris*, Stephen Swift, Annette Payne, and Xiaohui Lui

School of Information Systems, Computing and Mathematics, Brunel University,
Uxbridge, Middlesex UB8 3PH, UK
Eleftherios.Panteris@brunel.ac.uk
http://www.ida-research.net

Abstract. Biology has been revolutionised by microarrays and bioinformatics is now a powerful tool in the hands of biologists. Gene expression analysis is at the centre of attention over the last few years mostly in the form of algorithms, exploring cluster relationships and dynamic interactions between gene variables, and programs that try to display the multidimensional microarray data in appropriate formats so that they make biological sense. In this paper we propose a simple yet effective approach to biochemical pathway analysis based on biological knowledge. This approach, based on the concept of signature and heuristic search methods such as hill climbing and simulated annealing, is developed to select a subset of genes for each pathway that fully describes the behaviour of the pathway at a given experimental condition in a bid to reduce the dimensionality of microarray data and make the analysis more biologically relevant.

1 Introduction

Systems biology is a newly established field, attempting to describe biology at an organisation level by multidisciplinary research[1]. Microarray research is part of the toolbox used to define parts of the system and a lot of interest has focused on gene expression analysis. Computer science and informatics are part of this field and a lot of emphasis is given in microarray data analysis and data storage, as well as in distribution and display of data in terms of clustering programs and large databanks. Another side of systems biology that has flourished is the network modelling side; the application of mathematical models to try to describe biochemical pathways and biological processes in general[7].

All these multidisciplinary approaches aspire to eventually combine and produce functional descriptive models of biological systems that can be used among others to predict drug response and aid in cancer prevention and treatment.

Gene expression analysis[5] has attracted a large amount of attention over the last few years mostly in the form of algorithms exploring dynamic relationships between gene variables and programs that try to display the multidimensional microarray data in appropriate formats so that they make biological sense [16]. Due to the multidimensionality of the microarray experimental data, this has

M.R. Berthold et al. (Eds.): CompLife 2005, LNBI 3695, pp. 12–23, 2005.

proven a challenging task and there still a lot to be desired from the current work [2]. At the same time, the modelling community has a growing interest in the complexity of biochemical pathways and various modelling methods exist that try to predict how such pathways behave [14]. We have taken a heuristic approach to pathway analysis using the idea of signatures for each pathway.

We opted for an algorithm based on hill climbing [13] and simulated annealing [9] to mine for the signatures in all the 108 pathways from *Escherichia coli* . Both algorithm versions were effective in finding biologically relevant signatures and the results are promising that this is a valid way forward in the field.

Section 2 is the motivation behind this study, portraying the reasons why we used a novel interpretation of biochemical pathways for our problem. Section 3 explains the data treatment and their sources, and section 4 describes the signature mining process and its algorithm. Section 5 deals with the biological verification of the results. Section 6 summarises the findings and proposes future directions of work.

2   Motivation

Proteins are the building and functional blocks of the body and interact with each other in set ways, aptly named pathways. All processes that allow an organism to function are organised in pathways that work together to initiate and maintain the organism's response to internal and external stimuli, hence keeping it alive. Examples include the metabolic pathways in humans, responsible for decomposing and absorbing nutrients from food consumption.

As a large number of genomes have already been sequenced, interest turns to the functions and dynamics of the identified genes, as well as their means of influence to the physiology of the respective organisms. Many genes code for enzymes that catalyse metabolic reactions producing energy and various molecules that constitute the core activities of the cell.

Understanding the mechanisms involved in metabolic regulation has important implications in both biotechnology and in the pharmaceutical industry. The identification and validation of drug targets depends critically on knowledge of the biochemical pathways in which potential target molecules operate within cells. For this reason, the study of biochemical pathways is the focus of numerous drug discovery researchers and is central to the strategy of many biopharmaceutical and genomic companies. What is presented here is in essence the first part of a framework that utilises microarray data from genes with fluctuating expression levels to describe the state of the biochemical pathway they belong to, at any given experimental condition.

There is intense research going on in the areas that constitute systems biology, with researchers using very different methods to solve similar problems [1]. From the biological point of view, most researchers use methods that offer some, but not all the functionality a biologist would like to have, often with rather complex implementations. If pathway analysis and visualisation is going to be performed by biologists alone, it should be straightforward and with very few intermediate steps, so researchers can focus on the biological significance of the findings and not the programming implementation of the methods. So far this is not what is available in the research community, and the software available in public and commercial tools are mostly web based, but they do not provide all the functionality they should [3,6,11].

These constraints infiltrate the relationship biologists have with other sciences and computer science in particular. Ma and Zeng have shown in their paper [12] that modelling biochemical pathways is not straightforward, and mistakes can be made.

2.1   Biological Reasoning

We used two terms associated with pathways in the hope that they will help us describe them better; 'flow' of a pathway and 'rate of production'. Flow describes the functionality of the pathway. Disruption of 'flow' is done by removing or totally altering a gene from the pathway, by stopping the 'flow' the pathway seizes to function. Production rate on the other side describes how fast is the pathway produces the end product, the rate can be slow or fast and it is commonly regulated very well within the cell for all the pathways; there are pathways which their sole aim is to control another pathway and so on.

If genes from a pathway have a non variable expression across a large variety of microarray experiments that modify the environmental parameters of the organism, this possibly implies that these genes are not affected by these changes. If these genes are stable across most of the experiments that means that the genes are not essential to the regulation of the pathway but rather they provide the infrastructure, the structural network the pathway relies upon to function.

These genes probably code for structural proteins like membrane pumps, transmebrane proteins to form channels to the exterior of the cell etc, they are vital to the pathway and if altered in any way (mutated/deleted) the pathway ceases function, the 'flow' stops.

But what it is proposed here is that although these genes and their respective proteins are vital to the function of the pathway, they are not 'rate limiting' steps, but rather 'flow limiting' steps, so by eliminating them from the frame the focus will go onto the genes, that by being variable across the different experiments, are proving themselves to be 'rate controllers'. This set of genes probably code for enzymes or signalling molecules etc, that the environmental changes, mutations, inhibitor substances, heat, etc, have an effect upon, thus changing the production rate of the pathway. We are looking for a variation in individual gene expressions, not necessarily very high or very low numbers, but for considerable variations between experimental conditions. This implies to us that the genes in question are sensitive to changes to their immediate environment, thus by monitoring them there is a high chance that their expression variability reflects the true pathway behaviour at that experimental situation i.e. they are rate controllers and they control the production output of the pathway.

Using knowledge about biochemical pathways and their components, this study produces a practical picture of the behaviour of the whole genome of an organism based on microarray data and pathway data from major databases like KEGG [8]. By collecting numerous experiments from a given organism, *Escherichia coli* in this instance, for distinct environmental conditions and treatments, and then combining it with well established pathway information about genes and their biological contribution, we choose a subset of genes from each pathway, a 'signature', which is used to describe the behaviour of that pathway under the given condition.

A pathway's signature is a unique set of genes that can be monitored in any given microarray experiment to illustrate that pathway's behaviour. The signature is the collection of the 'true' expression indicators from the pathway. As mentioned above, they are the most 'expressively active' genes, in the sense that they are the more sensitive part of the pathway, the ones most responsive to external stimuli, i.e., the change in the environmental conditions affects them in such a way as to alter their expression in the cell. The rest of the genes that constitute the pathway are transcriptionally dormant in the sense that they do not respond readily to change, but they form the infrastructure as mentioned above.

Current pathway analysis methods of expression data, which include all the current clustering techniques, require all the genes of a pathway to be taken into account, and may lead to the erroneous conclusion that the activity of a pathway has remain unchanged. For example, if more genes in a pathway are transcriptionally dormant than transcriptionally active, the more numerous dormant ones mask the true picture of a change in the activity of that pathway.

Monitoring the signature of a pathway in all subsequent microarray experimental data would provide an immediate description of the behaviour of the pathway and subsequently of the whole organism in a global pathway /signature network. In essence, we aim to reduce the dimensionality of microarray data to provide a biologically relevant picture of the whole organism immediately, before resorting to clustering methods.

Our key emphasis lies on the utilisation of pathway knowledge to group all the scattered genes in a microarray dataset as pathways and monitor the pathway's behaviour as a whole, rather than genes individually. It is a different concept that aims to help biologists in pathway analysis, by portraying microarray data in a pathway orientated view, with genes grouped not only by expression similarity but also biologically.

Furthermore, it offers a simplified view of these pathways by using a specific subset of genes to portray the behaviour in each experiment. This offers new options to biologists who could group or 'cluster' the pathways according to behaviour in an experiment, thus finding interesting connections not easily observed in gene clustering techniques and visualizations.

3   Data Treatment

The datasets are from the Gene Expression Omnibus (GEO)[18] data repository at NCBI. Specifically they come from Escherichia coli and represent three different experimental conditions in 51 experiments in total. The variety of conditions are exploited to find the most sensitive genes under these conditions, since the larger the number of experimental conditions and number of experiments, the more fine tuned the dataset is. There are global experiments containing the majority of the *Escherichia coli* genes. For our purpose, the experimental data, representing 51 microarray experiments, were normalised to Standard Deviation of 1 and Mean of 0 so that they can be compared together. No further normalisation was necessary since the data were already normalised to log ratios when they were released in GEO.

The genes are chosen according to their variability in expression and have to be above a certain empirically defined global threshold, as used in microarray analysis [15,4] to be considered as statistically significant. The threshold is empirically selected depending on the dataset used and is considered for each time point independently and the selection process is repeated for every experiment. The KEGG *Escherichia coli* files were taken from the KEGG portal [8]. By combining the two, a list of important genes was assembled and these were used as the input of the algorithm.

4 *Signature* Mining Algorithm

Finding the best selection of genes in each pathway that represent that pathway's behaviour is problematic because each gene can be a member of several pathways and we needed to find a way to choose genes that represent each pathway out of the 108 of *Escherichia coli*.

Essentially we tried to find a way to move genes from one pathway to another based on their similarity of expression for the whole of the 51 experiments not just one experiment. Here we suggest an algorithm based on the concept of signature and heuristic research methods such as hill climbing [13] and simulated annealing [9].

Let G be the set of n genes, $G = 1, \cdots, n$, let be the n by T gene expression matrix for the n genes where the *ith* row of X, xi, is the gene expression profile for gene i. xij is defined as the *jth* element of the vector xi.

Let the pathway list P be a list of $m > 0$ lists where is the *ith* element of P, where $|p_i| > 0$. A signature s_i of a pathway p_i is defined as where $|s_i| > 0$. The list of signatures is denoted as S, where $|S| = m$. s_{ij} is defined as the *jth* element of the list s_i. How close two expression profiles a and b are, is defined as follows:

$$d(a,b) = \sqrt{\sum_{i=1}^{T}(x_{ai} - x_{bi})^2} \tag{1}$$

$$D \in \Re^{n \times n}, where D_{ij} = d(i,j) \tag{2}$$

```
(1)  INPUT:
     Euclidean Distance Matrix (D)
     (eq.2),Filtered Pathway list of lists
     (PA) (108 pathways long).
(2)  ITERATIONS
     For w=1:ITER do
(3)  RANDOM SELECTION with REMOVAL or
     REPLACEMENT
     Randomly chose a pathway and randomly
     chose a gene position from the pathway.
     If gene is present:
         REMOVE
     Else
         REPLACE gene back to position.
     End
(4)  GET EUCLIDIAN DISTANCES FROM ALL PATHWAYS
         For i=1:length of PA do
             OBTAIN all unique distances
             between the pathway genes from D
             for comparison.
         End
(5)  GET F1 (SIMILARITY FUNCTION)(eq. 4)
     For i=1:length of PA do
         F1(i)= SUM(distances of all genes
         from P(i))
     End
         Store F1(i)
(6)  GET F2 (SIZE FUNCTION)(eq. 5)
         For i=1:length of PA do
             F2(i)= length of P(i)
     End
         Store F2(i)
(7)  GET F3(EVALUATION FUNCTION)(eq. 6)
     For i=1:length of PA do
         F3(i)=SUM(F1(i)/F2(i))
         F3new(w)=F3new(w)+F3(i)
     End
(8)  a)   EVALUATION 'Hill Climb'
          If F3new(w) < F3old
     SET as F3old
          Elseif F3new(w) > F3old
     RESET to previous value
          End
     End
     b)   EVALUATION 'Simulated Annealing'
          If F3new(w) < F3old
     SET as F3old
          Elseif F3new(w) > F3old
     Accept with probability pr (eq 7)
     SET as F3old
             Else RESET to previous value
             End
          End
     End
(9)  OUTPUT:Signature list for all (108)
     pathways
```

Fig. 1. Algorithm. Signature Mining. Step 8: a) is for the hill climbing and b) for the simulated annealing.

The n by n symmetric matrix D contains all of the pairwise similarities between genes. Note that the larger d (a,b) is, the more dissimilar the genes a and b are. How close together the genes within a signature are is defined as follows:

$$FS(s_i) = \sum_{a=1}^{|s_i|-1} \sum_{b=a+1}^{|s_i|} d(s_{ia}, s_{ib}) \tag{3}$$

This is the sum of all pairwise differences between the elements of a signature. Equation 4 represents how well fitted the signatures are, and equation 5 represents how many genes have been allocated from each pathway. To 'mine' the signatures for each pathway we need to find a set S where F_1 is minimised and F_2 is maximised:

$$F_1 = \sum_{i=1}^{m} FS(s_i) \tag{4}$$

$$F_2 = \sum_{i=1}^{m} | s_i| \tag{5}$$

$$F_3 = \frac{F_1}{F_2} \tag{6}$$

The algorithm fitness F_3 is represented in equation 6 and needs to be minimised for the optimum solution, i.e. the smallest signature possible with the genes best describing the pathway.

$$pr = e^{-\Delta f}, \Delta f = \frac{f(old) - f(new)}{\theta t} \tag{7}$$

$$\varepsilon = \theta_o c^{iter} \tag{8}$$

$$c = (\frac{\varepsilon}{\theta_o})^{\frac{1}{iter}} \tag{9}$$

The signature mining algorithm takes as input a Euclidean distance comparison matrix of all the genes from all the pathways, and a pathway list of lists from KEGG of all the pathways and their genes. To mine the appropriate genes for each signature, we decided to randomly remove or replace a gene from a pathway and use a hill climbing or simulated annealing technique to evaluate the solution. The evaluation is based on a similarity and a size function, requiring minimisation of their fraction to progress. The algorithm is described below.

The simulated annealing step is defined with the above equations [9]. In equation 7 the probability that a worst solution is accepted is related to the difference between the solutions Δf and the starting temperature θ_o. The average probability will be $0.368(e^{-1})$ and this gets smaller as the temperature reduces as used in [17].

Since it is not possible to run the algorithm infinitely we choose the minimum temperature ε (equation 8) and in turn this helps calculate the decay constant c (equation 9) by which the probability of accepting a worse solution is reduced in each iteration. We used $\theta_o = 1000$, and $\varepsilon = 0.01$.

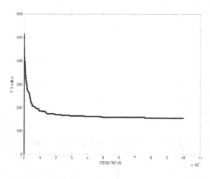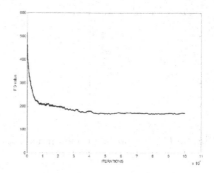

Fig. 2. Convergence plots of the Algorithm based on hill climbing on the left and the algorithm based on simulated annealing on the right. The y axis F_3 value refers to the evaluation function F_3 (equation 6).

The convergence of the algorithms is shown below (Fig.2), with the number of iterations being 100000. The convergence graph shows that the algorithm with the hill climbing method performs well: it sharply drops for the first 10000 iterations and then slowly stabilises to the minimum evaluation value possible and from the 11000 iteration onwards the slope levels up to almost a straight line. On the contrary the simulated annealing option affects the algorithm in that it slowly starts to stabilise from 40000 iterations onwards.

The performance of the algorithm based on hill climbing and simulated annealing is similar in that both versions have similar minimum values. However, the improvement of using simulated annealing in the process is visible when we look at the data and the difference the two versions produce in their analysis in the next section.

5   Application to *Escherichia coli* Data

From *Escherichia coli*, a pathway was chosen to illustrate the biological validity of the method. The Phenylalanine, Tyrosine and Tryptophan biosynthesis pathway, as defined in the KEGG pathway database, was chosen with focus on Tryptophan. The Tryptophan production is regulated from a specific operon that contains five genes, B1260, B1261, B1262, B1263, B1264.

Khodursky et al[10] have done a microarray experiment of *Escherichia coli* under Tryptophan starvation and observed a very specific response from the Tryptophan operon genes. These genes are activated in the absence of Tryptophan and induce its production. By starving the organism in their experiments, they monitored the activation of the pathway.

Using the signature mining algorithm we obtained a signature for each of the algorithm implementations, based on hill climb or simulated annealing, for the specific pathway that describes the behaviour of the pathway according to Khodursky et al [10]. The importance of the signature lies in the fact that we

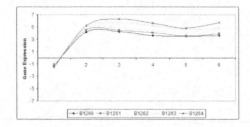

Fig. 3. The Tryptophan operon activation from the Khodursky *et al* [10] dataset

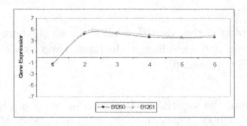

Fig. 4. The Signature genes present in the dataset. The genes describe the activation of the operon during Tryptophan starvation. The same genes were 'mined' with both versions of the algorithm.

Table 1. Distribution of signature genes per branch of the Phenylalanine (PHE), tyrosine (TYR) and tryptophan (TRYP) pathway. Genes B0928 and B2021 take part in both phenylalanine and tyrosine sides of the pathway.

Pa	TYR	TRYP	TRYP	TYR	PHE	TYR	TRYP	PHE	PHE
HC	B2329	B1260	B1261	B4054	B1713	B2600	B1704	B0928	B2021
SA	B2329	B1260	B1261			B2600	B0754	B0928	B2021

used the GEO dataset, explained above, to find the signature that describes the pathway in the Khodursky *et al* [10] dataset.

The Phenylalanine, Tyrosine and Tryptophan biosynthesis pathway includes genes from the biosynthesis of these three amino acids. These three processes are grouped together in the KEGG database due to the chemical similarity these amino acids have, hence the pathway contains 26 genes from all three processes.

Both algorithms gave the same genes from the tryptophan pathway as signature genes, B1260 and B1261. But the difference in the signature mining algorithms is evident in their ability to produce a pathway signature that represents the pathway better with the minimum number of genes.

As one can see in Table 1 where all three branches of the specific pathway are shown, the algorithm based on hill climbing (HC) gave 9 genes with some of them not very similar gene expression patterns. Whereas the algorithm based on simulated annealing (SA) produced 7 genes that are closer in terms of gene expression hence we can say that they describe the pathway better (data not shown).

Khodursky *et al* [10] are interested only in Tryptophan starvation so their dataset contains only the signature genes B1260 and B1261. The starvation response of *Escherichia coli* is to activate the genes that produce Tryptophan [10]. The response can be observed in Fig. 3 where the gene expression of the genes that constitute the tryptophan operon is plotted in six part starvation time course. The organism is placed in an environment without Tryptophan at the start of the experiments (see Fig. 2) and gene expression measurements are taken at 20 minutes intervals. It is obvious from the graph (Fig. 2) that the genes are highly upregulated moments after the starvation initiation.

Both our signatures have two out of five genes from the tryptophan operon. As it can be seen from Fig. 3 where they are sufficient to portray the behaviour of the pathway during the experiments. As mentioned above briefly, the signature was 'mined' from the GEO dataset and applied to the Khodursky *et al*[10] dataset. This is an important fact to stress, since that shows that we can find genes that are controlling the expression of the pathway using different datasets and then use only these genes to monitor the experiment at hand. The more extensive the mining dataset the more precise the signatures of the pathways will become.

The results obtained so far have provided early evidence that signature mining can be an effective way of analysing biochemical pathways. Once these pathways are chosen the signature mining algorithm can be applied across experimental conditions and datasets with ease.

The biological relevance of the signature mining algorithm is extensive, especially in the biochemical and pharmaceutical community, since it allows the researcher to observe the behaviour of a specific pathway in a clear and definitive way that does not involve genes that do not affect the pathway's regulation. Its relevance is obvious in drug related research where it could help in monitoring the changes of all the pathways of the organism when a drug is tested. In essence, it allows the researcher to have a snapshot of the whole organism processes in an easy and transparent way, easy to understand and use.

6   Concluding Remarks

This paper presents a novel interpretation in systems biology of biochemical pathways and the information that can be gathered from microarray experiments coupled with transparency of method and biological knowledge.

We have shown that a specially selected sub group of genes from a pathway, a signature, can describe its behaviour under a given experimental condition. Two different versions of the same simple yet effective algorithm based on hill climbing and simulated annealing were created to 'mine' for the appropriate genes for each pathway based on the pathway's behaviour across a large set of experiments of varying conditions. Both algorithms were able to select signatures for all 108 pathways, of which one was used as an example here. The algorithm based on simulated annealing was more specific providing a smaller signature for the pathway used.

Biological knowledge and verification was important in the design and subsequent application of the algorithm and as the preliminary results have clearly shown this interpretation of biochemical pathways to be an interesting way of explaining microarray data used for pathway analysis. Future work will include improvement of the algorithm run time and functionality. Ideally the algorithm should have picked all five genes that form the operon, and a more pathway specific evaluation function is currently being tested, with promising results, to solve that problem.

Furthermore, the algorithm will be included in a framework for microarray datasets for full exploitation of microarray data in relation to pathway analysis and pharmaceutical research. Application of the algorithm will not be restricted only to *Escherichia coli* but to other organisms with specific pharmaceutical concerns and ultimately to human data, with a continuation of the framework steps to include gene networks and interactions with protein-protein networks, offering a solid solution in that area of systems biology.

Acknowledgements

The authors would like to thank the reviewers for their useful insights and comments about the manuscript.

References

1. Aggawal, K., Lee, K.H.,; Functional genomics and proteomics as a foundation for systems biology.' Briefings in functional genomics and proteomics Vol 2, No 3, 175-184 2003
2. Claverie, J.; Computational methods for the identification of differential and coordinated gene expression. Human Molecular Genetics. Vol 8, No 10 1821-1832 1999
3. Dahlquist, K.D., Salomonis, N., Vranizan, K., Lawlor, S.C., Conklin, B.R.,; GenMAPP, a new tool for viewing and analyzing microarray data on biological pathways, Nature Genetics 31(1):19-20 2002
4. Duggan, D.J., Bittner, M., Chen, Y., Meltzer, P. Trent, J.; Expression profiling using cDNA microarrays. Nature Genetics. 21, 10-14 1999
5. Eisen, M., Spellman, P.T., Botstein, D., Brown, P.O, ; Clustering Analysis and display of genome wide expression patterns. PNAS 95. 14863-14868 1998
6. Goesmann, A., Haubrock, M., Meyer, F., Kalinowski, J., Giegerich, R.; PathFinder: reconstruction and dynamic visualization of metabolic pathways. Bioinformatics. 18: 124-129 2002
7. Huang, S.; Back to the biology in systems biology: What can we learn from biomolecular networks? Briefings in functional genomics and proteomics. Vol 2, No 4, 279-297 2004
8. Kanehisa, M., Goto, S.; KEGG: Kyoto Encyclopedia of Genes and Genomes. Nucleic Acids Research. 28, 27-30 2000
9. Kirkpatrick, S., Gelatt, Jr., C.D., Vecchi, M.P.; Optimization by Simulated Annealing, Science. 220, No. 4598, 671-680 1983

10. Khodursky A, B., Peter B, J.,, Cozzarelli N. R., Botstein D., Brown P.O., Yanofsky C.; DNA microarray analysis of gene expression in response to physiological and genetic changes that affect tryptophan metabolism in Escherichia coli. PNAS. USA. Oct 24;97(22):12170-5 2000

11. Kolpakov, F.A., Ananko, E.A., Kolesov, G.B., Kolchanov, N.A.; GeneNet: a gene network database and its automated visualization. Bioinformatics 14: 529-537 1998

12. Ma, H., Zeng, A., ; Reconstruction of metabolic networks from genome data and analysis of their global structure for various organisms. Bioinformatics 19: 270-277 2003

13. Michalewicz, Z., Fogel, D. B.; How To Solve It: Modern Heuristics, Springer, 1998

14. Papin, J., Price, N., Wiback, S., Fell, D., Palsson, B.; "Metabolic pathways in the post genome era," Trends in Biochemical Sciences., Vol. 28, 250-258 2003

15. Schena, M., Shalon, D., Heller, R., Chai, A., Brown, P.O., Davies, R.W.; Parallel Human Genome Analysis: Microarray Based Expression Monitoring of 1,000 Genes. PNAS 93. 10614-10619 1996

16. Slonim, D. K.; From patterns to pathways: gene expression data analysis comes of age. Nature Genetics. 32, 502 - 508 2002

17. Swift, S., Tucker, A., Vinciotti, V., Martin, N., Orengo, C.,Liu, X., Kellam, P.,; Consensus clustering and functional interpretation of gene expression data. Genome Biology 5: R94 2004

18. Edgar, R.,Domrachev M.,Lash A. E.; Gene Expression Omnibus: NCBI gene expression and hybridization array data repository. Nucleic Acids Research, Vol. 30, No. 1 207-210 2002

Recurrent Neuro-fuzzy Network Models for Reverse Engineering Gene Regulatory Interactions

Ioannis Maraziotis, Andrei Dragomir, and Anastasios Bezerianos

Department of Medical Physics, Medical School,
University of Patras, 26500 Rio, Greece
{imarazi, adragomir}@heart.med.upatras.gr,
bezer@patreas.upatras.gr

Abstract. Understanding the way gene regulatory networks (complex systems of genes, proteins and other molecules) function and interact to carry out specific cell functions is currently one of the central goals in computational molecular biology. We propose an approach for inferring the complex causal relationships among genes from microarray experimental data based on a recurrent neuro-fuzzy method. The method derives information on the gene interactions in a highly interpretable form (fuzzy rules) and takes into account dynamical aspects of genes regulation through its recurrent structure. The gene interactions retrieved from a set of genes known to be highly regulated during the yeast cell-cycle are validated by biological studies, while our method surpasses previous computational techniques that attempted gene networks reconstruction, being able to retrieve significantly more biologically valid relationships among genes.

1 Introduction

Large scale monitoring of gene expression activity opened the way for investigating complex biological processes at molecular level [1]. The emergence of the gene expression data posed new challenges to the data analysis research community due to both its large dimensionality and the complexity of information it contains, thus requiring novel data analysis and modeling techniques. Initial efforts targeted the inference of functional information for genes of unknown functionality, by means of clustering techniques [2-3], while other approaches used supervised learning techniques for discriminating between different sample classes (e.g. healthy vs disease tissue diagnosis) [4]. However, reconstructing and modeling gene networks remains one of the central problems in functional genomics.

The activity of genes is regulated by proteins and metabolites, which are produced by proteins. But proteins are also gene products, thus genes can influence each other (induce or repress) through a chain of proteins and metabolites. At genetic level, it is thus legitimate, and indeed common, to consider gene-gene

M.R. Berthold et al. (Eds.): CompLife 2005, LNBI 3695, pp. 24–34, 2005.

interactions, and these lead to the concept of gene networks. Gene networks ultimately attempt to describe how genes or groups of genes interact with each other and identify the complex regulatory mechanisms that control the activity of genes in living cells. The reconstructed gene interaction models should be able to provide biologists a range of hypotheses explaining the results of experiments and suggesting optimal designs for further experiments.

The reconstruction of gene networks based on expression data is hampered by peculiarities specific to this kind of data, therefore the methods employed should be able to handle under-constrain data, must be robust to noise (since experimental data obtained from microarrays are measurement noise-prone) and should provide interpretable results. Recently, there have been several attempts to describe models for gene networks. Boolean networks have been used due to their computational simplicity and their ability to deal with noisy experimental data [5]. However, the Boolean networks formalism assumes that a gene is either on or off (no intermediate expression levels allowed) and the models derived have inadequate dynamic resolution. Other approaches, reconstructing models using differential equations, turn out as computationally expensive and very sensitive to imprecise data [6], while Bayesian networks based models, although attractive due to their ability to deal with stochastic aspects of gene expression and noisy measurements, have the disadvantage of minimizing dynamical aspects of gene regulation [7].

Our approach uses a recurrent neuro-fuzzy method [8] to extract information from microarray data in the form of fuzzy rules, bringing together the advantages of computational power and low-level learning common to neural networks, and the high-level human-like reasoning of fuzzy systems. The dynamic aspects of gene regulatory interactions are considered by the recurrent structure of the neuro-fuzzy architecture we adopt, while the online learning algorithm drastically reduces the computational time.

Interactions among a set of target genes found to be highly involved in the yeast cell cycle regulation from previous biological studies [9] are studied and knowledge in the form of IF-THEN rules is inferred. The algorithm is trained on a subset of experimental samples and the inferred relations are tested for consistency on the remaining samples.

2   Methods

The computational approach we are using is that of a multilayer recurrent self-organizing neuro-fuzzy inference network (RSONFIN) [8]. Being a hybrid neuro-fuzzy architecture it is able to overcome drawbacks specific to pure neural networks, which function as black boxes, by incorporating elements specific to fuzzy reasoning processes, where each node and weight has its own meaning and function. In addition, its recurrent structure manages to bring the same advantages of a pure recurrent neural network (in terms of computational power and time prediction), while succeeding to extend the application of the classic neuro-fuzzy networks to temporal problems.

A recurrent neural network naturally involves dynamic elements in the form of feedback connections used as internal memories. Unlike the feed-forward neural network whose output is a function of its current inputs only and is limited to static mapping recurrent neural networks perform dynamic mapping using their ability to store information of the past (e.g. prior system states). From this point of view a recurrent fuzzy neural structure seems like the best choice to deal with time - series data, just like the ones used in the present work and which are concerned with the measurement of genes expression in a number of experiments throughout time.

In contrast to other neuro-fuzzy network architectures, where the network structure is fixed and the rules should be assigned in advance, there are no rules initially in the architecture we are considering; all of them are constructed during on-line learning. Two learning phases, the structure and the parameter learning phase are used to accomplish this task. The structure learning phase is responsible for the generation of fuzzy if-then rules as well as the judgment of the feedback configuration, and the parameter learning phase for the tuning of free parameters of each dynamic rule (such as the shapes and positions of membership functions), which is accomplished through repeated training from the input - output patterns. The way the input space is partitioned determines the number of rules. Given the scale and complexity of the data, the number of possible rules describing the causal relationships is kept under constraint by employing an aligned clustering-based partition method for the input space and by allowing a scene in which rules with different preconditions may have the same consequent part [8]. At the same time, the minimum number of fuzzy sets we allow is set to three, the lowest for which, in the case of gene expression data, we obtain adequate resolution.

The neuro-fuzzy system we are employing consists of a six-layer network, including a feedback layer (see Fig.1). In the first layer each node represents an input linguistic variable. No calculation takes place in this layer and each node transfers the value of an input variable to the next layer.

Each node in the second layer corresponds to one linguistic label (e.g. high, medium, low etc.) represented by a membership function, which has the following Gaussian form:

$$y^{(2)} = e^{\left\{ -\left(\varphi_i^{(2)} - m_{ij} \right) \big/ \sigma_{ij}^2 \right\}} \tag{1}$$

where m_{ij} and σ_{ij} are the center and width, respectively, of the membership function of the ith variable and the jth term, while by $y_k(n)$ and $\varphi(n)$ we represent the input and output of the kth node and of the nth layer, throughout the paper.

In the third layer, each node represents a fuzzy logic rule and performs a precondition matching of a rule:

$$y^{(3)} = y^{(6)} \cdot e^{-[D_i(\mathbf{x}-\mathbf{m}_i)]^T [D_i(\mathbf{x}-\mathbf{m}_i)]} \tag{2}$$

where m_i is the center of the ith rule and D_i is a diagonal matrix containing the widths of the same rule.

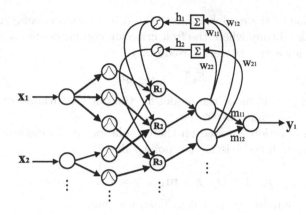

Fig. 1. Six-layer recurrent neuro-fuzzy network

The nodes of the fourth layer (output nodes) perform the following fuzzy OR operation to integrate the fired rules which have the same consequent part:

$$y^{(4)} = \sum_i \varphi_i^{(4)} \tag{3}$$

The fifth layer is responsible for the defuzzification operation. Nodes in this layer together with the links attached to them perform the following task:

$$y^{(5)} = \frac{\sum_i \varphi_i^{(5)} c_{ij}}{\sum_i \varphi_i^{(5)}} \tag{4}$$

where c_{ij} is the center of the i_{th} membership function of the jth output variable.

Finally, there is the feedback layer, which calculates the internal variable h_i and its firing strength. The firing strength of the internal variable contributes to the matching degree of a rule node in layer 3.

$$h_j = \sum_i y_i^{(4)} w_{ij} \tag{5}$$

$$y^{(6)} = \frac{1}{1 + e^{-h_i}} \tag{6}$$

where w_{ij} is the weight connecting the ith feedback layer node with the jth node of the 4th layer.

We have adapted the original model in order to achieve both higher accuracy, as well as a more parsimonious structure. The modifications concern on one hand the decision for the creation of a new rule and on the other hand the merging of similar rules based on the similarity of the fuzzy sets of the input and output variables.

The extension of the model as far as the rule creation is concerned, consists of a two criteria scheme, where the first criterion computes the overall error E_k of the network for a new pattern based on:

$$\|E_k\| = \|y_k^d - y_k\| \tag{7}$$

where y_k is the prediction of the model for the output variable while y_k^d is the actual value for the same pattern.

The second criterion is based on the calculation of the distance di between the pattern x and all the existing till point rules:

$$d_k(j) = D_j \|\mathbf{x} - \mathbf{m}_j\|, \ j = 1, 2, \cdots, N_R \tag{8}$$

where N_R is the number of the rules. Then we find:

$$d_{\min} = \arg \min \left(d_k \left(j \right) \right) \tag{9}$$

If both criteria below are true then a new rule is created:

$$\begin{aligned} \|E_k\| &> \lambda_{error} \\ d_{\min} &> \lambda_{dist} \end{aligned} \tag{10}$$

where $\lambda_{error} = 0.47$ and $\lambda_{dist} = 0.54$ are empirically set thresholds. The operation above occurs during the structure-learning phase of the model.

Regarding the merging of fuzzy rules, we have defined a procedure that takes place during the parameter learning phase. Usually the fact that two rules have different consequents but similar antecedents means that those two rules conflict each other. Therefore we decide whether we can combine two rules by evaluating the similarity of their antecedent part. Given for example two rules R_i and R_j their antecedent parts are the fuzzy sets A_{i1}, A_{i2}, A_{in} and A_{j1}, A_{j2}, , A_{jn} respectively, where A_{kl} is the kth fuzzy set of the lth fuzzy rule, the similarity for the antecedents can be computed as:

$$S\left(A_i, A_j\right) = \min_l \left\{ S\left(A_{il}, A_{jl}\right) \right\}, l = 1, 2, \cdots, n \tag{11}$$

if $S(A_i, A_j)$ exceed a value l_s (empirically l_s might be set between 0.6 and 0.8) then those fuzzy sets are considered to be similar enough and thus the rules that they constitute can be combined towards the creation of a new rule R_c. For the evaluation of the center and width of each one of the fuzzy sets A_{cl} of the new rule we use the average of the fuzzy sets A_{il} and A_{jl}, $c_{cl} = (c_{il} + c_{jl})/2$, $w_{cl} = (w_{il} + w_{jl})/2$. The same technique is used for the two consequents.

For the similarity measure of two fuzzy sets we use the formula previously derived in [10], which concerns bell - shaped membership functions. It should be pointed out that the value of l_s has an initial value that monotonically decreases in time so that higher similarity between two fuzzy sets is allowed in the initial stage of learning.

In order to recover functional relationships among genes and build models describing their interactions we employ transcriptional data from microarray experiments. A simple representation of the data would be that of a matrix with rows containing gene expression measurements over several experimental conditions/ samples. Gene expression patterns give an indication of the levels of activity of genes in the tissues under study. A number of network models that match the number of genes is constructed, each model having as output node a selected gene and input nodes the remaining genes from the dataset. Models attempt to describe the behavior of the gene selected as output based on possible interactions from the input genes and stores the knowledge on the derived gene-gene interactions in a pool of fuzzy rules. It must be noted at this point that although we are employing in the current study a data set with reduced number of genes, our method can be easily extended to the analysis of data sets containing thousands of genes, by introducing a clustering scheme prior to the models construction. Representative genes can be selected from the clusters as in [11] and subsequently fed to the neuro-fuzzy network.

For each model we have derived an error criterion, in order to test the model accuracy on a given data set, by using an overall error measure based on the difference between the prediction of the model for the output variable (i.e. gene) - y_i and its real value through all the samples of the data set - y_i^d:

$$E = \frac{1}{n} \sum_{i=1}^{n} \left[1 - \frac{|y_i| - |y_i^d|}{|y_i^d|} \right] \tag{12}$$

where n is the number of samples in the data set.

The number of rules governs the expressive power of the network. Intuitively, an increase in the number of rules describing the gene interactions in a certain model results in a decrease in the error measure of (1), accounted by the respective model. However, pursuing a low error measure may result in the network becoming tuned to the particular training data set and exhibiting low performance on the test sets (the well known overfitting problem). Therefore, the problem becomes one of combinational optimization: minimize the number of rules, while preserving the error measure at an adequate level by choosing an optimal set of values for the model parameters (overlap degree between partition clusters, learning rates, rule creation criteria).

The final rule base extracted from a certain dataset, which provides the plausible hypotheses for gene interactions, is built by selecting the rules that are consistent with all the models.

3   Results

To test the validity of our approach, we have used a dataset containing gene expression measurements during the cell-cycle of the budding yeast *Saccharomyces cerevisiae*. The experiments were performed by Spellman *etal*. [12] and consisted on 59 samples collected at different time points of the cell-cycle. The experiments are divided into three subsets, which were named according to the synchronization method used for the yeast cultures: *cdc*15 *arrest* (24 samples), *cdc*28 *arrest* (17 samples) and *alpha-factor* (18 samples). Missing values were filled in using an estimation method based on Bayesian principal component analysis (BPCA), which estimates simultaneously a probabilistic model and latent variables within the framework of Bayes inference [13].

We have chosen a subset of 12 genes (see Table 1) on the basis that they were identified from previous biological studies to be highly regulated during the yeast cell-cycle, their protein products playing key roles in controlling the cyclic biological processes [9]. The samples of the $cdc15$ subset were used as training data for the neuro-fuzzy network and the inferred fuzzy rules were tested for consistency on the other two experimental data subsets ($alpha$ and $cdc28$). The choice of the training set was determined by the larger number of samples in the $cdc15$ experimental set and therefore a larger number of training instances.

We have used the recurrent neuro-fuzzy approach to create twelve multi-input, single output models. Each model describes the state of an output gene based on the temporal expression values of the remaining eleven genes. As already described in the methods section, the structure of each model is determined by the on-line operation of the algorithm using the $cdc15$ subset as training set and repeated training of the algorithm is subsequently used to fine tune the membership functions as well as to determine the precondition and consequent parts of the rules.

Each one of the networks constructed describes the time series of the output gene through a number of fuzzy rules. Table 1 presents the set of rules describing gene SWI4. Columns describe rules. The numbers correspond to membership function values for each input variable - gene, expression level, i.e. 3 is high, 2 is medium and 1 is low (e.g. the second column contains the rule: "If SIC1 is low AND CLB5 is medium AND AND MBP1 is high AND CDC6 is low THEN SWI4 is medium").

Table 1. Rules describing the state of gene SWI4 based on the eleven remaining genes of the original dataset $cdc15$

Input Genes	Rules		
	Rule 1	Rule 2	Rule 3
SIC1	3	1	2
CLB5	3	2	1
CDC20	2	1	3
CLN3	3	1	2
SWI6	2	3	1
CLN1	3	2	1
CLN2	3	2	1
CLB6	3	2	1
CDC28	2	1	3
MBP1	2	3	1
CDC6	3	1	2
Prediction of SWI4	3	2	1

Table 2 presents the prediction errors on all three data subsets (computed as in Eq.1), as a relative measure of the models accuracy. The error scores shown in Table 2, emphasize the correlation in qualitative behavior between the fit and prediction and it has been used for the determination of the networks parameters that best match the prediction of all the twelve models considered.

Table 2. Gene prediction errors for the training and testing data subsets

Name of Predicted Gene	Prediction Errors		
	$cdc15$ data set	$cdc28$ data set	$alpha$ data set
SIC1	0.4129	0.4383	0.4400
CLB5	0.4842	0.3700	0.3965
CDC20	0.5608	0.5669	0.4135
CLN3	0.3179	0.3286	0.4129
SWI6	0.3336	0.1620	0.1588
CLN1	0.5371	0.4386	0.5765
CLN2	0.4374	0.4077	0.5188
CLB6	0.4725	0.5769	0.4753
SWI4	0.2429	0.4312	0.4959
CDC28	0.2058	0.2474	0.2141
MBP1	0.5958	0.2172	0.2382
CDC6	0.4446	0.3486	0.3441

In a parallel experiment, we have used the network models derived from the training set to predict time series for the gene expression patterns. Figure 2 shows the predicted time series for the expression ratio of genes CLN1 and SWI6 on both test data sets. It may be noticed that, while the former represents a gene with worst error fits, still the model is able to predict trends of expression values. In the case of the latter gene, the predicted patterns consistently fit the real expression data, fact which proves the accuracy of the derived models.

In the final step of our analysis, the rule base describing the gene interactions derived from the twelve models is used to build a gene network model (see Fig. 3) in the form of a graph in which each node represents a gene and the presence of an edge between two nodes indicates an interaction between the connected genes (either activation - represented by arrows, or repression - represented by closed circles). From a total of 36 rules derived by the 12 network models 19 were kept as being consistent throughout the models, all of them being consistent with the interactions retrieved by biological studies [9, 14].

Biologically accurate interactions have been successfully extracted using our method, such as the positive regulation of CLB6 by MBP1, the inhibition of CLB5 by CDC20, the positive regulation of CLN1 by CLN2 as well as the negative regulation of the same gene by CDC20. CLN3 activates by phosphorylation the transcription factor group SBF (formed by SWI4 and SWI6), which in turn

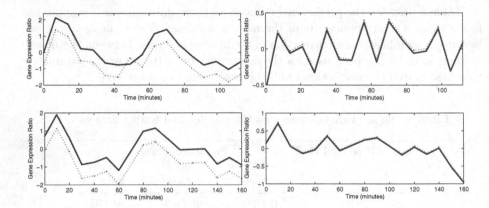

Fig. 2. Predictions of gene expression patterns. The two plots on the left represent the predictions for gene CLN1, while the ones on the right for the gene SWI6. For both genes the upper plot corresponds to the *cdc*28 data subset while the lower plot corresponds to the alpha data subset (solid lines represent actual expression, while the dotted ones the prediction of the model).

activates the cyclin CLN2. Other successfully found interactions include those between CLB5 and CLB6, while MBP1 (part of the transcription factor group MBF) is found to activate the cyclin CLB6. These are relations that besides their biological confirmation have been determined by other approaches like [15], [16]. But there are cases like the positive regulation of SWI4 by MBP1 that both the supervised learning analysis of Ref. [15], as well as the linear fuzzy approach of Ref. [16] failed to extract. It should be noted that the last relation described was successfully extracted despite the fact that the MBP1 transcription varies within a small range, which could have lead to a potential error produced by the model.

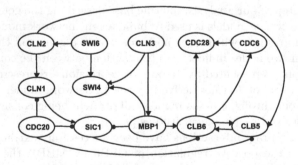

Fig. 3. Graph of the derived gene network interactions. The figure provides a schematic of the rules extracted using all the models in our application.

The results prove the accuracy and efficiency of our method in successfully capturing the interactions among the genes considered, despite the noise inherited from the microarray hybridization and the small amount of samples in the data.

4 Conclusion

The paper presents a method to extract causal interaction relationships among genes and tackle the inverse problem of gene regulatory networks reconstruction from microarray expression data. To this goal, we have adapted a recurrent neural fuzzy architecture that is able to achieve the task in a fast and comprehensive manner. The self-organizing structure of the method helps retrieving the optimal number of relationships underlying the data, while its recurrent part is able to take into account dynamic aspects of gene regulation. To our knowledge it is the first application of a neural fuzzy approach for the problem of gene regulatory networks reconstruction.

While our approach follows the current ideology regarding gene networks reconstruction - focusing attention on specific subsystems that are easier to analyze and feasible in terms of collecting necessary experimental data - it is able to supply starting points for deciphering the multiple complex biological systems. The inferred information provides biological insights, which can be used by biologists to design and interpret further experiments. The results prove the solid performance of a hybrid neuro-fuzzy approach, which is able to extract from a certain dataset more biological meaningful relations than other computational approaches. All the retrieved causal relationships among genes are in complete accordance with the known biological interactions in the yeast cell-cycle. Although the current study refers to a relatively small subset of genes, the method could be adapted to process larger sets, eventually even entire genome-scale expression data, from which subsets of genes involved in specific regulatory processes might be identified through suitable gene selection methods.

Acknowledgments

The authors would like to thank the European Social Fund (ESF), Operational Program for Educational and Vocational Training II (EPEAEK II) and particularly the Program IRAKLEITOS for funding the above work.

References

1. DeRisi, J.L., Iyer, V.R., and Brown, P.O.: Exploring the metabolic and genetic control of gene expression on a genomic scale. Science **278** (1997) 680-686
2. Mavroudi, S., Papadimitriou, S., and Bezerianos, A.: Gene expression data analysis with a dynamically extended self-organizing map that exploits class information. Bioinformatics **18** (2002) 1446-1453

3. Eisen, M.B., Spellman, P.T., Brown, P.O., and Botstein, D.: Cluster analysis and display if genome-wide expression patterns. Proc. Natl. Acad. Science **95** (1998) 14863-14868

4. Golub, T.R., Slonim, D.K., Tamayo, P. et al.: Molecular classification of cancer: class discovery and class prediction by gene expression monitoring. Science **286** (1999) 531-537

5. Golub, T.R., Slonim, D.K., Tamayo, P. et al.: Molecular classification of cancer: class discovery and class prediction by gene expression monitoring. Science **286** (1999) 531-537

6. Liang, S., Fuhrman, S., and Somogyi, R.: :REVEAL, a general reverse engineering algorithm for inference of genetic network architectures. Pac. Symp.Biocomputing (2000) 18-29

7. Tegner, J., Yeung, M.K., Hasty, J., and Collins, J.J.: Reverse engineering gene networks: integrating genetic perturbations with dynamical modeling. Proc. Natl. Acad. Science **100** (2003) 5944-5949

8. Friedman, N., Linial, M., Nachman, I., and Pe'er, D.: Using Bayesian networks to analyze expression data. J. Comp. Biology **7** (2000) 601-620

9. Juang, C.-F., and Lin, C.T.: A recurrent self-organizing neural fuzzy inference network. IEEE Trans. Neural Networks **10** (1999) 828-845

10. Juang, C.-F., and Lin, C.T.: A recurrent self-organizing neural fuzzy inference network. IEEE Trans. Neural Networks **10** (1999) 828-845

11. 9. Li, F., Long, T., Lu, Y., Ouyang, Q., and Tang, C.: The yeast cell-cycle network is robustly designed. Proc. Natl. Acad. Science **101** (2004) 4781-4786

12. Lin, C.T., and Lee, C.S.G.: Reinforcement structure/parameter learning for neural-network-based fuzzy logic control systems. IEEE Trans. Fuzzy Syst **2** (1993) 46-63

13. Guthke, R., Moller, U., Hoffmann, M., Thies, F., and Topfer,. S.: Dynamic network reconstruction from gene expression data applied to immune response during bacterial infection. Bioinformatics **8** (2005) 1626-1634

14. Spellman, P.T., Sherlock, G., Zhang, M.Q., Iver, V.R., Anders, K., Eisen, M.B., Brown, P.O., Botstein, D., Futcher, B.: Comprehensive identification of cell cycle-regulated genes of the yeast Saccharomyces cerevisiae by microarray hybridisation. Mol Biol Cell **9** (1998) 3273-3297

15. Oba, S., Sato, M., Takemasa, I. et al.: A Bayesian missing value estimation method for gene expression profile data. Bioinformatics **19** (2003) 2088-2096

16. Futcher, B.: Transcriptional Regulatory Networks of the yeast cell-cycle. Curr. Opin. Cell Biol. **14** (2002) 676-683

17. Soinov, L., Krestyaninova, M., and Brazma, A.: Towards reconstruction of gene networrks from expression data by supervised learning. Genome Biology **4** (2003) R6.1-R6.10

18. Sokhansanj, B., Fitch, P., Quong, J., and Quong, A.: Linear fuzzy gene network models obtained from microarray data by exhaustive search. BMC Bioinformatics **5** (2004) 1-12

Some Applications of Dummy Point Scatterers for Phasing in Macromolecular X-Ray Crystallography

Alexandre Urzhumtsev[1], Natalia Lunina[2], Pavel Afonine[3],
and Vladimir Y. Lunin[2]

[1] Département de Physique, Université H.Poincaré Nancy 1,
54506 Vandoeuvre-lès-Nancy, France
`Alexander.Ourjoumtsev@stmp.uhp-nancy.fr`
[2] Institute of Mathematical Problems of Biology, Russian Academy of Sciences,
142290 Pushchino, Russia
`{Lunina, Lunin}@impb.psn.ru`
`http://www.impb.ru/lmc`
[3] Lawrence Berkeley National Laboratory, 1 Cyclotron Road, BLDG 64R0121,
Berkeley, CA 94720, USA
`PAfonine@lbl.gov`

Abstract. The purpose of the X-ray macromolecular crystallography is to determine the electron density distribution $\rho(\mathbf{r})$ of the crystal and interpret it by atoms. $\rho(\mathbf{r})$ may be calculated by a Fourier series with complex coefficients. Their magnitudes are available from X-ray-diffraction experiment, however an accurate calculation of $\rho(\mathbf{r})$ is often impossible due to absence of estimates for corresponding arguments (phases) or their insufficient accuracy. To define or improve the phase estimates a model composed from 'dummy' scatterers may be used. The number and size of these scatterers depend on problem. At a conventional resolution the scatterers similar to carbon atoms are used for phase improvement. When phase information is not available models composed from a small number of large scatterers presenting whole molecular domains may be used. In another extreme case, at a subatomic resolution, scatterers presenting partial atomic charges may be used to model the density deformation.

1 Introduction

The goal of macromolecular X-ray crystallography is to find the coordinates and some other parameters like atomic displacement parameter, ADP, of atoms that compose the crystal under study. A traditional way to do it is to measure the intensities of X-rays diffracted by the electrons of the corresponding crystal, then reconstruct the electron density distribution $\rho(\mathbf{r})$ and then interpret it in terms of atomic model (see Table 1 for basic definitions). The magnitudes $F(\mathbf{s})$ of the complex Fourier coefficients $\mathbf{F}(\mathbf{s})$ of the function $\rho(\mathbf{r})$ may be derived directly from these intensities but the arguments, or *phases* $\varphi(\mathbf{s})$ of $\mathbf{F}(\mathbf{s})$ are lost in the diffraction experiment. This makes the direct computing of $\rho(\mathbf{r})$ as a Fourier

M.R. Berthold et al. (Eds.): CompLife 2005, LNBI 3695, pp. 35–45, 2005.
© Springer-Verlag Berlin Heidelberg 2005

synthesis impossible; the loss of phases is called the *phase problem*, its solution is called *phasing*. Another problem is that the experimental set $\{F^{obs}(\mathbf{s})\}$ of magnitudes is always limited in *resolution* (see Table 1):

$$for \quad \mathbf{s} \in S(\{F^{obs}(\mathbf{s})\} \text{ are available}), \quad |\mathbf{s}| = 1/d_{min}.$$

This limit d_{min} is defined by the experimental conditions, in particular by the wavelength of the X-rays used, and by the properties of the crystal, its internal disorder. In macromolecular crystallography the conventional resolution of the diffraction data set, 2-3 Å and lower, is insufficient to visualize individual atoms (minimal interatomic distances are about 1.0-1.5 Å) even when the phase estimates became known.

The uncertainty in the phases may be essentially reduced supposing that the electron density distribution is not an arbitrary function but one that may be presented by a sum of contributions from individual atoms (*atomicity* of the crystal). This atomicity plays the key role in phasing the data of small-molecules crystals. In macromolecular crystallography, it was shown a while ago [36] that even an incomplete and inexact atomic model helps in phasing and in further model improvement. At the same time, to build such an initial atomic model some relatively good phase values for structure factors of a resolution at least 2.5-3.0 Å are required (except a special case when a model of a homologous structure solved previously can be used). This situation presents a methodological loop for a number of structural studies, and some break is needed to cut it.

This article discusses an approach (Section 2) to solve this problem by constructing an auxiliary model composed from artificial scatterers called *dummy atoms*. Generally speaking, these dummy atoms have no relation to chemical atoms (for a recent review see [32]). This idea of a description of a molecule by such purely mathematical entities at a conventional resolution of 2-3 Å was extended then to extreme cases of very low and ultra-high resolution shown in Sections below.

2   Dummy Atoms and Phase Improvement

Interpretation of Fourier maps calculated with $F^{obs}(\mathbf{s})$ and the phase estimates $\varphi^{est}(\mathbf{s})$ is based on the fact that its peaks correspond to individual atoms when the phases are estimated relatively well and the resolution of the data set is high enough, about 1 Å. When the phase quality and resolution are not so high, the peaks correspond to whole atomic groups and no individual chemical atom can be distinguished. However, the principle of atomicity is still valid. This suggests building an auxiliary atomic model [4] where the 'atoms' positioned in the peaks of the map have no chemical meaning and are used only as an intermediate tool to calculate new phase values $\varphi^{mod}(\mathbf{s})$.

This nonlinear transformation of a Fourier map of a 2-3 Å to an atomic model gives a possibility to calculate $\varphi^{mod}(\mathbf{s})$ for all structure factors available and not only for those used for the initial map calculation. The main question is

Table 1. Crystallographic notion

Crystallographic notion	Mathematical object	Comments
$\rho(\mathbf{r})$, electron density distribution	3-dimensional periodic non-negative function	$\rho(\mathbf{r})$ in the crystal is generated mostly by atoms of the macromolecule plus by the solvent between the molecules
$\mathbf{F}(\mathbf{s}) = F(\mathbf{s})\ \exp i\varphi(\mathbf{s})$, structure factor	complex Fourier coefficient of the function $\rho(\mathbf{r})$	diffraction experiment gives the magnitudes $F^{obs}(\mathbf{s})$ for a set S={$\mathbf{s}$} of Fourier coefficients
$\mathbf{s} = (h,k,l)$, Miller indices	integer 3D index of the Fourier coefficient	
$d(\mathbf{s})$, resolution of the structure factor	$1/|\mathbf{s}|$	the smaller is $d(\mathbf{s})$, the higher is the resolution
d_{min}, resolution of the set of structure factors	min $d(\mathbf{s})$ for the set S of structure factors	d_{min} corresponds to the minimal size of a detail that can be distinguished at the corresponding Fourier synthesis
R-factor	$\dfrac{\sum |F^{obs}(\mathbf{s}) - F^{mod}(\mathbf{s})|}{\sum F^{obs}(\mathbf{s})}$	$F^{mod}(\mathbf{s})$ are structure factors calculated from some model; $F^{mod}(\mathbf{s})$ are considered to be optimally scaled to $\mathbf{F}^{obs}(\mathbf{s})$
R-free-factor		R-factor calculated for structure factors excluded from the refinement (control data set)
refinement	search for the values of model parameters that minimize the R factor or another crystallographic criterion	the criterion has a lot of local minima; a good enough starting point is necessary for a success of the refinement

whether $\varphi^{mod}(\mathbf{s})$ are better than $\varphi^{est}(\mathbf{s})$ and whether the addition of new structure factors with predicted phase values $\varphi^{mod}(\mathbf{s})$ indeed improves the quality of the starting image. (Crystallographers call these two processes of improving the phase estimates as *phase refinement* and *phase extension*, respectively). When an atomic model is built, one does not only 'interpret' the starting synthesis but gets a possibility to refine the parameters of this model to find a better one. This model improvement does not mean that the dummy atoms move toward the positions of 'chemical atoms' but simply that they reproduce better the structure factors. The atomic parameters are used only to calculated structure factors and are thrown away after this.

These models [23,25] are composed from dummy atoms, completely virtual objects allowed to be at any distance to each other and even to be superimposed. This feature is crucial to approximate large density peaks corresponding to atomic groups. An'optimized protocol of refinement of such a model [3,33] is another key point of the procedure. In order to extract the phase information from the refined model, a maximum-likelihood approach to estimate parameters of the corresponding phase probability distribution has been developed [23]. The whole procedure does not require high-resolution data neither for construction of the starting model nor for refinement.

It is worthy of noting that till the end of 70th crystallographers used mechanical models where each atom had a clear chemical meaning. The first dummy-atom models [4] were similar to those mechanical models in the sense that there were a number of restrictions on a mutual position of dummy atoms. These restrictions made the models not optimal for the phasing. On the contrary, another important feature of macromolecules to be branched chains of chemically linked atoms was successfully used [38,8] to improve dummy-atom models. With a progressive improvement of phases and corresponding Fourier maps, dummy atom models evolve into mixed models [25] when a part of a crystal cell may be already interpreted in terms of chemical atoms. Currently, when data quality and resolution are reasonably high, the full model of *chemical* atoms may be built automatically [19,27] starting from such dummy-atom models.

3   Dummy Atoms and *Ab Initio* Phasing

The term *ab initio* phasing is reserved in crystallography for the procedure of the initial determination of phases using a single set of $F^{obs}(s)$ and information of a general character, like atomicity of molecules and connectivity of atomic chains.

Ab initio phasing is a search for the exact phase values $\varphi^{exact}(s)$ for the given set of structure factors. In general, the less is dimensionality of the search space (the less is the number of structure factors with unknown phases) the more chances to find the correct solution in a limited time. For example, one may start phasing by selecting a few structure factors with largest values of magnitudes thinking that basically these Fourier coefficients form the image. We have shown previously that such images may be hardly interpretable [22]. Another possibility is to start phasing from lowest-resolution reflections at the resolution of 10, 20 Å or lower, depending on the size of the object. Of coarse, corresponding images do not show structural details but the obtained molecular shape and packing may serve as a starting point for further phase (image) improvement as well as may be the objects of interest themselves.

Dummy atoms models, already proven to be useful for phase improvement, are also important mathematical tools for such an *ab initio* phasing. It is clear that the models used for *ab initio* phasing of low-resolution reflections may be composed only from a very limited number of dummy atoms in order to avoid overfitting the data.

3.1   Low-Resolution Models

The simplest model of a globular molecule is a sphere with a uniform or Gaussian-distributed density inside [30,34,13,5]. More generally, a molecule of an arbitrary shape may be modeled by a few large spheres; that gives the name Few-Atoms Model (FAM) for this construction. These large spherical scatterers may be considered as a new kind of dummy atoms, of a much larger size, and again without any formal restriction on their mutual positions. The most trivial idea is that the FAM giving the lowest R-factor between $F^{mod}(\mathbf{s})$ and $F^{obs}(\mathbf{s})$ indicates the position of the molecule.

The multiple studies were done both for the simplest model composed from a single dummy atom and for models composed from several spheres. For both these groups of models it was observed that the positions of such dummy atoms giving the minimal R-factor value do not necessarily coincide with the center of the composing molecules or their principal domains [21]. In fact, this observation is in line with the basic idea of dummy atoms: do not try to interpret such models "chemically" but use them only as an intermediate tool to obtain the phases. Therefore, the goal of a modified procedure may be changed to a search for the FAMs that reproduce the best possible crystal structure factors regardless the chemical meaning of the positions of corresponding large dummy atoms. One may hope that a low R-factor means also a closeness of corresponding calculated phases to unknown searched values. If it is the case, R-factor might be used as a criterion to search for the correct phase set $\varphi^{exact}(\mathbf{s})$. Multiple checks with both calculated and experimental data show that this basic hypothesis generally is not held either.

3.2   Few Atoms Model Phasing

More detailed analysis of FAMs provides one with an important observation that if we generate a huge number of FAMs and take not a single best (by the R-factor value) FAM but several hundreds of best models, a typical distribution of corresponding phase sets may be described as follows.

Let's define a distance between 2 sets of phases as a closeness of two Fourier syntheses (for example, a least-squares distance between them) calculated with $F^{obs}(\mathbf{s})$ and these phase sets, respectively. With this measure, $\varphi^{mod}(\mathbf{s})$ calculated from the best FAMs are grouped in a very small number of clusters (usually 2 or 3) with one of them close to the correct solution. We already mentioned that the phase set for the FAM with the lowest R-factor does not necessary correspond to the correct cluster. Therefore, we cannot tell in advance which cluster is the correct one and we process all of them. For each cluster the phase sets are averaged giving for each structure factor its mean phase value and corresponding figure of merit [24]. This figure of merit reflects the dispersion of phase values among the sets. The 2-3 possible phase sets found from such averaging may be then analyzed one by one applying other criteria. Otherwise, the final choice may be done at later steps after eventual phase extension of all variants to a higher resolution. Multiple models composed from a smaller dummy atoms generated inside alternative envelopes may be used as such a selective statistical tool [20,28].

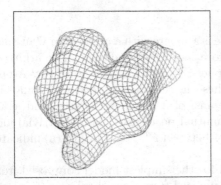 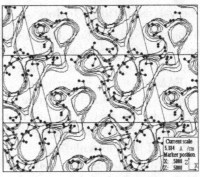

Fig. 1. Left: *ab initio* phased map at approximately 35 Å resolution for the ribosomal 50S particle from *T. thermophilus*, data by A. Yonath. Right: *ab initio* phased map at 14 Å resolution for myoglobin superimposed with Cα-atoms of the model.

The procedure [21,20] has been successfully applied to a number of experimental data sets, both with a known answer and for unknown structures (*e.g.*, Fig. 1).

4 Dummy Bond Electron Modeling

It is known from the study of small molecule crystals that in the Fourier maps of the resolution of 0.5-0.7 Å one may observe not only individual atoms but even the redistribution of the electron density from the isolated 'spherical atoms'. This redistribution is caused by formation of interatomic bonds; as a review, see [12]. This type of information is crucial for understanding the molecular mechanism, and several years ago Lecomte and his collaborators have started attempts of macromolecular studies at a subatomic resolution (for a review see [15]). Due to these density deformation effects conventional models of spherical atoms cannot reproduce well enough corresponding images and diffraction data as they do in conventional studies at the resolution of about 1-3 Å. By this reason, crystallographers switch to a multipolar model where the electron density distribution of an individual atom is no more spherical but is presented by a combination of spherical harmonics [6] (see also ref. in [12,18]). This approach is very efficient for studies of small molecules, at the same time its transfer to macromolecular studies is not so straightforward as it could be thought.

First, when one uses from the beginning the model composed from multipolar atoms, he imposes the presence of the bond density. Therefore, density peaks at the bonds are no more a feature of the experimental data but that of the model and the corresponding maps are no more an experimental proof of such a density (see *e.g.* [26]). Second, the multipolar models are described by about 30 parameters per atom instead of 10 parameters per atom for a model of spherical atoms with anisotropic ADPs. This increase in the number of parameters is not crucial for crystals of small molecules diffracting regularly to 0.5-0.6 Å; at such a resolution the number of $F^{obs}(\mathbf{s})$ is sufficient to refine multipolar models.

Currently for macromolecular crystals most of data sets available at a subatomic resolution have the resolution of 0.8-0.9 Å. At this resolution the number of $F^{obs}(\mathbf{s})$ is significantly smaller that at 0.5-0.6 Å, the multipolar models overfit the data and their refinement becomes unjustified. Third, the computational algorithms for multipolar modeling become too time consuming due to a huge growth of the number of parameters and structure factors at such resolutions.

Resuming the problem of macromolecular studies at subatomic resolution, we need a model that does not have many more parameters in comparison with the conventional models, for which the parts corresponding to density deformation can be easily removed and for which fast refinement algorithms can be applied or developed.

4.1 Dummy Bond Electrons

Alternatively to multipolar models that decompose the electron density into non-spherical 'atomic bricks', we may *complete* our conventional spherical-atom model by next-level 'mathematical' entities, new dummy atoms. These correcting scatterers may be at any moment removed from the model allowing to check if indeed they should be included or not. As previously, we do not care about physical interpretation of these new dummy atoms but use them as a mathematical tool to reproduce corresponding structure factors as good as possible.

The Fourier maps give the most of deformation density in the form of roughly spherical peaks at the given position near the 'chemical' atoms. Each such a complementary peak may be modeled by a Gaussian, that we call Dummy Bond Electron (DBE). Each DBE may require 2-3 parameters to be refined (shape of the peak and eventually its position along a given direction) resulting in total in 12-13 parameters per atom and not about 30 as in a multipolar model. In fact, in a number of studies of small molecular crystals a modeling of the deformation density peaks by point scatterers was tried previously (for example, [9,11,14,29]) but without a clear illustration of its advantages. We have proven the efficiency of DBE models in studies of macromolecular crystals applying various statistical criteria, *e.g.* R-free [10], that is possible here due to a very large number of experimental structure factor magnitudes available.

The electron density has been calculated by quantum-chemistry methods [31] for various individual amino-acid and nucleic-acid residues, the deformation density peaks were approximated by a Gaussian function forming a sort of a library (some similarity may be noted with the calculated database of multipolar models [35]). Then, for a given macromolecular model any set of DBE may be added at known positions with respect to heavy (non-hydrogen) atoms. We have shown also that fast refinement algorithms are easily applicable to DBE models [2] at a subatomic resolution.

4.2 Analysis of Density Deformation Maps and DBE Validation

In order to understand which molecular regions need DBE corrections, a series of tests with calculated and experimental data [7,37,17] has been carried out. They confirmed that even at the resolution of 0.9 Å the deformation density may be

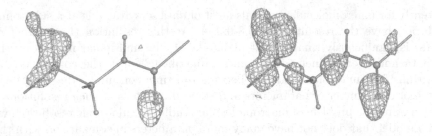

Fig. 2. Deformation density map calculated at the resolution of 0.9 Å for the protein RD1 [17]. The phases are calculated from a conventional model refined at 0.9 Å (left) and 0.6 Å (right).

Table 2. Comparison of R-factors for different models. Refinement of a multipolar model for Leu-enkephaline at 0.9 Å is unreasonable due to a too large number of parameters with respect to the number of data.

Crystal, model	resolution	Ndata	Nparam	R	Rfree
Leu-enkephaline	0.90 Å	1887			
- spherical anisotr.			474	2.91	3.70
- multipolar			1677	-	-
- DBE			658	2.17	3.17
Leu-enkephaline	0.56 Å	6894			
- spherical anisotr.			474	9.08	9.74
- multipolar			1677	7.90	8.63
- DBE			658	8.12	8.76
Antifreeze protein RD1	0.62 Å	106652			
- spherical anisotr.			3713	13.7	15.5
- DBE			5469	12.5	14.1

clearly observed. At the same time, as it has been pointed out previously [16], ADPs of bonded atoms (where DBE is supposed to be inserted) should not be too high. More interestingly, we noted [1] that these deformation density peaks are seen at the maps of 0.9 Å resolution if the model was refined at the resolution of 0.6 Å but they surprisingly disappear if the refinement was done only at 0.9 Å. In fact, refinement at the resolution of 0.9 Å or lower artificially increases ADPs values, and these too large values "hide" the deformation density. At the same time refinement at a higher resolution makes ADPs values more correct (Fig. 2).

Our tests confirmed that including the DBEs improves the quality of the models (Table 2). Interestingly, this improvement was observed even at a higher resolution while initially the DBE model was aimed to work at about 0.9 Å. As expected, the presence of DBEs during refinement corrects ADPs of heavy atoms and the phases calculated from the model. This improvement conserves the deformation-density peaks in Fourier maps of a resolution of at 0.9 Å even when the DBEs are excluded from the model at the stage when the structure factors are computed.

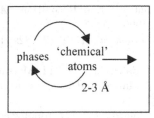
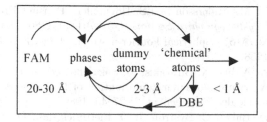

Fig. 3. Schematic presentation of the classical situation (left) and the 'cut' of this loop using dummy atoms (DA) at different scales (right). One FAM-scale DA corresponds to a molecular domain or to a whole molecule; one DBE-scale DA corresponds to a part of an atomic density.

5 Conclusion

This article shows the variety of applications of models composed from artificial scatterers (dummy atoms) that are irrelevant to chemical atoms. The original idea is based on a similarity of dummy atoms and chemical atoms in size. At the same time, in the direct-phasing FAM method each dummy atom is comparable in size with an atomic domain or the whole molecule while DBEs simulate small parts of the electron density of individual atoms (Fig. 3). All together, these techniques clearly illustrate the importance of a 'different-scale atomicity' in macromolecular crystallography and usefulness of mathematical abstractions for structural modeling.

A part of this work were supported by the CNRS - RAS collaboration and by PICS-3198/RFBR-05-01-22002. AU thanks Pole "Intelligence Logicielle" and CRVHP, LORIA, Nancy for financial support. VYL and NL were supported by RFBR grant 03-04-48155. AU is member of GdR 2417 CNRS. The authors thank E.Vernoslova, T.Petrova, T.Skovoroda, A.Podjarny and N.Muzet for contribution to different parts of the work presented in this article, and N. Messai for help in preparation of the Latex version of the manuscript.

References

1. P. V. Afonine, V. Y. Lunin, N. Muzet, and A. Urzhumtsev. On the possibility of the observation of valence electron density for individual bonds in proteins in conventional difference maps. *Acta Crystallographica Section D*, 60(2):260–274, Feb 2004.
2. P. V. Afonine and A. Urzhumtsev. On a fast calculation of structure factors at a subatomic resolution. *Acta Crystallographica Section A*, 60(1):19–32, Jan 2004.
3. R. C. Agarwal. A new least-squares refinement technique based on the fast Fourier transform algorithm. *Acta Crystallographica Section A*, 34(5):791–809, Sep 1978.
4. R. C. Agarwal and N. W. Isaacs. Method for obtaining a high resolution protein map starting from a low resolution map. *Proc. Natl. Acad. Sci. USA*, 74:2835–2839, 1977.

5. K. M. Andersson and S. Hovmöller. Phasing Proteins at Low Resolution. *Acta Crystallographica Section D*, 52(6):1174–1180, Nov 1996.

6. M. Atoji. Spherical Fourier method. *Acta Crystallographica*, 11(12):827–829, Dec 1958.

7. A. Aubry, N. Birlirakis, M. Sakarellos-Daitsiotis, C. Sakarellos, and M. Marraud. A crystal molecular conformation of Leucine-enkephalin related to the morphine molecule. *Biopolymers*, 28:27–40, 1989.

8. D. Baker, C. Bystroff, R. J. Fletterick, and D. A. Agard. *PRISM*: topologically constrained phased refinement for macromolecular crystallography. *Acta Crystallographica Section D*, 49(5):429–439, Sep 1993.

9. R. Brill. On the influence of binding electrons on X-ray intensities. *Acta Crystallographica*, 13(3):275–276, Mar 1960.

10. A. T. Brunger. Free R value: a novel statistical quantaty for assessing the accuracy of crystal structures. *Nature*, pages 472–475, 1992.

11. P. Coppens and M. S. Lehmann. Charge density studies below liquid nitrogen temperature. II. Neutron analysis of *p*-nitropyridine *N*-oxide at 30 K and comparison with X-ray results. *Acta Crystallographica Section B*, 32(6):1777–1784, Jun 1976.

12. P. Coppens, Z. Su, and P. Becker. *International Tables for Crystallography*, volume C, chapter Analysis of charge and spin density, pages 706–727. Kluwer Academic Publishers, Dordrecht-Boston-London, 1999.

13. G. W. Harris. Fast *ab initio* calculation of solvent envelopes for protein structures. *Acta Crystallographica Section D*, 51(5):695–702, Sep 1995.

14. E. Hellner. A simple refinement of density distributions of bonding electrons. I. A description of the proposed method. *Acta Crystallographica Section B*, 33(12):3813–3816, Dec 1977.

15. C. Jelsch, B. Guillot, A. Lagoutte, and C. Lecomte. Advances in protein and small-molecule charge-density refinement methods using *MoPro*. *Journal of Applied Crystallography*, 38(1):38–54, Feb 2005.

16. C. Jelsch, V. Pichon-Pesme, C. Lecomte, and A. Aubry. Transferability of Multipole Charge-Density Parameters: Application to Very High Resolution Oligopeptide and Protein Structures. *Acta Crystallographica Section D*, 54(6 Part 2):1306–1318, Nov 1998.

17. T. P. Ko, H. Robinson, Y. G. Gao, C. H. C. Cheng, A. L. DeVries, and A. H. J. Wang. The refined crystal structure of an Eel pout type III antifreeze protein RD1 at 0.62-A resolution reveals structural microheterogeneity of protein and solvation. *Biophysical J.*, 84:1228–1237, Jan 2003.

18. K. Kurki-Suonio. *Structural studies of crystals: on the occasion of the 75-th birthday of Academician B.K.Vainshtein*, chapter Alternative philosophy of charge density, pages 46–64. Nauka, Fizmatlit, Moscow, 1996.

19. V. S. Lamzin and K. S. Wilson. Automated refinement of protein models. *Acta Crystallographica Section D*, 49(1):129–147, Jan 1993.

20. V. Y. Lunin, N. L. Lunina, T. E. Petrova, A. G. Urzhumtsev, and A. D. Podjarny. On the *Ab Initio* Solution of the Phase Problem for Macromolecules at Very Low Resolution. II. Generalized Likelihood Based Approach to Cluster Discrimination. *Acta Crystallographica Section D*, 54(5):726–734, Sep 1998.

21. V. Y. Lunin, N. L. Lunina, T. E. Petrova, E. A. Vernoslova, A. G. Urzhumtsev, and A. D. Podjarny. On the *ab initio* solution of the phase problem for macromolecules at very low resolution: the few atoms model method. *Acta Crystallographica Section D*, 51(6):896–903, Nov 1995.

22. V. Y. Lunin, N. L. Lunina, and A. G. Urzhumtsev. Seminvariant density decomposition and connectivity analysis and their application to very low resolution macromolecular phasing. *Acta Crystallographica Section A*, 31:23–32, 1999.
23. V. Y. Lunin and A. G. Urzhumtsev. Improvement of protein phases by coarse model modification. *Acta Crystallographica Section A*, 40(3):269–277, May 1984.
24. V. Y. Lunin, A. G. Urzhumtsev, and T. P. Skovoroda. Direct low-resolution phasing from electron-density histograms in protein crystallography. *Acta Crystallographica Section A*, 46(7):540–544, Jul 1990.
25. V. Y. Lunin, A. G. Urzhumtsev, E. A. Vernoslova, Y. N. Chirgadze, N. A. Neveskaya, and N. P. Fomenkova. Phase improvement in protein crystallography using a mixed electron density model. *Acta Crystallographica Section A*, 41(2):166–171, Mar 1985.
26. P. Main. A theoretical comparison of the beta, gamma' and 2Fo-Fc syntheses. *Acta Crystallographica Section A*, 35(5):779–785, Sep 1979.
27. A. Perrakis, R. Morris, and V. S. Lamzin. Automatic protein model building combined with iterative structure refinement. *Nature Structural Biology*, 6:458–463, 1993.
28. T. E. Petrova, V. Y. Lunin, and A. D. Podjarny. *Ab initio* low-resolution phasing in crystallography of macromolecules by maximization of likelihood. *Acta Crystallographica Section D*, 56(10):1245–1252, Oct 2000.
29. U. Pietsch. X-ray bond charge in GaAs and InSb. *Phys.Stat.Sol.(b)*, 103:93–100, 1981.
30. A. D. Podjarny, B. Rees, J. C. Thierry, J. Cavarelli, J. C. Jesior, M. Roth, A. Lewitt-Bentley, R. Kahn, B. Lorber, J.-P. Ebel, R. Giege, and D. Moras. Yeast tRNAAsp - Aspartyl-tRNA Synthetase Complex: Low Resolution Crystal Structure. *J. Biomol. Struct. and Dynamics*, 5:187–198, 1987.
31. D. Sanchez-Portal, P. Ordejon, E. Artacho, and J. Soler. Density-functional method for very large systems with LCAO basis sets. *Int. J. Quant. Chem*, pages 453–461, 1997.
32. A. Urzhumtsev and V. Y. Lunin. Dummy atom models in macromolecular crystallography. *Crystallography Reviews*, 10(4):319–343, 2004.
33. A. G. Urzhumtsev, V. Y. Lunin, and E. A. Vernoslova. *FROG* – high-speed restraint–constraint refinement program for macromolecular structure. *Journal of Applied Crystallography*, 22(5):500–506, Oct 1989.
34. N. Volkmann, F. Schlunzen, A. G. Urzhumtsev, E. Vernoslova, A. Podjarny, M. Roth, E. Pebay-Peyroula, Z. Berkovitch-Yellin, A. Zaytsev-Bashan, and A. Yonath. On ab-initio phasing of ribosomal particles at very low resolution. *Joint CCP4 and ESF-EACBM Newsletter on Protein Crystallography*, 49(1):129–147, Jan 1995.
35. A. Volkov, Y. A. Abramov, and P. Coppens. Density-optimized radial exponents for X-ray charge-density refinement from *ab initio* crystal calculations. *Acta Crystallographica Section A*, 57(3):272–282, May 2001.
36. K. D. Watenpaugh, L. C. Sieker, J. R. Herriott, and L. H. Jensen. Refinement of the model of a protein: rubredoxin at 1.5 Å resolution. *Acta Crystallographica Section B*, 29(5):943–956, May 1973.
37. R. Wiest, V. Pichon-Pesme, M. Benard, and C. Lecomte. Electron distributions in peptides and related molecules. Experimental and theoretical study of Leu-enkephalin trihydrate. *J.Phys.Chem*, 98:1351–1362, Jan 1994.
38. C. Wilson and D. A. Agard. *PRISM*: automated crystallographic phase refinement by iterative skeletonization. *Acta Crystallographica Section A*, 49(1):97–104, Jan 1993.

BioRegistry: A Structured Metadata Repository for Bioinformatic Databases

Malika Smaïl-Tabbone, Shazia Osman, Nizar Messai, Amedeo Napoli, and Marie-Dominique Devignes

UMR 7503 LORIA, BP 239, 54506 Vandœuvre-lès-Nancy, France
{smail, osman, messai, napoli, devignes}@loria.fr
http://www.loria.fr/equipes/orpailleur

Abstract. One of the major challenges in the post genomic era consists in exploiting the vast amounts of biological data stored in the numerous heterogeneous biological databases distributed worldwide. Most research projects in bioinformatics start with data retrieval from selected sources. However, identifying appropriate data sources is not trivial and requires the representation of the knowledge about data sources. We present here the BioRegistry project which aims at providing means to represent and exploit knowledge associated with biological databases. As a first step, a repository structure has been designed to organise metadata associated with databases consisting of five metadata categories: database identification, topics covered, quality information, access/availability, and tracking of the metadata. The BioRegistry model and its relationships with the DCMI (Dublin Core Metadata Initiative) are described. Prototypes with various functionalities to feed, maintain and exploit the repository are presented.

1 Introduction

Biological datasets have tremendously grown in size and complexity in the past few years. Genome sequences, biomolecule structures, expression arrays, proteomics represent terabytes of data which are stored under various formats in distributed heterogeneous databases. More than 700 such databases have been listed at the beginning of the current year [1]. The extraction of knowledge from all these data is a crucial challenging task which ultimately gives sense to the tremendous data production effort with respect to domains such as evolution and disease understanding, biotechnologies, systems biology, pharmacogenomics, etc.

Knowledge discovery in databases (KDD) is a well-known process [2] that starts with two important steps: data selection from appropriate databases and data integration. In the biological domain, these tasks are hampered by various difficulties in terms of (i) identifying and characterising the relevant databases, (ii) designing data models to integrate the complex and distributed data. This paper deals with the first set of difficulties. We present here the BioRegistry project as a resource for cataloguing biological databases and facilitating relevant source discovery by querying and/or browsing.

M.R. Berthold et al. (Eds.): CompLife 2005, LNBI 3695, pp. 46–56, 2005.

After a short survey of the biological data integration context, we will explain the rationale of the project and present the model that has been designed to organise information about biological databases. We will then describe one attempt to automatically import the database descriptions from an existing resource. Implementation of various functionalities around the BioRegistry catalog will be presented and discussed in the perspective of future exploitation of this resource.

2   State of the Art

2.1   Biological Data Integration

Access to biological data in databases obviously necessitates, as a first step, the identification of relevant data sources. For example, the apparently simple query: "Which genes from the human X chromosome are preferentially expressed in the brain?" deals with both mapping and expression data which may or may not be contained in a single source at a given time. Most probably more than one data source can be found for each part of the query. The user may select one source because of a given quality criteria (e.g. manual revision of the data or update frequency) or availability information (e.g. access constraints). Once the relevant data sources have been selected, the user will need help for querying multiple data sources and getting integrated results.

Querying heterogeneous data sources and biological data integration have appeared as challenging problems in bioinformatics in 1995 [3,4,5]. Since then, numerous solutions have been proposed either through unified query interfaces (SRS, ENTREZ), data warehouses (GUS), database federations (SEMEDA [6], DISCOVERYLINK) or mediation architectures (TAMBIS, TINet, [7]). Web services are being developed today to standardise interactions with databases [8,9,10], thus allowing programs to automatically retrieve data from databases along with user-defined scenarios.

However, the choice of relevant data sources, given a user need, remains a major bottleneck, still poorly addressed by the expert himself. Who can claim to know the characteristics of all available biological databases at a given time? How can one express the criteria that will lead to the selection of the most relevant databases for a given query?

A few integrated architectures have dealt with the latter problem and modules capable of relating appropriate databases to user queries or sub-queries have been developed. In the mediation system TAMBIS [11] for instance, a knowledge base has been created to automatically associate query concepts and databases relative to these concepts. The TAMBIS ontology (TaO), which represents concepts in molecular biology and bioinformatics, is used to express both user query and source metadata in the same formalism so that queries can be automatically directed to matching sources. However, a dozen of databases only is taken into account by the system, so the usage of TAMBIS is rather limited.

A similar situation exists in the BioMediator architecture [12] in which a knowledge base contains the mediation schema represented as a hierarchy of concepts and a hierarchy of relations between concepts, annotations to explain how relations between data sources are obtained and maintained, and a catalog describing for each available data source the elements of the mediation schema they contain. Like in TAMBIS only a small number of sources, those that can be queried by the system, are described in the knowledge base. Other examples such as (BIS [13], BioDataServer [14], HKIS[15]) also illustrate that automatic source-query matching in mediation platforms yet only addresses a small number of pre-selected sources. Today, the exploitation of all available data sources still requires manual interaction between a user and a catalog of databases.

2.2   Existing Biological Databases Catalogs

The 2005 inventory of molecular biology databases published in NAR [1] is organised according to a pre-established hierarchy grouping together the databases according to a category list[1]: Nucleotide Sequence, RNA sequence, Protein sequence, Structure, etc. For each source, a summary paper is available with authors, citations, description and URL. Querying capabilities are still rather limited.

Thematic web portals such as the BioMed Central Database Gateway[2], the BioNetbook[3] at Pasteur Institute, the German site "bioinformatik"[4], Amos Bairoch's links at SwissProt[5], etc. provide access to numerous databases and resources. The classification provided by the portal may guide the user for selecting possible relevant databases. Manual exploration of the database sites and documentations is then necessary to refine the selection.

The DBCAT catalog, created and maintained by INFOBIOGEN [16], is probably the most structured catalog for molecular biology databases available so far (more than 500 databases). This flat file repository of structured metadata stores, for each database, information such as *Source Name, Domain covered by the source, Citation, Update Frequency, access URLs*. Another catalog named BioCAT has been designed in a similar manner for bioinformatic tools and is maintained at EBI in the frame of a collaboration between EBI and INFOBIO-GEN. In both resources, querying is possible through each field of the semi-structured format. However, apart from the Domain value, most field domains are open thus limiting the querying capabilities.

2.3   Rationale for the BioRegistry Project

This brief survey reveals the limits of existing solutions to the problem of identifying relevant data sources given a query or a user-need. To one extent (section 2.1), sophisticated integration models are designed to carry out this

---

[1] http://www3.oup.co.uk/nar/database/c/
[2] http://databases.biomedcentral.com/search
[3] http://www.pasteur.fr/recherche/BNB/bnb-en.html
[4] http://wwww.bioinformatik.de/cgi-bin/browse/Catalog/Databases/
[5] http://www.expasy.org/alinks.html

task. However, model complexity hampers large scale instantiation and the resulting systems poorly reflect the diversity of biological databases. To the other extent (section 2.2), users are faced with simple portals or catalogs, which give access to a large number of databases but offer quite poor query possibilities. More satisfying solutions should combine extensive representation of available databases and advanced discovery capabilities.

Inspiration may come from the closely related field of web services. In a web service architecture, the task of locating a relevant web service for a given application ("matchmaker" service) is usually performed inside a web service registry. The three well-known bioinformatic web service projects: MyGrid, Bio-Moby and Semantic Moby, have reported attempts to enrich the basic model of web service registry (UDDI) in order to augment the discovery capabilities (discussed in [17]). For instance, the MyGrid project has enriched the UDDI registry service with the ability of storing semantic metadata about the services it contains and has experimented with searches over this store driven by reasoning engine technology [18]. The main issue is then how to have all service providers registering their services with appropriate metadata and how to spread this augmented version of the registry service [19].

In the case of biological databases, not enough web services are yet deployed to allow retrieval of any desired data from any available database. We thus decided to create a biological databases registry called "BioRegistry", in which various metadata attached to biological databases are organised in a flexible and structured manner, enabling knowledge modelling about biological databases and advanced discovery capabilities. Various aspects of this work such as metadata valuation and exploitation using existing ontologies may reveal useful for web service registries.

3   The BioRegistry Model

Metadata (data about data) describe the content, quality, condition, and other characteristics of data. They play an important role in indexing, documentation and retrieval tasks. In 1995, an international committee of experts has proposed a standard model to describe metadata relative to web resources: the Dublin Core Metadata Initiative or DCMI [20]. This standard is composed of a core set of 15 elements including: *title, creator, subject, description, publisher, contributor, date, type, format, identifier, source, language, relation, coverage, and rights* [6].

Although the DCMI metadata model is intended to remain very simple and general, it provides two mechanisms that allow making more precise statements. Firstly, DCMI provides several "element refinements". For example *"created"* (dcterms:created) refines *"date"* (dc:date) to represent a date of creation. Secondly, DCMI defines several "encoding schemes" such as "vocabulary encoding schemes" which specify that a value is a term from a controlled vocabulary, or "syntax encoding schemes" that specify that a value is formatted in accordance

---

[6] http://dublincore.org/documents/dcmi-terms/

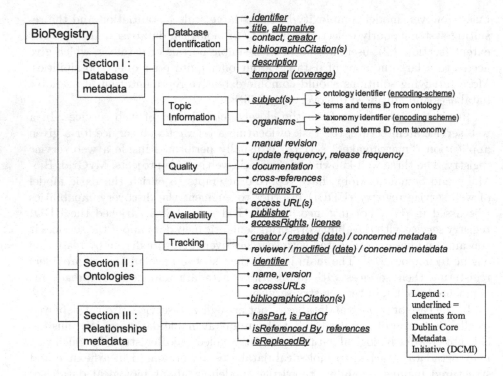

Fig. 1. Schematic representation of the BioRegistry metadata model

with some set of rules (e.g. date in the W3CDTF format YYYY-MM-DD). Nevertheless, describing resources in a particular domain still requires introducing some extensions. For instance, the Federal Geographic Data Committee (FGDC[7]) has built and approved in 1998 the Content Standard for Digital Geospatial Metadata. The complexity and specificities of biological databases also lead us to investigate which metadata could be attached to these databases and to propose a hierarchical model for organising these metadata. The BioRegistry metadata model (schematised in Figure 1) contains 3 sections. The first represents metadata associated with biological databases. The second describes ontologies and/or controlled vocabularies from which metadata terms can be extracted. The third is dedicated to relationships between databases. In this paper, we will mostly comment on the first section of the BioRegistry model. Choices of relevant metadata were performed by taking into consideration user needs. Five categories have been identified:

- Database identification: many DCMI elements have been used here: *identifier, title, alternative, creator, bibliographicCitation, description, temporal (coverage), created (date), modified (date)*.
- Topics covered by the database: this category is divided into two parts, the subjects covered by the data sources (DCMI element *subject*) and the

---

[7] http://www.fgdc.gov/fgdc/fgdc.html

organisms concerned (for example the Rat Genome Database contains data concerning the Rat organism).

- Database quality: in this category, many useful items are absent from the DCMI but crucial in the biological domain to document the quality of a database with respect to entry *revision* (manual or automatic), the existence of *documentation* and *cross-references* to other databases, *update and release frequencies*. The DCMI element refinement *conformsTo* was used to specify standard compliance (for example the MIAME standard for expression data)
- Database availability: this category contains the DCMI *publisher* element together with the various URL providing access to the database and description of access constraints for academic or industrial communities (free, registration required, fees).
- Metadata tracking: this category is aimed at tracking the possible modifications brought to metadata by the reviewers of the BioRegistry repository.

According to DCMI recommendations, standard data types are involved wherever possible (for example dates and time ranges at format W3CDTF). Most importantly, existing controlled vocabularies and/or domain ontologies are used to fill metadata fields where appropriate.

The field on *subjects* for instance contains terms extracted from the biomedical thesaurus MeSH, maintained by NLM[8]. This thesaurus was chosen because it is widely used to index scientific literature, it presents a broad coverage of many biological domains and is regularly updated to take into account changes in the topics addressed by scientific papers. It should be mentioned also that it is already present as a DCMI encoding scheme. However, more focused vocabularies/ontologies may also be used in the future. Concerning the field on *organisms* the NCBI taxonomy[9] of living organisms has been chosen since this taxonomy is also used to annotate biological sequences.

Besides the *Metadata* section, the BioRegistry model contains a section on *Ontologies* for describing and referencing the ontologies. Reference to the appropriate vocabulary/ontology is then associated with each term present in the fields on *subjects* and *organisms* as for DCMI encoding schemes. New vocabularies/ontologies can be added if needed.

The third section of the BioRegistry model will contain metadata representing relationships between databases. Here again some DCMI fields can apply: *hasPart, isPartOf, isReferencedBy, references, isReplacedBy, etc.*.

The BioRegistry has been implemented as an XML schema available at http://bioinfo.loria.fr/Members/devignes/Bioregistry/SchemaBioregistry. The hierarchical structure of the model is efficiently represented in the schema formalism. In addition, the schema specification allows one to define types and constraints on the metadata to enter into the BioRegistry, which may in turn facilitate editing of the BioRegistry content.

---

[8] http://www.nlm.nih.gov/mesh/
[9] http://www.ncbi.nlm.nih.gov/Taxonomy/

4   Populating the BioRegistry

In the first stage of the work, the inclusion of several databases in the BioRegistry repository has been performed manually. Examples of collected metadata are visible on the BioRegistry web page[10] for about 14 databases. To accelerate the process, an automatic procedure was designed to incorporate metadata from the DBCAT catalogue into the BioRegistry model. For some DBCAT fields the correspondence with BioRegistry elements is obvious. This is the case for the BioRegistry *Title, Contact, bibliographicCitation, Description, update and release frequencies, accessURL* elements for which values were directly imported from the corresponding fields in the DBCAT file. In order to fill the BioRegistry topic information subsection, several algorithms are designed to further exploit DBCAT content. The constraint here is to translate the DBCAT information into controlled vocabulary terms (DCMI encoding-schemes). The main application field of a database is represented in the *Domain* field of the DBCAT catalogue (DNA, RNA, Protein, Genomic, Mapping, Protein Structure, Literature, Miscellaneous). This leads us to convert these metadata (as well as their misspelled, synonymous or multilingual forms) into a few MeSH terms (Table 1) to be entered in the *subjects* subsection of the BioRegistry repository. Additional MeSH terms are also retrieved from MedLine as those indexing the publications referred to in the DBCAT citation field. Analysis of the results actually reveal that this latter procedure yield quite abundant noise. Some filters should be included before entering all MeSH terms into the BioRegistry document.

Since the DBCAT catalogue does not contain any field related to the organisms concerned with the data in a given database, the DBCAT *Description* field is parsed to retrieve any matching terms with the NCBI taxonomy. Retrieved terms are entered in the field on *organisms* of the BioRegistry repository. This procedure reveals to be very helpful in extracting appropriate organism names as long as these are mentioned in the DBCAT *Description* field.

The DBCAT catalogue has not been updated since 2001. To avoid entering obsolete hyperlinks in the BioRegistry repository, each of the URLs extracted from the DBCAT files is tested before writing it into the BioRegistry file.

Automatically created XML files (one per database, i.e. 509) are currently being manually checked and curated thanks to an editor, developed as a java application (BioRegistry Metadata Editor) and capable of checking the schema-specified constraints. Once curated and validated, individual XML files can be imported into the BioRegistry repository.

Additional automatic or semi-automatic procedures to populate and update the BioRegistry will be developed in the future. Exploitation of the Nucleic Acids Research 2005 compilation of molecular biology databases maintained at NCBI [1] is envisaged. Alert and survey mechanisms have to be designed to detect any change or new release in existing databases as well as new databases appearing on the web.

---

[10] http://bioinfo.loria.fr/Members/devignes/Bioregistry/presentationBioregistry/

Table 1. Correspondence between DBCAT Domain values and MeSH terms to be entered into the BioRegistry repository

Domain	Derived values	MeSH term	TermID	Tree Number
DNA	Adn Dna	**DNA**	D004247	D13.444.308
RNA	Rna	**RNA**	D012313	D13.444.735
Protein	PROT Prot Proteins PROTEIN	**Proteins**	D011506	D12.776
Genomic	GENOMIC GENOMICS Pathway maps	**Genomic**	D023281	G01.273.343.350
Mapping	MAP	**Chromosome Mapping**	D002874	E05.393.183
Protein Structure	Protein structure (3D)	**Protein Conformation**	D011487	G06.184.603.790.709
Literature	LIT Lit Litterature	**Information Services**	D007255	L01.453
Miscellaneous	Misc MISC	**None**		

5   Querying the BioRegistry

A first exploitation of the BioRegistry is form-based querying, triggering structured information retrieval of the metadata. This task is highly analogous to an information retrieval problem in which databases, instead of documents, would be searched for, and where indexation would be based on metadata reflecting information about the databases rather than on the data extracted from documents. In addition to the topics addressed by the databases, user queries may involve other criteria such as data quality (documentation, update frequency, manual revision, etc.) or data availability (access constraints, etc.). The BioRegistry should allow the biologist to formulate a multi-criteria query combining various metadata categories and to recover a sorted list of data sources with metadata matching more or less the query. A similarity calculation measure for matching attribute-value pairs will be used to perfom the sorting of the BioRegistry sources with regard to the user-query. This measure will be built according to the local-global principle which consists in defining local similarity measures on the different metadata fields (or attributes) and choose an aggregation (or amalgamation) function to define a global similarity measure. In particular, for an ontology-based metadata (i.e., *subjects* and *organisms*), the local similarity measure will take into account the hierarchical or taxonomic links between the terms [21].

Browsing through the BioRegistry repository is an alternative to form-based querying in the process of database discovery. The structured organisation of metadata in the BioRegistry model allows easy extraction of various sets of databases and/or metadata, thus offering numerous possibilities to create customised views over the biological databases. Once a given set of databases and metadata has been selected (for example the "subjects" of all the databases in the repository, or the metadata associated with only the databases dealing with "human" organism), methods such as formal concept analysis [22,23] can be adopted to visualise the sharing of metadata across the databases. An attempt to represent the BioRegistry content in the frame of formal concept analysis, inspired by the work [24] in the field of information retrieval, has been published elsewhere [25,26]. In both approaches, controlled vocabularies and ontologies, used to fill the fields on subjects and organisms, can be exploited as a means to query re-formulation and/or refinement in order to improve the recall as in [27,28].

6 Discussion

The metadata model described here for biological databases is the core component of the BioRegistry project. The first objective fulfilled by this component is to facilitate and optimise the selection of relevant databases to query in a given context. Efforts are underway to populate this repository in the most exhaustive and updated manner. Contacts with scientific and technical information (STI) institutions such as INIST (http://www.inist.fr/) in France and NCBI in the USA have been made. An international committee should be set up to propose this description of biological databases metadata as a standard to the bioinformatic community. Ideally in the future, any person involved in the construction or maintenance of a biological database should be able to fill in a BioRegistry submission form online in order to enter his database into the repository.

The next objective of the BioRegistry project is to offer a mediation possibility to relevant databases and to assist users in the design and execution of scenarios/workflows. This will require (i) implementing and exploiting the third section of the BioRegistry model concerning relationships between databases, (ii) enriching the BioRegistry model with a description of the programming interface required for invoking a database. Ultimately, this will enable the BioRegistry project to take into account biological web services.

Acknowledgments

This work was funded by grants from Region Lorraine (PRST Intelligence Logicielle). We are grateful to G. Vayssex for permission to export DBCAT data.

Special thanks to Marie Jacquot, Nadine Mercier and Hanane Moustain for developing the DBCAT to BioRegistry migration application, and to Mickael Lambotte for the BioRegistry Metadata Editor. N. Messai benefited from a fellowship co-financed by Région Lorraine and Communauté Urbaine Grand Nancy.

References

1. Galperin, M.Y.: The Molecular Biology Database Collection: 2005 update. Nucleic Acids Research **33** (2005) National Center for Biotechnology Information and National Library of Medicine and National Institutes of Health.
2. Frawley, W.J., Piatetsky-Shapiro, G., Matheus, C.J.: Knowledge discovery in databases: An overview. In: Knowledge Discovery in Databases. AAAI/MIT Press (1991) 1–30
3. Davidson, S.B., Overton, G.C., Buneman, P.: Challenges in Integrating Biological Data Sources. Journal of Computational Biology **2** (1995) 557–572
4. Karp, P.D.: A strategy for database interoperation. Journal of Computational Biology **2** (1995) 573–586
5. Markowitz, V.M.: Heterogeneous molecular biology databases. Journal of Computational Biology **2** (1995) 537–538
6. Kohler, J., Philippi, S., Lange, M.: SEMEDA : ontology based semantic integration of biological databases. Bioinformatics **19** (2003) 2420–2427
7. Eckman, B.A., Kosky, A.S., Leonardo A. Laroco, J.: Extending traditional query-based integration approaches for functional characterization of post-genomic data. Bioinformatics **17** (2001) 587–601
8. Buttler, D., Coleman, M., Critchlow, T., Fileto, R., Han, W., Pu, C., Rocco, D., Xiong, L.: Querying Multiple Bioinformatics Information Sources: Can Semantic Web Research Help? SIGMOD Record **31** (2002) 59–64
9. Wroe, C., Stevens, R., Goble, C., Roberts, A., Greenwood, M.: A suite of DAML+OIL Ontologies to Describe Bioinformatics Web Services and Data. International Journal of Cooperative Information Systems **12** (2003) 197–224
10. Oinn, T., Addis, M., Ferris, J., Marvin, D., Greenwood, M., Carver, T., Matthew, Pocock, Wipat, A., Li, P.: Taverna : a tool for the composition and enactment of bioinformatics workflows. Bioinformatics **20** (2004) 3045–3054
11. Goble, C.A., Stevens, R., Ng, G., Bechhofer, S., Paton, N.W., Baker, P.G., Peim, M., Brass, A.: Transparent Access to Multiple Bioinformatics Information Sources. IBM Systems Journal **40** (2001) 532–551
12. Shaker, R., Mork, P., Brockenbrough, J., Donelson, L., Tarczy-Hornoch, P.: The biomediator system as a tool for integrating biological databases on the web. In: Proceedings of the Workshop on Information Integration on the Web (held in conjunction with VLDB 2004), Toronto (2004)
13. Lacroix, Z., Boucelma, O., Essid, M.: The biological integration system. In: WIDM '03: Proceedings of the 5th ACM international workshop on Web information and data management, New York, NY, USA, ACM Press (2003) 45–49
14. Freier, A., Hofestädt, R., Lange, M., Scholz, U., Stephanik, A.: Biodataserver: A sql-based service for the online integration of life science data. In Silico Biology **2** (2002) 5
15. Boulakia, S.C., Lair, S., Stransky, N., Graziani, S., Radvanyi, F., Barillot, E., Froidevaux, C.: Selecting biomedical data sources according to user preferences. Bioinformatics **20** (2004) i86–i93

16. Discala, C., Benigni, X., Barillot, E., Vaysseix, G.: DBCAT: a catalog of 500 biological databases. Nucleic Acids Research **28** (2000) 8–9

17. Lord, P., Bechhofer, S., Wilkinson, M.D., Schiltz, G., Gessler, D., Hull, D., Goble, C., Stein, L.: Applying semantic web services to Bioinformatics: Experiences gained, lessons learnt. In Sheila A. McIlraith, Dimitris Plexousakis, F.v.H., ed.: The Semantic Web ISWC 2004: Third International Semantic Web Conference, Hiroshima, Japan, November 7-11, 2004. Proceedings. Volume 3298., Springer-Verlag GmbH (2004) 350–364

18. Lord, P., Wroe, C., Stevens, R., Goble, C., Miles, S., Moreau, L., Decker, K., Payne, T., Papay, J.: Semantic and personalised service discovery. In Cheung, W., Ye, Y., eds.: WI/IAT 2003 workshop on Knowledge Grid and Grid Intelligence, Halifax, Canada (2003) 100–107

19. Oinn, T., Addis, M., Ferris, J., Marvin, G., Greenwood, M., Carver, T., Wipat, A., Li, P.: Taverna, lessons in creating a workflow environment for the life science. In: Proceedings of GCF Workflow Workshop, Berlin (2004)

20. Dekkers, M., Weibel, S.: State of the dublin core metadata initiative. D-Lib Magazine **9** (2003)

21. Bergmann, R.: Highlights of the european inreca projects. In: Proceedings of the 4th International Conference on Case-Based Reasoning. (2001) 1–15

22. Ganter, B., Wille, R.: Formal Concept Analysis. Mathematical Foundations, Springer-Verlag (1999)

23. Carpineto, C., Romano, G.: Concept Data Analysis: Theory and Applications. John Wiley & Sons (2004)

24. Carpineto, C., Romano, G.: Order-theoretical ranking. Journal of the American Society for Information Science **51** (2000) 587–601

25. Messai, N., Devignes, M.D., Napoli, A., Smal-Tabbone, M.: Treillis de concepts et ontologies pour l'interrogation d'un annuaire de sources de données biologiques (bioregistry). In: 18ème Congrès INFORSID 2005, Grenoble (2005)

26. Messai, N., Devignes, M.D., Napoli, A., Smal-Tabbone, M.: Querying a bioinformatic data sources registry with concept lattices. In: Proceedings of the 13th International Conference on Conceptual Structures (ICCS '05) Conceptual Structures: Common Semantics for Sharing Knowledge, Kassel, Germany (2005)

27. Safar, B., Kefi, H., Reynaud, C.: OntoRefiner, a user query refinement interface usable for Semantic Web Portals. In: Proceedings of Application of Semantic Web technologies to Web Communities, Workshop ECAI'04, Valencia, Spain (2004) 65–79

28. Messai, N.: Treillis de Galois et ontologies de domaine pour la classification et la recherche de sources de données génomiques. Rapport de dea informatique de lorraine, UHP-Nancy 1 (2004)

Robust Perron Cluster Analysis for Various Applications in Computational Life Science

Marcus Weber and Susanna Kube

Zuse Institute Berlin (ZIB), Germany

Abstract. In the present paper we explain the basic ideas of Robust Perron Cluster Analysis (PCCA+) and exemplify the different application areas of this new and powerful method. Recently, Deuflhard and Weber [5] proposed PCCA+ as a new cluster algorithm in conformation dynamics for computational drug design. This method was originally designed for the identification of almost invariant subsets of states in a Markov chain. As an advantage, PCCA+ provides an indicator for the number of clusters. It turned out that PCCA+ can also be applied to other problems in life science. We are going to show how it serves for the clustering of gene expression data stemming from breast cancer research [20]. We also demonstrate that PCCA+ can be used for the clustering of HIV protease inhibitors corresponding to their activity. In theoretical chemistry, PCCA+ is applied to the analysis of metastable ensembles in monomolecular kinetics, which is a tool for RNA folding [21].

1 Introduction

The application and improvement of cluster algorithms plays an important role in several areas of computational life science. Given a number of N objects $q \in \Omega$ with certain features, we are interested in identifying objects with similar behaviour in order to combine them into N_C clusters. For this purpose, we want to construct membership functions $y_i : \Omega \rightarrow [0, 1], i = 1, \ldots, N_C, N_C \ll N$, which form a partition of unity. Then, each object in Ω can be assigned to the clusters with certain weights given by the values of the membership functions. A cluster can be considered as a vector that remains almost invariant under the action of a matrix T, i.e.

$$T y_i \approx y_i. \tag{1}$$

In molecular dynamics, T is the discretised version of a spatial transition operator [14] and clusters are conformations for which the large scale geometric structure is conserved. In this case, the matrix T contains transition probabilities between different conformations. In general, T must be a row stochastic matrix. For example, it can result from the normalisation of a symmetric matrix whose entries represent some pairwise similarity measure, e.g. a covariance matrix.

Equation (1) is similar to an eigenvalue problem for an eigenvalue near $\lambda = 1$. A perturbation analysis shows that the space of eigenvectors of T corresponding

M.R. Berthold et al. (Eds.): CompLife 2005, LNBI 3695, pp. 57–66, 2005.

to eigenvalues near $\lambda = 1$ indicates a partition of Ω into the clusters we are looking for [4]. In Robust Perron Cluster Analysis, the space spanned by the membership functions y_i equals the space of the N_C first eigenvectors of T. In this case, the number N_C of clusters equals the number of discrete eigenvalues of T near $\lambda_1 = 1$. If each object is uniquely assigned to a cluster, then a rearranging of the rows and columns of T results in an almost block diagonal matrix. Therefore, the identification of clusters can also be seen as a detection of the almost block diagonal structure of T.

There are several other spectral methods which can be applied to reduce the dimensionality of given data. For example, Principle Component Analysis (PCA) and Independent Component Analysis (ICA) use the eigenvectors of a covariance matrix to compute a set of important directions within the data. However, they fail to separate non-overlapping data sets. An illustrative example can be found in [7]. PCCA+ was especially designed to identify spatially separated clusters and is close to Laplacian projection methods used in graph partitioning [20] [18], for example the relaxation of the *normalised cut* minimisation problem used by Shi and Malik [16] and the Multicut Algorithm by Meila and Shi [11]. The main differences between Robust Perron Cluster Analysis and these methods are:

- The results of Perron Cluster Analysis are given in terms of almost characteristic functions, i.e. fuzzy sets.
- These functions are a simple linear transformation of the eigenfunctions of the operator $\mathcal{T}$.
- There is a detailed perturbation analysis for the PCCA+ approach based on Markov chain theory, which provides robustness of this method.

2   Robust Perron Cluster Analysis Approach

The basis for Robust Perron Cluster Analysis is a stochastic matrix $T \in I\!R^{N \times N}$ with an eigenvalue cluster near 1. The clusters we are looking for are represented by vectors y_i, $i = 1, \ldots, N_C$, combined into a nonnegative matrix $Y \in I\!R^{N \times N_C}$. In order to meet the partition-of-unity constraint, Y has to be row stochastic, see also [3]. Since Y should fulfil

$$T y_i \approx y_i,$$

the idea of PCCA+ is to construct Y as a linear transformation of the matrix $X \in I\!R^{N \times N_C}$, which contains the N_C first eigenvectors of T corresponding to eigenvalues near $\lambda_1 = 1$, see [5]. Therefore, the task for PCCA+ is to find a corresponding transformation matrix $\mathcal{A} \in I\!R^{N_C \times N_C}$, such that

$$Y = \mathcal{A}X$$

is a nonnegative, row stochastic matrix. Since there are many feasible solutions, one searches for a solution which maximises the functional

$$\sum_{i=1}^{n} \frac{\langle y_i, T y_i \rangle_\pi}{\langle y_i, e \rangle_\pi} \to \max,$$

where $e = (1, \ldots, 1)$ is a constant vector, y_i is the ith column of Y and $\langle \cdot \rangle_\pi$ is a π-weighted inner product with the unique invariant row vector which meets $\pi = \pi T$. If the stochastic matrix T is the discretisation of a transition operator, then this is equivalent to the maximisation of metastability [5]. If the stochastic matrix is constructed based on a geometrical cluster problem (see below), then this optimisation problem minimises the overlap between different clusters. Instead of solving a constrained optimisation problem, another approach tries to find an optimal initial guess $\mathcal{A}$ wrt. the maximisation problem without regarding the non-negativity constraint for Y [20]. The smallest entry of Y, the so-called minChi-indicator, measures the feasibility of the initial guess as a solution of the clustering. This is also applied in order to determine N_C, i.e. the correct number of clusters. The minChi-indicator is used for the geometrical cluster problems shown in this paper.

For an application of Robust Perron Cluster Analysis in conformation dynamics see [5]. Now, we will give some other application examples for PCCA+.

3   Graph-Based Spectral Clustering via PCCA+

Suppose we want to cluster $N_o \in \mathbb{N}$ objects, each of them described by $N_f \in \mathbb{N}$ features given by real numbers. That means we have to apply PCCA+ to an $N_o \times N_f$ real valued object-feature-matrix X. As input for PCCA+, we need an $N_o \times N_o$ diagonalisable stochastic matrix T which measures the similarity between objects in some sense. For this purpose, T is constructed out of a symmetric nonnegative matrix $W \in N_o \times N_o$ by scaling its rows to row sum 1, see [20]. The symmetric matrix W can be seen as weight matrix for an undirected graph where each object is represented by a vertex. The pairwise similarities between these vertices are expressed by weights of the corresponding edges. One example for computing this weight matrix can be taken from our analysis of gene expression data [20] in cooperation with the Max Planck Institute for Molecular Genetics. With some parameter $\beta > 0$, the weight $W(i, j)$ of the edge between object i and object j is defined as

$$W(i, j) = \exp(-\beta \, d^2(i, j)),$$

where $d(i, j)$ denotes the standard Euclidean distance between the ith and the jth row of X interpreted as vector in the N_f-dimensional space.

As an example, we examined the expression data of $N_f = 2000$ genes taken from $N_o = 50$ breast cancer patients. As preprocessing, we rescaled the features to zero mean and variance 1. After constructing T, we applied PCCA+ and got two clusters[1] y_1, y_2. Each patient $i \in \{1, \ldots, N_o\}$ was assigned to the cluster

---

[1] The minChi-indicator also allowed more than two clusters, $N_C = 2$ has been chosen in order to compare the results of PCCA+ with results from literature [20].

Survival fit, 2 clusters

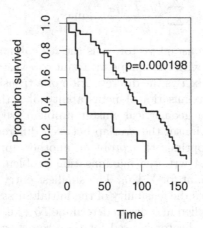

Fig. 1. Comparison of survival curves resulting from PCCA+ applied to gene expression data of breast cancer research

$k = 1, 2$, for which $y_k > 0.5$. In Figure 1 we compared the survival time of these two groups of patients and recognised a significant difference. The low p-value denotes the probability, that the difference of these two curves arises randomly. For a comparison with other clustering methods see [20].

A second example for a graph based clustering turns up in the research of HIV protease inhibitors. We examined data kindly provided by Martin Däumer and Rolf Kaiser from the Institute of Virology, Cologne University, and Joachim Selbig from the Department of Biochemistry and Biology at the University of Potsdam [1]. The aim of this project is to find out if structural similarities between different inhibitors imply functional similarities. In a first step it was examined how good $N_o = 7$ different protease inhibitors bind to $N_f = 2311$ different mutants of HIV protease which are described by their genotype. This

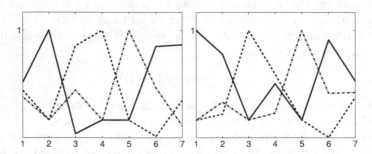

Fig. 2. Two clusterings of seven HIV protease inhibitors on the basis of 2311 HIV mutants. The computation of the activity coefficients differs between the two pictures. The three membership functions y_1, y_2 and y_3 are plotted as solid, dash and dash-dot line.

behaviour was measured by the activity coefficients. Our task was to identify those inhibitors with the same functional behaviour. For the computation of the 7×7-similarity matrix W, the pairwise correlation coefficients of the seven activity "vectors" have been shifted and normalised to the interval $[0, 1]$. Then the stochastic matrix T has been constructed and PCCA+ has been applied. The result was a 3-clustering of the protease inhibitors indicating different behaviour according to the HIV protease mutants. For a verification of the result, we used the fact that the activity coefficients can be computed in different ways. PCCA+ has been applied to these different activity coefficients and we always got similar results. In Figure 2 the results of two of the clusterings are shown. The x-axis shows the seven protease inhibitors. Their grades of membership, i.e. the curves for y_1, y_2 and y_3, are plotted in different line styles on the y-axis. Each HIV protease can be assigned to the cluster for which the corresponding grade of membership is maximal, i.e. for both experiments we get the result

$$\text{cluster}_1 = \{1, 2, 6, 7\}, \quad \text{cluster}_2 = \{3, 4\}, \quad \text{cluster}_3 = \{5\}.$$

Now it remains to examine the structural similarities between the different protease inhibitors which is still ongoing work. If it turned out that the structure of the inhibitors allows the same clustering, laboratory work could be done in a more tightly focused way.

4   Analysis of Metastable Ensembles in Monomolecular Kinetics

The understanding of transition pathways between different conformations of a molecule is an important issue in structural biology. Although the restriction of degrees of freedom to a few dihedral angles significantly reduces the complexity of the problem, this is still very difficult. Often, scientists are interested in single pathways, for example those over lowest energy barriers [2]. On the other hand, it is well known that molecular kinetics is not purely deterministic. All kinds of trajectories could appear, some with higher probability than others. Therefore, it seems natural to consider population probabilities. Starting with a given probability density in position space, we are interested in the evolution of the density to figure out intermediate states.

A description of molecular dynamics based on all conformations is unfeasible for large molecules. Therefore, we work with a set concept based on metastable conformations as introduced in [14]. First, we reduce the position space to a number of N states represented by basis functions [19] or boxes [15]. Then, we cluster states into metastable conformations by applying PCCA+ to the transition rate matrix Q. The infinitesimal generator Q of T^τ provides important chemical information concerning transition pathways of single molecules. Given an initial weighting x_A of the states, one can compute the corresponding weights and the spatial configuration density at each time step $t \in [0, \infty)$ via

$$\dot{x} = Q^\top x, \quad \text{with} \quad T^\tau = \exp(\tau Q). \tag{2}$$

This is the desired dynamic in configuration space, which is not based upon single molecules but upon ensembles.

It is easy to verify that the eigenvectors, which are essential for PCCA+, remain the same for the transition rate matrix Q. Assume, Q is diagonalisable by some nonsingular matrix X, i.e.

$$Q = X\Theta X^{-1} = X\mathrm{diag}(\theta_1, \ldots, \theta_p)X^{-1}.$$

Then

$$T^\tau = \exp(\tau Q) = X\exp(\tau\Theta)X^{-1} = X\mathrm{diag}(\exp(\tau\theta_1), \ldots, \exp(\tau\theta_p))X^{-1},$$

see [8]. Since $\exp(0) = 1$, an eigenvalue cluster of T^τ at 1 corresponds to an eigenvalue cluster of Q at 0. The number N_C of metastable sets is determined by this number of eigenvalues.

The entry $q_{ij}, i \neq j$, can be considered as the reaction rate of the monomolecular reaction

$$x_i \quad \rightharpoonup \quad x_j$$

where x_i stands representatively for the weight or "concentration" of state i. Equation (2) is not very interesting because the kinetics simply converges against the equilibrium distribution π. If one is interested in a simulation of a transition from metastable conformation A to a metastable conformation B and the corresponding transition behaviour, then (2) has to be solved as an initial value problem with initial distribution x_A and an absorbing end state given by the distribution x_B. Chemically, one would permanently eliminate conformation B out of the ensemble in order to push the reaction into the direction of this product. Mathematically this can be done by projection of x onto the orthogonal complement of the desired end point x_B before applying Q. Thus, the absorbing kinetics equation is:

$$\dot{x}(t) = Q^\top \left(x - \frac{\langle x, x_B \rangle}{\langle x_B, x_B \rangle} x_B\right), \quad x(0) = x_A. \tag{3}$$

The rate matrix Q can be obtained directly from the transition probability matrix T, but on the other hand, it offers a new approach to identify metastable conformations if the transition probability matrix is not available or difficult to compute. Furthermore, we are able to reduce our model not only to a set of basis functions whose number can be very large, but also to the few metastable sets which contain all important information about the system.

Example: n-Pentane. We present the application to the n-pentane molecule $CH_3(CH_2)_3CH_3$ which was modelled with Merck Molecular Force Field [9][10] at a temperature of $300K$. The rate matrix Q was calculated directly from the transition probability matrix T. T itself resulted from a conformation dynamics simulation with ZIBgridfree, a program package based on meshfree methods which was developed at Zuse-Institute Berlin, see [19],[12].

We found 9 eigenvalues of Q close to 1,

$$\lambda = \{1.0000, 0.9988, 0.9985, 0.9978, 0.9976, 0.9967, 0.9947, 0.9601, 0.9589\},$$

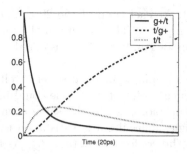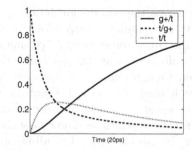

Fig. 3. Matlab [6] plot of a conformation kinetics simulation. *Left:* From g+/t conformation of pentane to the t/g+ conformation. *Right:* From t/g+ conformation of pentane to the g+/t conformation. Due to symmetry of pentane, both kinetics simulations should be equivalent. Differences result from unsymmetric approximations of transition probabilities.

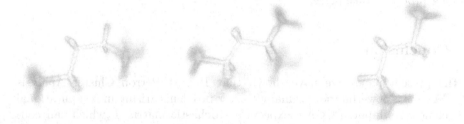

Fig. 4. Volume rendering of two conformations of pentane (left and right) and the corresponding transition macrostate (middle) in *amira/amiraMol* [17],[13]

followed by a gap to the 10th eigenvalue $\lambda_{10} = 0.8170$. This corresponds to 9 metastable conformations which can be distinguished according to the orientation of one of the two dihedral angles ($\pm$g and t denote the $\pm$ gauche and trans orientations):

$$\text{conformations} = \{-g/t, t/+g, -g/-g, t/t, t/-g, +g/t, +g/+g, -g/+g, +g/-g\}$$

The results for a $(g+/t) \rightharpoonup (t/g+)$ transition of pentane and the reverse experiment are shown in Figure 3. Only the concentrations of the conformations $(g+/t), (t/g+)$ and (t/t) are plotted. The corresponding Matlab algorithm needs less than 1 second CPU time for the computation of a 20ps reaction kinetics simulation with a 60×60-rate matrix Q, i.e. the numerical simulation of the "reduced model" is much faster than a full dynamics simulation of the same length. Figure 3 can be interpreted as follows. The conformational change from $(g+/t)$-pentane to $(t/g+)$ crosses the (t/t) conformation which can be seen as transition state. The transition from $(t/g+)$-pentane into $(g+/t)$-pentane is visualised in Figure 4. The left picture shows the start conformation $(t/g+)$, the right one the end conformation $(g +/t)$. At each step of the 20ps kinetics simulation, a similar density plot can be computed. The picture in the middle shows the transition state at 3.5ps simulation length. It can be considered as the mean conformation at this particular time.

Even though pentane is a very simple example, it illustrates very well the concept behind our method. From the chemical point of view, one could imagine that we start with a mixture of different molecules of the same chemical substance from which we know how the single molecules are distributed to the clusters. In this example, they all belong to the conformation $(g + /t)$. As time goes on, this distribution is driven towards equilibrium. Now, for example, suppose that molecules in a certain conformation are especially appropriate for a certain docking process, i.e. they do not contribute to the kinetics after this docking has taken place. This conformation is the target conformation of the reaction equation, here $(t/g+)$. The reaction kinetics calculation delivers information about the time scale of this process. Furthermore, it shows which other conformations are favoured in the meantime which can be of interest if several docking processes take place.

5   Conclusion

In the present paper, we have shown that Robust Perron Cluster Analysis (PCCA+) is a powerful tool for many cluster problems arising in computational life science. As input, PCCA+ expects a stochastic matrix T which can contain dynamics/kinetics information or similarity values from geometrical cluster problems. The aim of PCCA+ is to recover the almost block diagonal structure of T. The corresponding clustering is given in terms of a membership function for each of these "blocks". The number of almost-blocks in the matrix T need not to be known a priori. It is provided by the number of eigenvalues close to 1 or by the minChi-value. The property of the membership functions to be linear combinations of eigenfunctions allows their direct use in conformation kinetics. We prefer PCCA+ because it is easy to implement and has shown to be competitive with other clustering methods like Supervised Principal Component Analysis [20].

Acknowledgement. The authors especially want to thank Peter Deuflhard for various support of our work and for mathematical motivation. We also want to mention the cooperations with the group of M. Vingron at the Max Planck Institute for Molekular Genetics in Berlin, and the cooperation with J. Selbig at the University Potsdam and the Max Planck Institute for Molecular Plant Physiology. They provided us with application examples for PCCA+. Furthermore, we want to thank the group of P. Schuster from the Institute of Theoretical Chemistry at the University of Vienna for their support concerning conformation kinetics.

References

1. N. Beerenwinkel, B. Schmidt, H. Walter, R. Kaiser, T. Lengauer, D. Hoffmann, K. Korn, and J. Selbig. Diversity and complexity of HIV-1 drug resistance: A bioinformatics approach to predicting phenotype from genotype. *PNAS*, 99:8271–8276, 2002.
2. P. G. Bolhuis, C. Dellago, P. L.Geissler, and D. Chandler. Transition path sampling: throwing ropes over mountains in the dark. *Journal of Physics: Condensed Matter*, 12:A147–A152, 2000.
3. P. Deuflhard, W. Huisinga, A. Fischer, and Ch. Schütte. Identification of almost invariant aggregates in reversible nearly uncoupled Markov chains. *Lin. Alg. Appl.*, 315:39–59, 2000.
4. P. Deuflhard and Ch. Schütte. Molecular conformation dynamics and computational drug design. In J.M. Hill and R. Moore, editors, *Applied Mathematics Entering the 21th Century*. ICIAM 2003, Sydney, Australia, 2004.
5. P. Deuflhard and M. Weber. Robust Perron Cluster Analysis in Conformation Dynamics. In M. Dellnitz, S. Kirkland, M. Neumann, and Ch. Schütte, editors, *Lin. Alg. App. – Special Issue on Matrices and Mathematical Biology*, volume 398C, pages 161–184. Elsevier Journals, 2005.
6. TheMathWorks Inc. Germany. Matlab(R) 6.5.0, 1994–2005.
7. D. Gleich and L. Zhukov. Soft clustering with projections: PCA, ICA, and Laplacian. Technical report, California Institute of Technology, Computer Graphics Research, 2004.
8. G.H. Golub and C.F. van Loan. *Matrix Computations*. Johns Hopkins University Press, 3rd edition, 1996.
9. T.A. Halgren. *J. Am. Chem. Soc.*, 114:7827–7843, 1992.
10. T.A. Halgren. Merck molecular force field. *J. Comp. Chem.*, 17(I-V):490–641, 1996.
11. M. Meila and J. Shi. A random walks view of spectral segmentation. *AI and Statistics (AISTATS)*, 2001.
12. H. Meyer. Die Implementierung und Analyse von HuMfree–einer gitterfreien Methode zur Konformationsanalyse von Wirkstoffmolekülen. Master's thesis, Free University Berlin, February 2005.
13. J. Schmidt-Ehrenberg, D. Baum, and H.-C. Hege. Visualizing dynamic molecular conformations. In *IEEE Visualization 2002*, pages 235–242. IEEE Computer Society Press, 2002.
14. Ch. Schütte. *Conformational Dynamics: Modelling, Theory, Algorithm, and Application to Biomolecules*. Habilitation Thesis, Fachbereich Mathematik und Informatik, Freie Universität Berlin, 1999.
15. Ch. Schütte, A. Fischer, W. Huisinga, and P. Deuflhard. A direct approach to conformational dynamics based on hybrid Monte Carlo. *J. Comput. Phys., Special Issue on Computational Biophysics*, 151:146–168, 1999.
16. J. Shi and J. Malik. Normalized Cuts and Image Segmentation. *IEEE Transactions on Pattern Analysis and Machine Intelligence*, 22(8):888–905, 2000.
17. D. Stalling, M. Westerhoff, and H.-C. Hege. Amira - a highly interactive system for visual data analysis. In Christopher R. Johnson and Charles D. Hansen, editors, *Visualization Handbook*. Academic Press, November 2004.
18. D. Verma and M. Meila. A Comparison of Spectral Clustering Algorithms. Technical Report 03-05-01, University of Washington, 2003.

19. M. Weber. *Meshless Methods in Conformation Dynamics*. PhD thesis, Free University Berlin, 2005. In preparation.
20. M. Weber, W. Rungsarityotin, and A. Schliep. Perron cluster analysis and its connection to graph partitioning for noisy data. Technical Report ZR-04-39, Zuse Institute Berlin, 2004.
21. M. T. Wolfinger, W. A. Svrcek-Seiler, Ch. Flamm, I. L. Hofacker, and P. F. Stadler. Efficient computation of RNA folding dynamics. *J. Phys. A: Math. Gen.*, 37:4731–4741, 2004.

Multiple Alignment of Protein Structures in Three Dimensions

Evgeny Krissinel and Kim Henrick

European Bioinformatics Institute, Hinxton, Cambridge, CB10 1SD, UK
keb@ebi.ac.uk
http://www.ebi.ac.uk/msd-srv/ssm

Abstract. The paper describes the algorithm of multiple alignment of protein structures in 3D used in the EBI-MSD web service SSM (Secondary Structure Matching) located at URL given in the title. Structure alignment is known as a computationally hard procedure, with multiple alignment being considerably harder then a more conventional pairwise alignment. We base our approach on an efficient SSM algorithm for pairwise structure alignment, which allowed for multiple alignment of a considerably larger number of structures (up to 100), on comparison with alternative techniques, in real time.

1 Introduction

Comparison studies play an important role in structural biology. It is widely acknowledged that structural similarity is a clue for the identification of protein function and evolution. Often structural similarity is estimated by sequence identity, obtained in the course of sequence alignment, assuming that higher sequence similarity is a necessary condition for structures to be geometrically similar. Vast data on protein structures, accumulated in PDB over last decades, allow nowadays for a detail structure analysis. It was found (cf. Refs. [1,2,3,4]) that structural similarity is not a simple function of sequence identity. As appears, only 20% of identical residues in two chains is often sufficient for structures to be very similar.

This result implies that structure-related studies should use geometry-based tools whenever possible. A number of methods for the comparison of protein structures have been developed over last decade. Most of the effort was invested into algorithms for pairwise structure alignment [1, 5–18] but only a few techniques for the alignment of multiple structures in 3D have been reported [19–21].

In this paper, we describe the algorithm of multiple structure alignment employed in the EBI-MSD web-server SSM (found at URL given in the title). The server delivers both pairwise and multiple alignments of protein structures in 3D. The SSM's pairwise alignment algorithm was detailed in Ref. [1].

2 General Notes

Multiple structure alignment (MA) may be defined as identification of residues that occupy geometrically equivalent positions in all (more than 2) aligned struc-

M.R. Berthold et al. (Eds.): CompLife 2005, LNBI 3695, pp. 67–78, 2005.

tures. Geometrically equivalent residues are found in close proximity of each other, when structures are properly rotated and translated (superposed). Evidently, there are many different rotations and translations that put some of residues into superposition, so there are many different alignments. From that manifold, we focus on alignments which maximise a certain score function, which normally depends on the number of superposed residues and a measure of distance between the superposed structures.

Although the above definition is identical to that of pairwise alignment (PA), it is important to realize that, in general, multiple alignment *does not* reduce to the set of all-to-all pairwise alignments of given structures. Identification of geometrical equivalence is always a subject to certain criteria, and unless structural similarity is high, a small distortion of one structure may noticeably change the pairwise alignment. As a result, if ith residue of structure A, r_A^i, may be aligned to residue r_B^j of structure B, and the latter – to residue r_C^k of structure C, it does not necessarily mean that residues r_A^i and r_C^k may be also aligned. Only in the simplest case of highly similar structures, when geometrical equivalence of residues is established well within the used geometrical criteria, multiple alignment is given by the intersection of all-to-all pairwise alignments.

It follows from the above that multiple alignment almost always results in lower pairwise scores, so that structures appear more distant then would be concluded from the pairwise comparisons. On the other hand, MA is biased to spotting out structural features that are common for all aligned structures, therefore one may expect that multiple alignments are less affected by artefacts of employed geometrical criteria.

The problem of multiple structure alignment, just as that of PA, does not have an exact solution, and any solution is subject to accepted definitions of structural similarities and scores. There are no common agreements on the latter. All known methods (cf. Refs. [19,20,21]) use different techniques that try to improve a starting alignment chosen from initial pairwise all-to-all alignments, typically by chosing a pair of most similar structures and consecutive addition of closest structural neighbours to the alignment. In Ref. [21], MA is sought from initial PAs by Monte-Carlo moves representing indels. Neither of techniques guarantees convergence to optimal solution. We suggest an approach, which is based on iterative removal of structural elements that have least chances to get aligned, according to a heuristic score. After all non-aligneable structural elements are identified, the solution is refined by iterative multiple alignment of backbone C_α atoms.

3   Algorithm

3.1   Multiple Alignment of Structural Elements

In the following discussion, we define structural element as one or more secondary structure elements (SSE), found in a certain geometrical orientation to each other and ordered in the same way along aminoacid chain. We also assume that

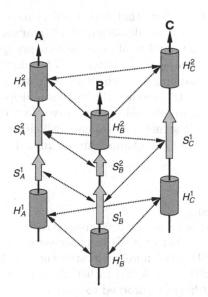

Fig. 1. Schematic of multiple alignment of structural elements ($H^k_{A,B,C}$ stand for helices and $S^k_{A,B,C}$ - for strands of chains A, B and C, respectively; arrows denote pairwise alignments of SSEs). Chains A and B may be unambiguously aligned, while there is an ambiguity of their alignment to chain C: any of strands $S^k_{A,B}$ may be aligned to S^1_C. Given that geometrically best pairwise alignments map S^2_A and S^1_B onto S^1_C, as shown in the Figure, the strands cannot be multiply aligned by a simple intersection of their pairwise alignments.

multiple alignment preserves the connectivity of structural elements, i.e. for any aligned pairs (S^i_A, S^j_B) and (S^m_A, S^n_B) (where superscripts denote the element's serial numbers and subscripts - chains) sign$(m - i)$ = sign$(n - j)$.

The problem of multiple alignment of structural elements is illustrated in Fig. 1. In this illustration, multiple alignment of helices may be unambiguously obtained as an intersection of their pairwise alignments, while strands do not seem to align because strands S^2_A and S^1_B in structures A and B, aligned to the only strand S^1_C in structure C, do not align to each other.

The above consideration, however, does not mean that one should not *try* to align strands in this example. Indeed, if pairs (S^2_A, S^2_B) and (S^2_A, S^1_C) were found as geometrically equivalent, one can assume that pair (S^2_B, S^1_C) could be also aligned, however with a lower pairwise score than that of pair (S^1_B, S^1_C). Similar reasonings lead to the conclusion that pair (S^1_A, S^1_C) could be aligned as well. It is not a rare situation in protein structure comparison that a particular structure element of one structure may be equivalenced with more than one element of another structure; should that be the case, solution with maximal pairwise score is chosen [1].

Therefore, it may be suggested, that, having the results of pairwise all-to-all alignments as a starting point, one possibly needs to remap those structural elements that may be connected by PA relations (dotted arrows in Fig. 1), but do not multiply align as an intersection of PAs. For the schematic in Fig. 1, one would need to choose from 4 remappings: (S^i_A, S^j_B, S^1_C), $i, j = 1, 2$, the one which maximises a defined MA score. Being apparently correct in general, this simple recipe has two main drawbacks. Firstly, remapping of structural elements changes the optimal orientation of structures and, as a result, pairwise scores for all structural elements also change. For example, remapping of strands in Fig. 1

might make some helices non-matching. This means that structural elements should be remapped gradually, one-by-one, with recalculation of all pairwise alignments after each remapping. Secondly, the number of possible remappings depends exponentially on the number of aligned structures. Extension of example in Fig. 1 onto 11 chains gives 2^{10} possible remappings if each chain, except one, has only two candidate strands for alignment. In practice, this makes multiple alignment of more than 10-15 structures computationally prohibitive. In this situation, our suggestion is to gradually exclude structural elements, which have least chances to get aligned, from consideration. The chances are estimated by a heuristic score, as described below.

Algorithm of multiple SSE alignment

1. Initialise an empty list $\mathcal{L}$ of excluded SSEs.
2. Calculate $N(N-1)/2$ pairwise alignments between all N given structures.
3. For each SSE $\notin \mathcal{L}$, calculate the total number of SSEs in other structures it is aligned to, P_x^i (i stands for the SSE serial number in structure x). If $P_x^i = N - 1$ for all i and x, then all SSEs not found in list $\mathcal{L}$ have been multiply aligned and algorithm quits. Otherwise, proceed to step 4.
4. For each SSE $\notin \mathcal{L}$ with $P_x^i < N - 1$, calculate the alignment score Q_x^i. We define this score as a sum of Q-scores in all pairwise alignments for the given SSE (cf. Eqs. (8,10) in Ref. [1]):

$$Q_x^i = \sum_y \sum_j \frac{\left(N_{xy}^{ij}\right)^2}{\left(1 + \left(RMSD_{xy}^{ij}/R_0\right)^2\right) N_x^i N_y^j} \tag{1}$$

where y enumerates structures, j enumerates SSEs in a structure, N_{xy}^{ij} is the number of aligned residues in ith SSE of structure x and jth SSE of structure y, $RMSD_{xy}^{ij}$ - r.m.s.d. of aligned residues, N_x^i and N_y^j are the total numbers of residues in the SSEs. R_0 is an empirical parameter measuring the importance of r.m.s.d. versus the alignment length, chosen at 3 Å [1].

5. Identify the least Q_x^i and place ith SSE of structure x into list $\mathcal{L}$. If all SSEs of structure x are found in list $\mathcal{L}$, then multiple alignment does not exist and algorithm quits. Otherwise, proceed to step 6.
6. Recalculate $N-1$ pairwise alignments between structure x and other structures, with SSEs found in list $\mathcal{L}$ excluded from consideration, and return to step 3.

As seen from the above, the described algorithm may be implemented using any method for pairwise alignment. Using the similarity Q-score is an empirical element of the algorithm. The score was chosen on the ground of observation, described in Ref. [1], that it represents a considerably better measure for structural similarity than the more conventional r.m.s.d. and alignment length. For the pairwise alignments, we employ the SSM algorithm [1], being encouraged by its efficiency and quality quoted recently in an independent study [22]. Because SSM algorithm is based on matching SSEs, it allows for efficient removal of non-matching SSEs from consideration in step 6 above.

3.2   Multiple C_α Alignment

Multiple alignment of structural elements yields a list of geometrically equivalent SSEs in the given structures. These data can be used for identifying common substructures in general and may be sufficient in some studies. A detail analysis of structural similarity requires structure alignment on the level of individual residues, including those not contained in SSEs. Below we describe an algorithm for multiple alignment of residues represented by their C_α atoms.

The algorithm follows the ideas of SSM algorithm for pairwise C_α alignment (SSM-PA), described in Ref. [1]. Using SSE alignment as an initial guess for the superposition of structures, SSM algorithm looks for pairs of C_α atoms which may be mapped onto each other such as to maximise a score function. Obtained alignment is then used for the calculation of improved superposition and the whole process is iterated until alignment does not change.

The SSM-PA algorithm may be adapted to multiple alignment after corresponding changes in its part that maps C_α atoms and redefinition of the score function. Below we discuss these changes and summarise the algorithm. In what follows, a_i stands for ith C_α atom of chain A and $|a_i, b_j|$ is distance between two atoms. We will also refer to groups of atoms, all from different chains, as $\mathcal{G}_{i,j,k...} = \{a_i, b_j, c_k ...\}$. A group is considered as mapped, if all atoms in the group are found to be in geometrically equivalent positions.

SSM-PA defines a pair of atoms (a_i, b_j) as mappable if they belong to compatible SSEs (see details in Ref. [1]) and $|a_i, b_j| \leq |a_i, b_m|$ and $|a_i, b_j| \leq |a_n, b_j|$ for any unmapped atoms a_n and b_m. This definition allows to identify the pair unambiguously and efficiently. Having sorted all pairs by increasing distance prior the mapping, SSM-PA builds optimal C_α alignment by mapping pairs one-by-one starting from top of the list. Each new pair is checked for the connectivity conflict with all previously mapped pairs, that is, for any two mapped pairs (a_i, b_j) and (a_n, b_m) the equality $\mathrm{sign}(n - i) = \mathrm{sign}(m - j)$ should hold.

In order to find mappable atoms in more than two chains, one has to define a distance measure for groups of atoms, $|\mathcal{G}|$. A few distance measures may be proposed, for example,

$$|\mathcal{G}| = \sqrt{\frac{1}{2N(N-1)} \sum_{x,y \in \mathcal{G}} |x,y|^2} \tag{2}$$

$$|\mathcal{G}| = \max_{x,y \in \mathcal{G}} |x,y| \tag{3}$$

$$|\mathcal{G}| = \max_{x \in \mathcal{G}} |x, \bar{g}| \tag{4}$$

where $\bar{g}$ is a central-mass atom in the group. One may see that a straightforward use of these or similar distance measures for mapping groups of atoms results in the evaluation of a large number of groups, even after introducing a reasonable distance cut-off. It may be shown that computation complexity of such algorithm is proportional to $N!$, which makes it unfeasible for $N > 5 - 8$ structures. Therefore, we suggest a simplified procedure for the identification of mappable groups of atoms.

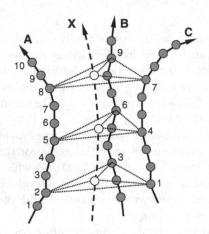

Fig. 2. Schematic of multiple C_α alignment. A fragment of superposed chains is shown in the Figure. Mapped groups of atoms are connected by dotted lines, chain X represents the consensus structure. Suppose that structure B is the closest one to X in pairwise score, then mapped groups $\{a_2, b_3, c_1\}$, $\{a_5, b_6, c_4\}$ and $\{a_8, b_9, c_7\}$ are identified as atoms b_3, b_6, b_9 and atoms from chains A and C closest to them. See text for details.

Introduce *consensus* structure X made of atoms placed in mass centers of the mapped groups (cf. Fig. 2). Next, find structure A^* that is closest to X in pairwise score (initially this structure may be defined as one with minimal sum of pairwise scores to other structures). Now one can identify mappable groups as those made from atoms a_i^* and atoms mappable to them in all other structures, chosen as in pairwise SSM alignment procedure [1], outlined above.

The proposed approach may be viewed as a simplified version of the central star method used in multiple sequence alignment (cf. Ref. [23]). While sequence alignment may be done in one pass, structure alignment involves recalculation of structure superposition after each alignment, which recalculation may change the choice of structure A^*. We found in a number of trial studies that this approach is a good approximation to the full-metric solution. Both approaches give identical answers for the alignment of structures with pronounced similarity, and a moderate number of differences (few percent of aligned residues) in case of dissimilar structures.

As noted in Ref. [1], not all mappings improve the alignment score. After all possible mappings are done, the algorithm should try to improve the alignment score by unmapping the groups with large distance measure $|\mathcal{G}|$. We define the alignment score as

$$Q = N_{align}^2 / \left\{ \left[1 + (D_\mathcal{G}/R_0)^2 \right] N_{min} N_{max} \right\} \tag{5}$$

where N_{align} is number of aligned groups, N_{min} and N_{max} are minimal and maximal number of residues in the aligned chains, R_0 is the same empirical parameter as in Eq. (1). $D_\mathcal{G}$ is calculated as r.m.s.d. of all mapped groups:

$$D_\mathcal{G} = \sqrt{\sum_{a_i^*} |\mathcal{G}_{...i...}|^2 / N_{align}} \tag{6}$$

where any of Eqs. (2-4) may be used for the calculation of $|\mathcal{G}_{...i...}|$. We use Eq. (2) because then Eq. (5) reduces to the pairwise Q-score [1] at number of structures $N = 2$.

Algorithm of multiple C_α alignment

1. Using the results of multiple SSE alignment, make initial superposition of structures and find structure A^* with least sum of pairwise Q-scores to other structures.
2. Calculate core C_α-alignment as an intersection of all pairwise alignment obtained in the last iteration of multiple SSE alignment.
3. Identify all mappable groups of atoms respecting to umapped atoms a_i^* as described above, and sort them by increasing the distance score $|\mathcal{G}_{...i...}|$. Starting from top of the list, map groups that do not have the connectivity conflict with all previously mapped groups.
4. Unmap groups in the reverse order until maximum value of Q-score, as defined by Eqs. (5,6), is reached.
5. Mapped groups of atoms represent a multiple alignment. If it does not differ from the one previously obtained then quit. Otherwise proceed to step 6.
6. Calculate consensus structure as mass centers of the mapped groups (see Fig. 2). Using algorithm for fast optimal superposition, described in Ref. [1], superpose all structures with the consensus structure.
7. Identify structure A^* which superposes with best pairwise score on the consensus structure, and proceed to step 2.

3.3 Implementation, Output Data and Scores

The described algorithm of multiple alignment of protein structures in three dimensions has been implemented as an additional function of the EBI-MSD web-server SSM, which also may be used as a standalone (off-line) application in in-house setups. The development is based on the new CCP4 Coordinate Library [24]. The output data include:

Alignment length: number of aligned groups of C_α atoms
Consensus r.m.s.d. and Q-score: r.m.s.d. and Q-scores of each structure alignment to consensus structure
Overall r.m.s.d. and Q-score: calculated as Eqs. (6) and (5), respectively
Superposition matrices: rotation-translation matrices of best structure superposition on consensus structure
Pairwise scores: $N \times N$ matrices of pairwise r.m.s.d., Q-score and sequence identity
SSE and C_α alignments: tables of aligned SSEs and residues.

All output data may be downloaded in XML or plain text format (and FASTA format for aligned sequences), superposed structures may be visualised using the Rasmol [25] software.

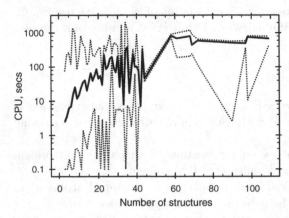

Fig. 3. CPU time as a function of the number of aligned structures obtained from the log of SSM server at EBI-MSD. All calculations were done on a single 1.2Ghz PC. Upper dotted line represents the maximum CPU time required, lower dotted line - minimum CPU time, and solid line gives the average. 95% of the data correspond to multiple alignment of up to 30 structures.

4 Results and Discussion

Fig. 3 represents data on the computational performance of the described algorithm, obtained from the log of SSM server at EBI-MSD for 2004-2005 year period. As may be seen from the Figure, calculation time is not a simple function of the number of aligned structures N. However, in the region of $3 \leq N \leq 30$, where most of the data have been collected, the average computation time has polynomial trend on N. In each particular case, calculation time also depends on the structure size (number of SSEs) and structural similarity: calculations are, on average, longer for larger and less similar structures. As may be seen from Fig. 3, these factors make a difference of more than 4 orders of magnitude.

The computational complexity of MA algorithm may be estimated as $O(N^2 n_m)$ times complexity of pairwise alignment, where n_m stands for the number of SSEs in the longest chain. Complexity of SSM-PA depends on structure topology and similarity and ranges from $O(nm)$ to $O(m^{n+1}n)$, where n, m are the numbers of SSEs in the aligned structures.

Figs. 4A,B present the typical results of multiple alignment. As seen from Fig. 4A, our MA algorithm is capable of discovering common substructures in different-fold structures, as defined by SCOP classification [28]. The β-sheet, common to all structures, was aligned with overall r.m.s.d. of 2.7Å and Q-score of 0.14, which implies a noticeable similarity. This similarity is present despite a rather low sequence identity of the aligned parts, which ranges from 0 for pair 1sar:A-1jqq:C to 0.14 for pair 1sar:A-1jy4:B.

Multiple alignment of same-family structures usually shows high structural similarity, as one would expect to obtain from SCOP classification. Fig. 4B demonstrates very clearly that structural differences occur only on protein surface, while internal parts match closely, forming a core of chain fold. The aligned parts were matched with overall r.m.s.d. of 1.55Å and Q-score of 0.53, which indicates a strong structural similarity. Sequence identity of the aligned structures in this example varies from 0.31 (4dfr:A-1dhf:A) to 1.0 (1ra8-5dfr).

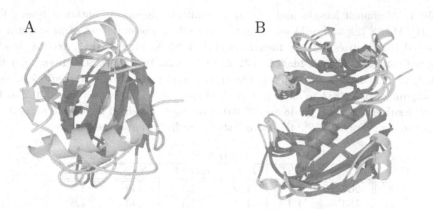

Fig. 4. Results of multiple alignment of A) different-fold structures 1sar:A, 1lqm:B, 1jqq:C and 1jy4:B (SCOP families d.1.1.2, d.17.5.1, b.34.2.1 and k.35.1.1, respectively), and B) same-family structures 4dfr:A, 1ra8, 5dfr, 1dhf:A, 1mvs:A, 1ia1:B and 1ia3:A (all belong to SCOP family c.71.1.1). Aligned parts are shown in dark grey. The pictures were obtained using Molscript [26] and Raster 3D [27] software.

Since there is no commonly accepted mathematical definition for multiple structure alignment, quality assessment of the results is difficult. A detail discussion of this question is outside the scope of present study. Table 1 shows a typical example of comparison of multiple alignments obtained from Combinatorial Extension [21], MASS [20] and SSM servers.

Visual inspection of the alignments reveals that the servers, in general, agree with each other. As seen from Table 1, SSM's alignments are somewhat longer than those from MASS at higher r.m.s.d. (alignment length in CE seems to be reported wrongly, see remarks in the Table caption). This fact means that, comparing to SSM, MASS is more willing to sacrify the alignment length in favour of lower r.m.s.d. The balance between N_{align} and $RMSD$ depends on empirical parameters (such as distance cut-off) used in particular algorithms, and, generally, is not an indicator of a method's quality or robustness. We discussed this question in details in Ref. [1].

Comparison of the servers performance may be done only with the following important remarks. Firstly, the run time depends drastically on the selection of aligned structures, which is demonstrated in Fig. 3 by a considerable difference between the maximal and minimal CPU time required for the alignment. Therefore, fair comparison may be done only using the averaged run times from the servers' logs, which are not available on-line. Secondly, in difference of SSM, CE-MA and MASS are not interactive servers. Instead, they deliver results by e-mail. We measured the response time of CE-MA as a difference between the "send" time tags of the e-mails confirming the submission and delivering the results. MASS does not confirm submission by e-mail, and we measured its response time as a difference between the actual time of delivery and delivery time for a MA of 3 identical structures, which is supposed to be very fast. All measurements were done in off-peak time period, without parallel submissions. The

Table 1. Alignment lengths and r.m.s.d. of multiple alignments obtained from CE-MA [21], MASS [20] and SSM servers (present study). The initial set of 24 structures contained PDB entries 4dfr:A, 1dyh:A, 1dyi:A, 1rb3:A, 2drc:B, 1ra3, 1re7:A, 1ra9, 1rx2, 5dfr, 1dg8:A, 1dg5:A, 1dhf:A, 1u70:A, 1dr2, 1hfq, 1u72:A, 1pd9:A, 1dyr, 1j3j:B, 1ia4:B, 1vj3:A, 1m78:A and 1t6t:2. The entries were picked from the results of pairwise alignment of 4dfr:A to all entries of PDB such that Q covers a range of 0.2 to 1. Then the subsets of 8, 12, 16, 18 and 20 structures were obtained by leaving every 3rd, and removing every 2nd, 3rd, 4th and 6th structure from the set, respectively.

N	CE		MASS		SSM	
	N_{align}	RMSD	N_{align}	RMSD	N_{align}	$RMSD^{‡}$
8	205*	1.2	130	1.1	146	1.5
12	183*	1.4	121	1.1	143	1.6
16	188*	1.5	118	1.1	140	1.5
18	187*	1.4	119	1.1	140	1.5
20	187*	1.4	118	1.1	140	1.5
24	187*†	1.4	39	1.1	77	1.5

*Alignment length, reported by CE, is apparently wrong because it exceeds chain lengths of individual structures (160 for 4dfr:A). † In 24-structure set, CE-MA omitted PDB entry 1t6t:2. ‡ Consensus r.m.s.d. is shown.

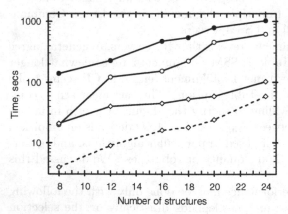

Fig. 5. Response time of CE-MA [21] (filled circles), MASS [20] (open circles) and SSM (diamonds) for producing alignments in Table 1. Dashed line shows CPU time of SSM. See text for details.

last factor, that affects the comparison, is the server's hardware. SSM-MA runs on a single 1.2Ghz Linux PC, and it is not likely that it may get a substantial advantage, if any, on the hardware basis.

Figure 5 shows comparison of response time, measured as described above, obtained from CE-MA, MASS and SSM for producing multiple alignments in Table 1. As seen from the Figure, SSM outperforms CE-MA by almost an order of magnitude for all data sets in Table 1. MASS seems to be 4 to 6 times slower than SSM except for the data set of 8 structures, when MASS is as fast as SSM.

5   Conclusion

We have described here the algorithm of multiple alignment of protein structures employed in the EBI-MSD web service SSM. The service has been launched in June 2002 and since then served tens of thousands requests yearly. Vast experience of using SSM proved its high efficiency and quality of the results [22]. Our MA algorithm is different from a few others avaliable in that it seeks a solution by gradual removal of structural elements that are less likely to get aligned, rather than by a progressive clustering of the most similar chains. We have shown in this paper that SSM-MA is capable to handle large sets of structures and in most instances the results are delivered in a few minutes time. We have also described the basic scores used in SSM-MA output. These scores are derived from those of pairwise structure alignment by generalisation on the many-structure case. Like in the case of PA (cf. Ref. [1]), our experience suggests that Q-score is a better measure of structural similarity than the traditionally used r.m.s.d. and alignment length. As found, SSM-MA is capable of picking similarities in remote structures from different SCOP folds and classes, which suggests usability of the method for structure classification and studying the structure-function relationships.

Acknowledgement. E.K. is supported by the research grant No. 721/B19544 from the Biotechnology and Biological Sciences Research Council (BBSRC) UK.

References

1. Krissinel, E. and Henrick, K.: Secondary-structure matching (SSM), a new tool for fast protein structure alignment in three dimensions. Acta Cryst. **D60** (2004) 2256—2268.
2. Chotia, C. and Lesk, A.M.: The relation between the divergence of sequence and structure in proteins. EMBO J. **5** (1986) 823–826.
3. Chotia, C.: One thousand families for the molecular biologist. Nature **357** (1992) 543–544.
4. Hubbard, T.J.P. and Blundell, T.L.: Comparison of solvent-inaccessible cores of homologous proteins – definitions useful for protein modelling. Protein Engng. **1** (1987) 159–171.
5. Holm, L. and Sander, C.: Protein structure comparison by alignment of distance matrices. J. Mol. Biol. **233** (1993) 123-138.
6. Orengo, C.A. and Taylor, W.R.: SSAP: Sequential Structure Alignment Program for protein structure comparison. Meth. Enzym. **266** (1996) 617-635.
7. Falicov, A. and Cohen, F.E.: A surface of minimum metric for the structural comparison of proteins. J. Mol. Biol. **258** (1996) 871-892.
8. Gerstein, M. and Levitt, M.: Using iterative dynamic programming to obtain accurate pairwise and multiple alignments of protein structures. In Proc. of the Fourth Int. Conf. Intell. Syst. Mol. Biol., Menlo Park, Calif.: (1996) AAAI Press, pp. 59-67.
9. Singh, A.P. and Brutlag, D.L.: Hierarchical protein structure superposition using both secondary structure and atomic representations. In Proc. Int. Conf. Intell. Syst. Mol. Biol. ISMB-97: (1997) AAAI Press, pp. 284-293.

10. Vriend, G. and Sander, C.: Detection of common three-dimensional substructures in proteins. Proteins **11** (1991) 52-58.
11. Mizuguchi, K. and Go, N.: Comparison of spatial arrangements of secondary structural elements in proteins. Protein Engng. **8(4)** (1995) 353-362.
12. Mitchell, E.M., Artymiuk, P.J., Rice, D.W. and Willett, P.: Use of techniques derived from graph theory to compare secondary structure motifs in proteins. J. Mol. Biol. **212** (1990) 151-166.
13. Alexandrov, N.N.: SARFing the PDB. Protein Engng. **9** (1996) 727-732.
14. Grindley, H.M., Artymiuk, P.J., Rice, D.W. and Willett, P.: Identification of tertiary structure resemblance in proteins using a maximal common subgraph isomorphism algorithm. J. Mol. Biol. **229** (1993) 707-721.
15. Shindyalov, I.N. and Bourne, P.E.: Protein structure alignment by incremental combinatorial extension (CE) of the optimal path. Protein Engng. **11(9)** (1998) 739-747.
16. Gibrat, J.-F., Madej, T. and Bryant, S.H.: Surprising similarities in structure comparison. Current Opinion in Structural Biology **6** (1996) 377-385.
17. Kleywegt, G.J. and Jones, T.A.: Detecting folding motifs and similarities in protein structures. Meth. Enzym. **277** (1997) 525-545.
18. Russell, R.B. and Barton, G.J.: Multiple protein sequence alignment from tertiary structure comparison. Proteins: Struct. Funct. Genet. **14** (1992) 309–323.
19. Shatsky, M., Nussinov, R. and Wolfson, H.J.: MultiProt - a Multiple Protein Structural Alignment Algorithm. In Lecture Notes in Computer Science: (2002) Springer Verlag, pp. 2452:235–250.
20. Dror 0., Benyamini H., Nussinov R. and H. Wolfson: Multiple structural alignment by secondary structures: algorithm and applications. Protein Science **12** (2003) 2492–2507.
21. Guda C., Lu S., Scheeff E.D., Bourne P.E. and Shindyalov I.N.: CE-MC: a multiple protein structure alignment server. Nucl. Acids Res. **32** (2004) W100–W103.
22. Kolodny, R., Koehl, P. and Levitt, M.: Comprehensive Evaluation of Protein Structure Alignment Methods: Scoring by Geometric Measures. J. Mo. Biol. **346** (2005) 1173–1188.
23. Gusfield, D. *Algorithms on Strings, Trees and Sequences.* Cambridge University Press, New York, (1997), pp 348–350.
24. Krissinel, E.B., Winn, M.D., Ballard, C.C., Ashton, A.W., Patel, P., Potterton, E.A., McNicholas, S.J., Cowtan, K.D. and Emsley, P.: The new CCP4 Coordinate Library as a toolkit for the design of coordinate-related applications in protein crystallography. Acta Cryst. **D60** (2004) 2250—2255.
25. Sayle, R. A., and Milner-White, E. J.: RasMol: Biomolecular graphics for all. Trends in Biochemical Sci. **20** (1995) 374-376.
26. Kraulis, P.J.: MOLSCRIPT: a program to produce both detailed and schematic plots of protein structures. J. Appl. Cryst. **24** (1991) 946-950.
27. Merritt, E.A. and Bacon, D.J.: Raster3D: Photorealistic Molecular Graphics. Meth. Enzymol. **277** (1997) 505-524.
28. Murzin, A.G., Brenner, S.E., Hubbard, T. and Chothia, C.: SCOP: a structural classification of proteins database for the investigation of sequences and structures. J. Mol. Biol. **247** (1995) 536-540.

Protein Annotation by Secondary Structure Based Alignments (PASSTA)

Constantin Bannert and Jens Stoye

Technische Fakultät, Universität Bielefeld, Germany

Abstract. Most software tools in homology recognition on proteins answer only a few specific questions, often leaving not much room for the interpretation of the results. We develop a software *Passta* that helps to decide whether a protein sequence is related to a protein with known structure. Our approach may indicate rearrangements and duplications, and it displays information from different sources in an integrated fashion.

Our approach is to first break each sequence of the Protein Data Bank (PDB) into Secondary Structure Elements (SSEs). Given a query sequence, our goal is then to 'explain' it by SSE sequences as good as possible. Therefore, we use the Waterman-Eggert algorithm to compute pairwise alignments of SSE sequences with the query. In a graph-based approach, we then select those alignments that reproduce the query in an optimal way. We discuss two examples to illustrate the potential (and possible pitfalls) of the method.

1 Introduction

The need to characterize and annotate the enormous amount of gene sequences emerging from the genome sequencing projects led to the development of many useful algorithms and tools. A common approach used is homology recognition, where a query sequence is compared to one or many already characterized sequences or structures. If a certain similarity between those can be found, we can assume the existence of a common ancestor, and hence, homology. Shi *et al.* [1] defined four major groups of homology recognition:

1. Methods that do pairwise sequence comparison, usually by computing pairwise alignments. They are able to detect closely related homologs, but often miss remote homologies.
2. Tools in the second group are also based on sequence comparison. However, they use multiple alignments of related sequences and compute profiles or probabilistic models from them to improve the detection of remote homologs.
3. The third group of methods uses structure- and sequence information.
4. Homology detection in the fourth group relies on structure information only. Methods in this group are usually threading methods.

BLAST [2], FASTA [3], and the Smith-Waterman algorithm [4] are well known examples from the first group of methods. They compute pairwise alignments, but are usually used to search a whole sequence database.

M.R. Berthold et al. (Eds.): CompLife 2005, LNBI 3695, pp. 79–90, 2005.
© Springer-Verlag Berlin Heidelberg 2005

PSI-Blast [5] can be seen as an enhanced BLAST. It is better suited to detect remote homologies, and since it computes and uses profiles of simililar sequences, it can be assigned to the second group. The same holds for HMM-based approaches, see [6] and references in [1].

The Jumping Alignment algorithm 'Jali' [7] also belongs to the second group of methods, but the concept already suggests a connection to protein structure. The query is aligned simultaneously to all sequences in a multiple alignment (usually derived from a protein family). However, only one sequence in the alignment, the 'reference sequence', contributes to the actual computation of the Jali score. The algorithm may change (or 'jump' between) the sequences, if the properties of the multiple alignment allow to do so. Structure may play a role, if the aligned family has modular properties, e.g., divides into two subfamilies. However, tests in [8] whether the jumps of the algorithm reflected the secondary structure of some protein families revealed only few examples where this was the case.

All approaches mentioned so far are strictly sequence-based alignment approaches. Since they process their information sequentially, they are not well suited for the detection of rearrangements and duplications. When used in a database search context, the results are usually presented as alignments of the query to the corresponding database hits. If the hit is only partial, it is not at once clear whether the unmatched part of the query bears similarities to other proteins in the database.

This was our motivation to develop a new software, called *Passta* (Protein annotation by secondary structure based alignments). Passta is a fragment-based alignment approach on secondary structure elements (SSEs) or, more precisely, SSE sequences. SSEs can be seen as the smallest structural entities in a protein, and we decided to use them, even though their structure is not fixed *in vivo*, but depends on the environment and other factors (see, e.g. [9]). Given a query sequence, the aim of Passta is to show how well it can be represented with SSEs found in sequences of the PDB. Further information is provided by linking the SSEs to the SCOP classification database [10,11], and by showing the position of the matched SSEs in their chain, which helps to find possible rearrangements and duplications. Each site in the query can only be aligned with one SSE at a time, but we display all such alignments simultaneously.

Two methods in the third of the groups listed above also use SSEs, MAP [12] and SEA [13]. Both use predicted SSEs. MAP derives a secondary structure 'map' from the SSEs to find the most likely fold from a database of domains with known structure. SEA ('SEgment Alignment') uses a graph-based approach to compare two protein sequences. For both proteins, SSEs are predicted with several secondary structure prediction methods and represented in two unweighted graphs. Ye *et al.* [13] then solve a network matching problem: They search for a path in each graph/network, such that the corresponding SSEs in both paths are maximally similar.

The main difference between their and our concept is that we represent residue-level alignments from many different proteins simultaneously in one graph and search for those that best explain the query, while Ye *et al.* use one graph for each target. Therefore, the SEA approach can not detect similarities to different database hits at the same time.

Other methods in homology recognition using SSEs are mostly in group four. They use the three-dimensional coordinates of the SSEs, mostly for vector representations in protein structure comparison. These methods are basically out of scope here, however, we would like to mention a recent study by Shih and Hwang [14] who investigated alternative/permuted alignments by structural comparison, where the SSEs were not required to be sequential. Their results indicate that this area is to some degree overlooked, and investigations here will be useful to improve our understanding of the organization and evolution of proteins.

We present the basic framework of Passta in Section 2 and illustrate its performance in Section 3. The first example we give is a plastocyanin sequence from *Oryza sativa* (rice), which serves as a proof of concept. The second one is the CASP6 target 'T0269' (see http://predictioncenter.llnl.gov/casp6/Casp6.html).

2 Material and Methods

Terminology: Based on the atomic coordinates of a structure determination experiment, each amino acid in a protein structure is assigned a secondary structure *state* (see e.g., [9], chapter 17). The standard here is the DSSP algorithm [15], which was also used to assign the secondary structure states to the databases in Section 2.1. These states can be grouped into *classes*. The DSSP states G, H, and I are helical states, B and E are strand states, and the remaining three (T, S, and blank) are random coil or loop states. A 'SSE' is a protein segment where all amino acid states are equal or at least belong to the same class.

2.1 Database Integration

Passta uses a relational database (Passta DB) that integrates information from three secondary source databases which themselves are all derived from the Protein Data Bank (PDB) [16].

The **Protein Topology Graph Library (PTGL)** [17] is based on the atomic coordinates of PDB proteins satisfying certain quality criteria. These proteins were decomposed into SSEs of known local structure, and their topology was stored. The aim of the PTGL is to provide this topology information to the user, but we currently use only the decomposed SSEs.

The **PDBFinder II database** (submitted) is an enhanced version of the PDBfinder database [18]. It provides extensive information for almost all proteins in the PDB. Most of this information could also be found in other databases as well, but here it is all in one place. We parse loops and coils from the PDBFinder II database.

The **SCOP database** [10,11] is a classification on protein domains. It defines four hierarchical levels: *family, superfamily, common fold*, and *class*. Two domains in the same family are closely related, indicated by a high percentage of sequence identity and structural similarity. Domains in the same superfamily are rather distantly related, but should have a common ancestor. Mostly, this means low sequence identity, but high structural similarity. Two domains have a common fold if their secondary structure elements have the same arrangement, i.e. topology. However, they are not necessarily related. The different folds are grouped into classes according to their main class of secondary structure. The ASTRAL database [19,20] can be seen as bridging the gap between SCOP and the PDB. Here, we use the 'SPACI' score that the ASTRAL consortium assigns to each protein domain in SCOP. It summarizes the quality of the structure determination experiment. We use it to determine a representative among sequence-identical SSEs and Chains.

Integration. After the integration of the source databases into the Passta DB, it contains tables and data for most proteins, chains, and SSEs available. Also, some precomputed information needed in the annotation approach ('PasstaRun', see Section 2.2) was stored.

The PDB and Chain ID fields are common in all of the source databases, so we used them to cross-index all information before storing it in the Passta database.

The SSEs with helical or strand conformation were taken from the PTGL. The coils were parsed from the PDBFinderII, because the PTGL does not provide them. To ensure consistency, we required the SSE sequences from the PTGL to map back to the PDBfinder II chain sequence. If this was not possible, we excluded the whole chain from the database. If more than twenty percent of a SSE sequence was made up of 'X's (amino acid unknown), we excluded this SSE as well. Some SSEs are not contigous, they contain a chain break, indicated by a gap character ('−'). We decided to split those SSEs into two of the same class. After the integration, the Passta DB contains 21572 proteins, 44048 chains, and almost 1.5 million SSEs. That is about 90 % of all possible data.

MaxScores. For each SSE in the database, we stored its maximal alignment (i.e., exact match) score under several substitution matrices. We use these values as rough estimates of alignment quality (see Section 2.2).

Redundancy. Many chain- and SSE sequences are not unique. Since redundant sequences slow down the search procedure, we mark them in order to exclude them from the search process. The procedure is basically the same for chain and SSE sequences: We select all equal sequences and compare the SPACI scores of their 'parent' proteins. The sequence with the best associated SPACI score is marked as being the representative, i.e. non-redundant. First, we do this on the level of chains. Then, we apply it to the SSEs from those chains that were marked non-redundant right before.

SCOP. We also integrated the SCOP classification into the database. For most SSEs we now know the structural domain it belongs to.

2.2   PasstaRun - Annotation Strategy

The annotation of the query with PDB SSEs is implemented in two stages. The first stage ('Pass One') is a filtering approach. It selects some candidate chains for use in the second stage ('Pass Two'), the annotation process itself. We used the blosum62 substitution matrix with gap costs of 12 and 2 for initiation and extension, respectively.

Pass One. Given a query R of length n, Pass One starts by selecting all non-redundant SSEs of length 6 or more from the database. Each SSE comes with its associated *MaxScore*. The Waterman-Eggert algorithm [21] computes pairwise, local, non-intersecting alignments of two sequences. It starts with the optimal, i.e. highest scoring alignment, then co- and suboptimal alignments are computed. We apply the Waterman-Eggert algorithm in Pass One to compute several alignments between each SSE sequence and R, until the ratio *score/MaxScore* drops below a predefined constant.

Let A be the set of all alignments found in this way. Each alignment $\alpha \in A$ is represented by a 5-tuple (b, e, c, p, s). Elements b and e, $1 \le b \le e \le n$, are the begin and end indices of the aligned SSE w.r.t. the query; $c \ge 1$ is a unique chain identifier; $p \ge 1$ is the position of the SSE in its chain; and s is the alignment score. For a given alignment $\alpha = (b, e, c, p, s)$, we will refer to the individual components of the 5-tuple as $b(\alpha) := b$, $e(\alpha) := e$, $c(\alpha) := c$, $p(\alpha) := p$, and $s(\alpha) := s$, respectively.

Our goal in Pass One is to find a set of good candidate chains for use in Pass Two. In fact, we only need to find a set of good alignments, since we know the chains that an aligned SSE sequence is contained in. We use a graph-based approach to solve this problem. We define a directed acyclic graph $G = (V, E)$, where the set of vertices V is made up of representations of all alignments in A, plus two other vertices; $head = (0,0,0,0,0)$ and $tail = (n+1, n+1, 0, 0, 0)$, such that $V = A \cup \{head, tail\}$.

An edge exists between two vertices $u, w \in V$, $u \ne w$, if and only if

1. u and w do not overlap (and u is before w), i.e. $e(u) < b(w)$, and
2. there is no alignment v between u and w, i.e. $\nexists v \in V : e(u) < b(v)$ and $e(v) < b(w)$.

A *path* P in G is a sequence of vertices $(v_1, v_2, \ldots, v_k)$ such that v_i and v_{i+1} are connected by an edge for all $1 \le i < k$. Any path from *head* to *tail* corresponds to a selection of non-overlapping alignments. The *weight* of a path $P = (v_1, v_2, \ldots, v_k)$ is given by $weight(P) := \sum_{i=1}^{k} s(v_i)$, it indicates the selection quality. to some degree. So, the problem to find good candidate chains transforms to a *single-source shortest path* problem from *head* to *tail*, which can be solved easily and efficiently (see e.g. [22]).

We compute all such optimal paths, i.e. where $weight(P)$ is maximal, and collect all chain IDs of their vertices for use in Pass Two. However, the SSEs we use in the alignments are all non-redundant. SSEs with an identical sequence may also exist in other chains. We identify those and collect their chain IDs as well.

Pass Two. The goal of the second pass is to annotate the query with the best selection of SSEs from those chains that passed Pass One. Since the result should be biologically feasible, we have placed certain constraints on the algorithm. Otherwise, Pass One and Pass Two are quite similar. We describe changes in the definitions and the algorithm where applicable. Pass Two first collects for each chain c in the list *all* SSEs, regardless of redundancy status or size (i.e., the complete chain). Then we recompute the set of alignments A. However, it makes no sense to align sequences of length one or two to the query. Therefore, we divided the alignment phase into an *align* and an *extend* part.

Align. Let $l(S)$ be the length of an SSE sequence S. If $l(S) = 2$, we compute all exact matches between S and the query R and insert the corresponding alignment(s) into A. If $l(S) \geq 3$, we use the Waterman-Eggert algorithm. We accept an alignment α between S and R whenever its *score/MaxScore* ratio is larger than a predefined constant.

Extend. Each time we insert an alignment α into A in the *align* phase, we look at the SSEs adjacent to S in its chain. If they exist and their length is less or equal to 4, we align them to R, allowing neither insertions nor deletions. If the score is larger than zero, we include the new alignment into A (for an example, see Fig. 1).

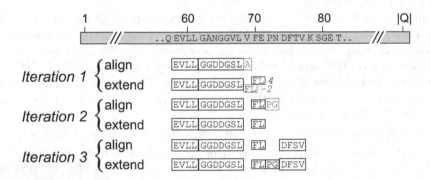

Fig. 1. Example to illustrate the alignment phase in Pass Two. The example corresponds to the result shown in Fig. 2. Alignments displayed with reduced size are from an 'extend' phase. *Iteration 1, align:* A local alignment of the SSE sequence 'GGDDGSLA' and the query is computed, yielding ('GGDDGSL', 'GANGGVL'). *Iteration 1, extend:* The SSE left of 'GGDDGSLA' is already aligned, so no extension is performed here. However, the right SSE is small enough, and we match it (a) at the end position of the **local** alignment and (b) at the position where the **global** alignment would have ended. Only the latter is accepted here because its score exceeds zero. *Iteration 2, align:* 'PG' is of length two, and there is no exact match with the query, so it is not aligned. *Iteration 2, extend:* No alignment was accepted in the alignment phase, therefore nothing is to be extended. *Iteration 3, align:* The SSE sequence 'DFSV' is aligned to the query. *Iteration 3, extend:* Now it is possible to extend with 'PG' to the left of 'DFSV'.

The directed acyclic graph G is built as described in Section 2.2. However, there are some differences: In Pass Two, we expand the edge definition. There is now also an edge between two vertices $u, w \in V$, if:

1. $c(u) = c(w)$ i.e., both SSEs come from the same chain,
2. $e(u) < b(w)$ (the no-overlap condition from Pass One), and
3. there is no other alignment v between u and w, where the SSE in v comes from the same chain as the one in u and w: $\nexists v \in V : c(u) = c(v) = c(w)$ and $e(u) < b(v)$ and $e(v) < b(w)$.

The definition of a *path* is the same as in Pass One. However, a path $P = (v_1, v_2, \ldots, v_k)$ in Pass Two can contain jumps and rearrangements. If the chains of two adjacent vertices in P are different, i.e. $c(v_i) \neq c(v_{i+1})$, we call this a *jump*. If they are equal but their positions are not consecutive, i.e. $c(v_i) = c(v_{i+1})$ and $p(v_{i+1}) - p(v_i) \neq 1$, we call this a *rearrangement*. Let $j(P)$ be the number of jumps in a path P, and $r(P)$ the number of rearrangements. We penalize jumps and rearrangements by two parameters, *jump cost* (jc) and *rearrangement cost* (rc). The weight of a path P is now given by

$$weight(P) = \sum_{i=1}^{k} s(v_i) - j(P) \times jc - r(P) \times rc.$$

This makes the annotation of the query with small chance hits from different chains highly unlikely, if the jump cost is chosen well. Finally, the alignments in the optimal path are visualized in a HTML page.

3   Results and Discussion

We present and discuss two examples that we annotated with Passta to illustrate some application possibilities. The complete and colored versions of the presented alignments can be found at `http://www.cebitec.uni-bielefeld.de/~ban nert/res/`*filename*, where *filename* is `Pcya36-6.htm` or `T0269-40-5.htm`.

3.1   Plastocyanin from Rice

In the first example, the query was a plastocyanin (PC) sequence with a length of 154 amino acids, from rice. We used a jumpcost value of 36. An excerpt of the resulting file is shown in Fig. 2. Passta aligned two PDB chains to the query, the raw score of the optimal path is 369. The annotation suggests that the first part of the alignment is similar to 1fsk_C, the heavy chain of an Immunoglobulin (IG) antibody from *Mus musculus* (mouse). There are two rearrangements (SSEs not consecutive) in the order of the SSEs from this chain. The second chain '1ag6' is a PC, from spinach (*Spinacia oleracea*). The annotation with '1ag6' is doubtlessly correct, however, given the high sequence similarity of the query to other PCs in the PDB, it is not surprising.

Chain	SPXID	Local Query-SSE alignments
1fsk C	NA	SVAKTTP
	21431	SVFP APGSAAQTNSMV SVFPAPGSAAMV LAPGSAAQ
1ag6	22857	EVLLGGDDGSL FLPGDFSV SGE IVFKNNAGF PHNVVFDEDEI PSGVD
Query	Len 154	MAALSSAAVTIPSMAPSAPGRRRMRSSLVVRASLGKAAGAAAVAVAASAMLAGGAMAQEVLLGANGGVLVFE PNDFTVKSGETITF KNNAGF PHNVVFDEDAVPSGVDV

Chain	SPXID	Position of aligned SSEs in their Chain
1fsk C	NA	30
	21431	31 32 33 31 32 33 32
1ag6	22857	2 3 4 5 6 7 8 9 10 11
Query	Len 154	MAALSSAAVTIPSMAPSAPGRRRMRSSLVVRASLGKAAGAAAVAVAASAMLAGGAMAQEVLLGANGGVLVFE PNDFTVKSGETITF KNNAGF PHNVVFDEDAVPSGVDV

Fig. 2. Passta alignment of Plastocyanin from rice (excerpt). The upper table shows the local alignments of the SSE sequences with the query. The lower table displays the position of the aligned SSEs in their PDB chain. While the SSEs of '1ag6' are consecutively aligned, those of '1fsk' are to some degree rearranged and duplicated. Some query segments and SSEs are shaded. Dark SSE segments are strand SSEs, light ones are helix SSEs. The intensity of the shaded segments in the query string corresponds to the score they contribute. The darker, the better the score of the segment.

What about the IG matched to the first part of the query? It is classified into another SCOP fold than '1ag6', namely 'Immunoglobulin-like beta-sandwich' instead of 'Cupredoxin'. The query PC in this experiment is 154 amino acids long. The length of the other PCs in the Passta database is only about 100 residues. A multiple alignment of the query and all non-redundant PCs from the PDB shows that the first 55 residues of the query are not matched by any other PC in the DB (see Fig. 3). An alignment of the query against the whole SCOP family 49504 'Plastocyanin/azurin-like' reveals that the first 20 positions of the query are unmatched by *any* other domain in the family, and that the sequence similarity within the first 50 residues is in general quite low (data not shown). Therefore, we could not find a close homolog matching this region. The IG is not even a remote homolog of the query, but according to Russell *et al.* in [12] it has a loose structural similarity to PCs.

In SCOP, structural similarities are classified into the same fold. The classification depends on the topology of the SSEs. Here, it matters whether a beta-sheet is parallel or antiparallel. The loose similarity observed in our example rather corresponds to the *architecture* level as defined in CATH [23]. The orientation of the SSEs is not important at this level. A loose structural similarity being more informative than a hit with absolutely no relationship to the query, we consider the annotation with the mouse IG as success.

3.2   CASP Target T0269

T0269 (PDB code '1vgs') is a thioredoxin peroxidase from the archaeon *Aeropyrum pernix*, with two domains and a length of 250 residues. In this experiment, we used a minimum length of 5 (instead of 6) for the non-redundant SSEs that were aligned in Pass One. The Passta alignment reached a raw score of 191 and used three chains, '1n8j_A', '1prx_A', and '1uth_A' (see Fig. 4). There are eight

```
PC03        ------------------------------------------------------------ETFTV
PC07        ------------------------------------------------------------ETYTV
PC08        ------------------------------------------------------------ANATV
PC09        ------------------------------------------------------------ANATV
PC16        ------------------------------------------------------------ANATV
PC04        ------------------------------------------------------------QTVAI
PC01        ------------------------------------------------------------ASVQI
PC12        ------------------------------------------------------------MIDV
PC14        -------------------------------------------------------------IDV
PC17        -------------------------------------------------------------LEV
PC02        -------------------------------------------------------------VEV
PC13        -------------------------------------------------------------VEV
PC05        -------------------------------------------------------------AEV
PC06        -------------------------------------------------------------AEV
PC18        -------------------------------------------------------------AEV
PLAS_ORYSA  MAALSSAAVTIPSMAPSAPGRRRMRSSLVVRASLGKAAGAAAVAVAASAMLAGGAMAQEV
PC19        ------------------------------------------------------------DATV
PC10        ------------------------------------------------------------AQIV
PC15        ------------------------------------------------------------AAIV
PC11        ------------------------------------------------------------AKV
                                                                       :

PC03        KMGADSGLLQFEPANVTVHPGDTVKWVNNKLPPHNILFDDKQVPG-ASKELADKLSHSQ-
PC07        KLGSDKGLLVFEPAKLTIKPGDTVEFLNNKVPPHNVVFDAALNPA-KSADLAKSLSHKQ-
PC08        KMGSDSGALVFEPSTVTIKAGEEVKWVNNKLSPHNIVFAADGV----DADTAAKLSHKG-
PC09        KMGSDSGALVFEPSTVTIKAGEEVKWVNNKLSPHNIVFAADGV----DADTAAKLSHKG-
PC16        KMGSDSGALVFEPSTVTIKAGEEVKWVNNKLSPHNIVFDADGV----PADTAAKLSHKG-
PC04        KMGADNGMLAFEPSTIEIQAGDTVQWVNNKLAPHNVVVEGQ----------PELSHKD-
PC01        KMGTDKYAPLYEPKALSISAGDTVEFVMNKVGPHNVIFDKVPAG-----ESAPALSNTK-
PC12        LLGADDGSLAFVPSEFSCSPGCKIVFKNNAGFPHNIVFDEDSIP---SGVDASKISMSEE
PC14        LLGADDGSLAFVPSEFSISPGEKIVFKNNAGFPHNIVFDEDSIP---SGVDASKISMSEE
PC17        LLGSGDGSLVFVPSEFSVPSGEKIVFKNNAGFPHNVVFDEDEIP---AGVDAVKISMPEE
PC02        LLGGDDGSLAFLPGDFSVASGEEIVFKNNAGFPHNVVFDEDEIP---SGVDAAKISMSEE
PC13        LLGGDDGSEAFLPGDFSVASGEEIVFKNNAGFPHNVVFDEDEIP---SGVDAAKISMSEE
PC05        LLGSSDGGLAFVPSDLSIASGEKITFKNNAGFPHNDLFDEDEVP---AGVDVTKISMPEE
PC06        LLGSSDGGLAFVPSDLSIASGEKITFKNNAGFPHNDLFDKKEVP---AGVDVTKISMPEE
PC18        KLGSDDGGLVFSPSSFTVAAGEKITFKNNAGFPHNIVFDEDEVP---AGVNAEKISQPE-
PLAS_ORYSA  LLGANGGVLVFEPNDFTVKSGETITFKNNAGFPHNVVFDEDAVP---SGVDVSKISQEE-
PC19        KLGADSGALEFVPKTLTIKSGETVNFVNNAGFPHNIVFDEDAIP---SGVNADAISRDD-
PC10        KLGGDDGSLAFVPSKISVAAGEAIEFVNNAGFPHNIVFDEDAVP---AGVDADAISYDD-
PC15        KLGGDDGSLAFVPNNITVGAGESIEFINNAGFPHNIVFDEDAVP---AGVDADAISAED-
PC11        EVGDEVGNFKFYPDSITVSAGEAVEFTLVGETGHNIVFDIPAGAPGTVASELKAASMDEN
            :*        : *   .  .*  : :       ** :.                  *
```

Fig. 3. Excerpt of a ClustalW alignment of the query ('PLAS_ORYSA') with all other non-redundant plastocyanins in the PDB

rearrangements altogether, six in the alignment sequence of '1prx_A' and two in '1uth_A'.

There is no SCOP classification available for '1uth_A', and Fig. 4 shows that the similarities of this chain to the query are few. Its use is probably due to a chance hit of SSE number 4. '1n8j_A' and '1prx_A' are classified into the same SCOP family ('Glutathione peroxidase-like'), which contains thioredoxins. '1n8j_A' is an alkyl hydroperoxide reductase from *Salmonella typhimurium*, '1prx_A' is a human peroxidase. So, both proteins are correctly chosen from the database.

However, we selected this example because of the interesting sequence of SSE alignments in '1prx_A'. There are six rearrangements here, but a closer look reveals that there are really only three sequences of SSEs. If SSE number 14 was removed between number three and four, the first sequence is (10-16),

Chain	SPXID	Local Query-SSE alignments
1n8j A	95401	EGRWSVPF YPADFT VSPTELGDVADHYEELQKLG
1prx A	33076	VKLIALSIDSVEDHLAWSKDINAY KLPFFPIID DRNRLAIL
1uth A	NA	
Query	Len : 250	MPGSIPLIGERFPEMEVTTDHGVIKLPDHYVSQGKWFVLFSHPADFTPVCTTEFVSFARRYEDFQRLGVDLIGLSVDSVFSHIKWKEWIERHIGVRIPFPIIADPQGTVARRLGLLHA

Chain	SPXID	Position of aligned SSEs in their Chain
1n8j A	95401	6 7 8 9 10
1prx A	33076	10 11 12 13 14 15 16
1uth A	NA	
Query	Len : 250	MPGSIPLIGERFPEMEVTTDHGVIKLPDHYVSQGKWFVLFSHPADFTPVCTTEFVSFARRYEDFQRLGVDLIGLSVDSVFSHIKWKEWIERHIGVRIPFPIIADPQGTVARRLGLLHA

EANT TV IID GRIR YPATTGR DSILRVVISLQLGDS RVA PVDWKDGD		
	IGEMYFMPP YVCM DHPSAK LKQFSE PAKL	
ESATHTVRGVFIVDARGVIRTMLYYPMELGRLVDEILRIVKALKLGDSLKRAVPADWPNNEIIGEGLIVPPPTTEDQARARMESGQYRCLDWWFCWDTPASRDDVEEARRYLRRAAEKPAKLLYEEARTHLH		

2 3 14 4 21 22 6 2324 25		
	4 15 16 30	
ESATHTVRGVFIVDARGVIRTMLYYPMELGRLVDEILRIVKALKLGDSLKRAVPADWPNNEIIGEGLIVPPPTTEDQARARMESGQYRCLDWWFCWDTPASRDDVEEARRYLRRAAEKPAKLLYEEARTHLH		

Fig. 4. Passta alignment of T0269 from *Aeropyrum pernix*. The residues 1-32 are unmatched. Then, five SSEs of '1n8j_A' are consecutively aligned to the query (number 6 to 10). From there on, '1prx_A' is used to annotate the query. After the alignment of SSE number 16, a second sequence SSEs is consecutively aligned, from number 2 to 6. However, the sequence is interrupted by number 14, and after number 4 a third sequence of consecutive SSE-alignments is started.

the second one (2-6), and the last one (21-25). Of course, since the alignments in the second sequence are very small, this could be just a coincidence. But it could also indicate some evolutionary event that took place in the past.

4   Conclusion and Outlook

Passta delivers a snapshot that may provide useful information: It shows how well a query sequence can be represented by PDB sequences, and at which positions. Since the classes of the chains that the aligned SSEs originate from are also displayed, some information on the secondary structure composition is available as well. Finally, the position information given for every aligned SSE w.r.t. its chain may indicate duplications, repeats or other evolutionary events.

Of course, some problems remain to be solved: Since Passta is presently based on pairwise sequence alignments, we can not expect it to find remote homologs in the 'twilight zone'. We also have to admit that some of the computed alignments are not very robust. Small variations of the jumpcost parameter can lead to large variations in the resulting alignment.

We plan to use a set of secondary structure specific substitution matrices as soon as possible. If the values for gap initiation and gap extension costs are wisely chosen, this should further improve the annotation quality of Passta.

Acknowledgments

We would like to thank Patrick May, Hans-Michael Kaltenbach, and Klaus-Bernd Schürmann for many interesting discussions; and the anonymous referee for some helpful remarks.

References

1. Shi, J., Blundell, T.L., Mizuguchi, K.: FUGUE: Sequence-structure homology recognition using environment-specific substitution tables and structure-dependent gap penalties. J. Mol. Biol. **310** (2001) 243–257
2. Altschul, S.F., Gish, W., Miller, W., Myers, E.W., Lipman, D.J.: Basic local alignment search tool. J. Mol. Biol. **215** (1990) 403–410
3. Pearson, W.R.: Rapid and sensitive sequence comparison with FASTP and FASTA. In Doolittle, R.F., ed.: Molecular Evolution: Computer Analysis of Protein and Nucleic Acid Sequences. Volume 183 of Meth. Enzymol. Academic Press, San Diego, CA (1990) 63–98
4. Smith, T.F., Waterman, M.S.: Identification of common molecular subsequences. J. Mol. Biol. **147** (1981) 195–197
5. Altschul, S.F., Madden, T.L., Schäffer, A.A., Zhang, J., Zhang, Z., Miller, W., Lipman, D.J.: Gapped blast and psi-blast: a new generation of protein database search programs. Nucleic Acids Res. **25** (1997) 3389–3402
6. Eddy, S.R.: Profile hidden Markov models. Bioinformatics **14** (1998) 755–763
7. Spang, R., Rehmsmeier, M., Stoye, J.: A novel approach to remote homology detection: Jumping alignments. J. Comp. Biol. **9** (2002) 747–760
8. Bannert, C.: Systematic investigation of jumping alignments. Technical Report **2003-05** (2003) http://www.cebitec.uni-bielefeld.de/~bannert/pubs.html.
9. Bourne, P.E., Weissig, H.: Structural Bioinformatics. Wiley Liss (2003)
10. Hubbard, T.J., Ailey, B., Brenner, S.E., Murzin, A.G., Chothia, C.: SCOP: A Structural Classification of Proteins database. Nucleic Acids Res. **27** (1999) 254–256
11. Murzin, A.G., Brenner, S.E., Hubbard, T., Chothia, C.: SCOP: A structural classification of proteins database for the investigation of sequences and structures. J. Mol. Biol. **247** (1995) 536–540
12. Russell, R.B., Copley, R.R., Barton, G.J.: Protein fold recognition by mapping predicted secondary structures. J. Mol. Biol. **259** (1996) 349–365
13. Ye, Y., Jaroszewski, L., Li, W., Godzik, A.: A segment alignment approach to protein comparison. Bioinformatics **19** (2003) 742–749
14. Shih, E., Hwang, M.J.: Alternative alignments from comparison of protein structures. Proteins **56** (2004) 519–527
15. Kabsch, W., Sander, C.: Dictionary of protein secondary structure: Pattern recognition of hydrogen-bonded and geometrical features. Biopolymers **22** (1983) 2577–2637
16. Berman, H.M., Westbrook, J., Feng, Z., Gilliland, G., Bhat, T.N., Weissig, H., Shindyalov, I.N., Bourne, P.E.: The protein data bank. Nucleic Acids Res. **28** (2000) 235–242
17. May, P., Barthel, S., Koch, I.: Ptgl - a web-based database application for protein topologies. Bioinformatics **20** (2004) 3277–3279

18. Hooft, R.W.W., Sander, C., Vriend, G.: The pdbfinder database: A summary of pdb, dssp and hssp information with added value. CABIOS **12** (1996) 525–529
19. Brenner, S.E., Koehl, P., Levitt, M.: The astral compendium for protein structure and sequence analysis. Nucleic Acids Res. **28** (2000) 254–256
20. Chandonia, J.M., Walker, N.S., Lo Conte, L., Koehl, P., Levitt, M., Brenner, S.E.: Astral compendium enhancements. Nucleic Acids Res. **30** (2002) 260–263
21. Waterman, M.S., Eggert, M.: A new algorithm for best subsequence alignments with application to trna-rrna comparisons. J. Mol. Biol. **197** (1987) 723–728
22. Cormen, T.H., Leiserson, C.E., Rivest, R.L., Stein, C.: Introduction to Algorithms, 2nd Ed. MIT Press / McGraw-Hill (2001)
23. Pearl, F.M.G., Lee, D., Bray, J.E., Sillitoe, I., Todd, A.E., Harrison, A.P., Thornton, J.M., Orengo, C.A.: Assigning genomic sequences to cath. Nucleic Acids Res. **28** (2000) 277–282

MAPPIS: Multiple 3D Alignment of Protein-Protein Interfaces

Alexandra Shulman-Peleg[1], Maxim Shatsky[1],
Ruth Nussinov[2,3], and Haim J. Wolfson[1,*]

[1] School of Computer Science, Raymond and Beverly Sackler Faculty of Exact Sciences, Tel Aviv University, Tel Aviv 69978, Israel
[2] Sackler Inst. of Molecular Medicine, Sackler Faculty of Medicine, Tel Aviv University, Tel Aviv 69978, Israel
[3] Basic Research Program, SAIC-Frederick, Inc, Lab. of Experimental and Computational Biology, Bldg. 469, Rm. 151, Frederick, MD 21702, USA
{shulmana, wolfson}@post.tau.ac.il

Abstract. A protein-protein interface (PPI) is defined by a pair of regions of two interacting protein molecules that are linked by non-covalent bonds. Recognition of conserved 3D patterns of physico-chemical interactions may suggest their importance for the function as well as for the stability and formation of the protein-protein complex. It may assist in discovery of new drug leads that target these interactions. We present a novel method, MAPPIS, for multiple structural alignment of PPIs which allows recognition of a set of common physico-chemical properties and their interactions without the need to assume similarity of sequential patterns or backbone patterns. We show its application to several biological examples, such as alignment of interfaces of G proteins with their effectors and regulators, as well as previously created clusters of interfaces.

Availability: The program and supplementary information, including colored figures, can be found at: http://bioinfo3d.cs.tau.ac.il/mappis/

1 Introduction

Association and dissociation of protein molecules are crucial for most of the cellular processes. A *protein-protein interface* (PPI) is defined by a pair of regions of two interacting protein molecules that are linked by non-covalent bonds. Interface structures contain the 3D information of the interactions created between pairs of binding sites. Comparison and understanding of the physico-chemical and geometrical nature of these interactions may assist in recognizing certain interface binding organizations, that are important for the formation and stability of protein-protein complexes [1,2]. Their recognition may assist in development of efficient drugs to prevent protein association or dissociation.

Sequence patterns have been widely used for comparison and annotation of protein binding sites [3]. However, there are numerous examples of functionally

---

* Corresponding authors.

M.R. Berthold et al. (Eds.): CompLife 2005, LNBI 3695, pp. 91–103, 2005.
© Springer-Verlag Berlin Heidelberg 2005

similar interfaces that do not exhibit such sequential patterns [4]. A sequence order independent structural alignment method was used by Keskin et al. [5] to classify all known PPIs according to their C_α patterns. However, representation by backbone atoms does not capture the physico-chemical nature of the interfaces, which is important for the interaction. Additional representations have been used and several methods have been developed for alignment between binding sites [6,4,7,8,9,10]. However, these align between single binding sites and do not consider the interactions with the corresponding binding partners. Recently, we have developed a method for alignment between a pair of PPIs [11,12].

Consider the classical problem of pattern detection modulo rigid (Euclidean) motion. Define two equally sized point sets as ϵ-congruent, if there is an Euclidean transformation and an associated one-to-one mapping, such that the maximal distance between a pair of matched superimposed points is below ϵ. For a pair of point sets A and B, the *Largest Common Point Set* (LCP) problem is the task of detecting the maximal size ϵ-congruent subsets $A' \in A$ and $B' \in B$, $|A'| = |B'|$. The optimal solutions are computationally expensive [13] therefore in practice approximation algorithms are required [14]. Extension of the problem to detect a common point set between a set of K structures has many important applications for the analysis of protein and drug molecules. However, even in 1D space for the case of exact congruence ($\epsilon = 0$) the problem is NP-Hard [15].

Here, we define a new optimization problem of detecting the highest scoring spatial pattern common to a set of PPIs. The scoring function considers physico-chemical properties and interactions shared by a set of PPIs. Our main motivation is similar to the multiple sequence alignment thesis, namely, that a feature common to a number of proteins is (probably) functionally more significant than a similar feature found only between a pair of proteins. We present a novel method, MAPPIS, for multiple structural alignment of PPIs, which optimizes the introduced scoring function. The computational problem involves two NP-Hard subproblems. The first problem is the selection of pairwise transformations for construction of a uniquely defined multiple alignment [15]. The second problem is, given a multiple superposition, detect the highest scoring common pattern comprised of PPI physico-chemical properties and interactions, i.e. the matching problem [16]. Applying a *branch-and-bound* method allows us to practically overcome the exponential nature of the first problem. To solve the second problem we apply a hierarchical greedy technique. The overall scheme guarantees an approximation to the optimal solution. Although each of these subproblems has been previously addressed [16], introduction of protein-protein interactions imposes a new algorithmic formalism and consequently a completely new program implementation. The method's running times are practical (a matter of seconds on a standard PC). We show its application to recognition of conserved interactions shared by interfaces of G proteins with their effectors and regulators as well the PPI clusters created by Mintz et al. [12].

2   The Largest Common Interface Problem

In this work we extend the multiple Largest Common Point set problem [15] to the Largest Common Interface problem. The input is K *protein-protein interfaces* $\{(A_i, B_i)\}_{i=1}^{K}$. We assume that correspondence between the interface sides (interacting protein chains) is given, i.e. we do not seek for an alignment between A_i and B_j[1]. We define a *protein-protein interface* (PPI) as a pair of interacting binding sites from two non-covalently linked protein molecules. Each binding site is represented by a set of *pseudocenters* [7] which are points in 3D space that represent centers of potential interactions: *hydrogen-bond donor, hydrogen-bond acceptor, mixed donor/acceptor, hydrophobic aliphatic and aromatic(pi) contacts*. These are extracted from the side-chains as well as the protein backbone. For example, the side chain of Arg is represented by 3 donors (nitrogen atoms) and an aliphatic pseudocenter (located at the center of mass of its 3 carbons), while the side-chain of Pro is represented by an aromatic pseudocenter located at the center of its ring. Only surface exposed pseudocenters that are within 4Å from the surface of the binding partner are considered (see Box 1).

Box 1: PPI Representation.
(a)   The   interface   surfaces
are   represented   as   dots   and
pseudocenters (from both backbone
and side chains) as balls. Hydrogen
bond donors are blue, acceptors
- red, donors/acceptors - green,
hydrophobic aliphatic - orange and

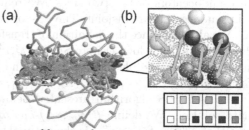

aromatic - white. (b) The interactions created by complementary pseudocenters of
an interface (represented by arrows). The bar presents the complementarity of the
properties. Specifically, hydrogen bond donors are complementary to acceptors,
while hydrophobic aliphatic and aromatic interact with similar ones. The interaction distance thresholds are 3.9Å [17] for hydrogen bonds and 8Å for the rest.

We define an *interaction* as a pair of close enough pseudocenters, one from each side of the interface, possessing *complementary* physico-chemical properties (see Box 1 (b)). According to our definition the number of properties complementary to a given pseudocenter may be larger than the number of real interactions in which it can participate. Exact definition of real interactions is not straightforward [17]. To partially overcome this problem we consider all possible interactions till the last stage of our method. When the final matching is computed, we select only high scoring matched interactions so that the number of interactions in which each pseudocenter can participate is not exceeded.

---

[1] This correspondence can be obtained from the biological data. Otherwise, it can be estimated by running twice the pairwise alignment between (A_1, B_1) - (A_i, B_i) and (A_1, B_1) - (B_i, A_i), for each $i \neq 1$.

Two superimposed pseudocenters are considered *similar* if they have the same physico-chemical properties, while the mixed donor/acceptor is similar to both. Two interactions $i = (a, b)$ and $i' = (a', b')$ are considered similar if the corresponding pseudocenters (a, a') and (b, b') are similar and $|a - T(a')| \leq \epsilon$ and $|b - T(b')| \leq \epsilon$, where T is an Euclidean 3D transformation.

Select one of the PPIs (e.g. the first) as the *pivot* interface. For K interfaces we define the similarity with respect to the *pivot* PPI. Denote by $IN = \{(a_1, b_1), ..., (a_K, b_K)\}$ a set of similar interactions iff $\forall i = 2...K$ $(a_1, b_1) \sim (a_i, b_i)$. We denote by $PC = (p_1, ..., p_K)$ a set of similar pseudocenters iff $\forall i = 2...K$ $(p_1) \sim (p_i)$. In the general case we are given similarity scoring functions, $S_{IN}(IN[i]) = S_{IN}((a_1, b_1), (a_i, b_i))$ and $S_{PC}(PC[i]) = S_{PC}(p_1, p_i)$. The particular scoring functions used in this work are defined in the Appendix. Given a set of matched interactions $\{IN_t\}$ and a set of matched pseudocenters for each interface side, $\{PC[A]_p\}$ and $\{PC[B]_l\}$, we define the scoring function of multiple alignment to be the minimum[2] of the scores between the pivot PPI and the rest of PPIs: $S = min_{i=2...K} S_i$, where S_i is defined as:

$$S_i = \sum_t S_{IN}(IN[i]_t) + \sum_p S_{PC}(PC[A][i]_p) + \sum_l S_{PC}(PC[B][i]_l).$$

In other words, S_i is the sum of scores of all the matched interactions and the pseudocenters of the two sides. We require that the matched interactions and pseudocenters are disjoint. The S_i score is based on a matching, i.e. IN and PC. We also introduce the definition of transformation based upper-bound score, which we will use below. Denote $S(T = (t^2, t^3, ..., t^K))$ to be the upper bound on the score for all possible matches of IN and PC after multiple superimposition of the set $\{(A_i, B_i)\}_{i=1}^K$, i.e. $\{t^i(A_i, B_i)\}_{i=1}^K$, where t^1 is identity.

The MAPPIS algorithm presented below guarantees to detect an approximate solution. We define $(\beta, \gamma_{in}, \gamma_{pc})$-*approximation* algorithm as follows. Assume that $S^{opt} = S_{IN}^{opt} + S_{PC}^{opt}$ is the optimal score with distance tolerance ϵ. The approximation algorithm guarantees to detect a solution with a score at least $\frac{1}{\gamma_{in}} S_{IN}^{opt} + \frac{1}{\gamma_{pc}} S_{PC}^{opt}$ with distance tolerance at most $\epsilon + \beta$.

3   MAPPIS Algorithm

Our goal is to maximize the scoring function S as defined above. First we generate a polynomial number of pairwise transformations between the pivot interface and all other interfaces. Then, we apply a *branch-and-bound* technique to effectively filter out the low scoring multiple alignments. Third, we apply the approximation method to solve the multiple matching problem. The overall scheme guarantees an approximation to the optimal alignment score.

The Transformation Search. For each pair of interfaces $(I_1 = (A_1, B_1), I_i = (A_i, B_i))$, $2 < i \leq K$ and I_1 is the pivot interface, we create a polynomial size set of 3D transformations that can superimpose one interface onto the

---

[2] In this definition the similarity score is measured by the distance of the outlier from the pivot.

other. A common alignment technique considers each pair, one from I_1 and one from I_i, of ϵ-congruent point triplets to create a 3D transformation. To guarantee a β approximation to the optimal alignment (including multiple superposition) we need to sample $(\frac{\epsilon}{\beta})^6$ alignments for each triplet pair [14].

The Combinatorial Stage of the Multiple Alignment. Here we adopted a very fast *branch-and-bound* technique that allows to apply effectively the *bound* criterion [16]. It works as follows. Denote S^* to be the maximal score found so far. Each query triangle (or any other set of features) from interface I_1 defines a set of possible transformations between I_1 and the rest of the PPIs, $CB = \{T^2, T^3, ..., T^K\}$, where $T^i = \{t_j^i\}$ is a set of transformations for PPI i. We require that $\forall t^i \in T^i$ $S(t^i) > S^*$, otherwise the transformation t^i can be rejected. A multiple alignment is a combination of $K - 1$ transformations, $(t^2, t^3, ..., t^K)$. We iteratively traverse a set CB in the following manner. Assume that we have created a vector of first m transformations $T = (t^2, t^3, ..., t^m)$. We try to extend it with a transformation t^{m+1}, $T^* = (t^2, t^3, ..., t^m, t^{m+1})$. Clearly, $S(T^*) \leq S(T)$. We can effectively estimate $S(T^*)$ without actually solving the matching problem. $S(T^*)$ is less than the sum of the maximal possible scores of elements (IN and PC) of I_1 which have at least one close element from each transformed point set $t^2(I_2)...t^{m+1}(I_{m+1})$. As we extend the vector T the number of such elements drops very quickly. If $S(T^*)$ drops below S^*, then we disregard the vector T^* and start to build another combination. Essentially, we continue with the vector T and try to add another transformation from T^{m+1}, and so on. The number of traversals may be exponential, however the practical running times are significantly lower due to the filtering (*bound*) step.

Multiple 3D-Pivot Matching. During the iterations from the previous stage, once we reach the end of the transformation set traversal we have a uniquely defined set of K transformations. At this stage we need to compute a set of matched interactions, IN, and a set of matched pseudocenters, PC, that maximize the score S. Therefore, we face another combinatorial problem. However, this optimization problem is NP-Hard even for 2 interfaces (optimization of only PC score, for 2 structures, is solvable by maximal weight bipartite matching algorithm which is polynomial). For two interfaces the matching problem is similar to the 3D 4-partite matching problem, where we look for the largest set of disjoint 4-tuples such that each 4-tuple consists of ϵ-close points. Two interfaces define four partitions (A_1, B_1, A_2, B_2). There are interaction edges between (A_1, B_1) and (A_2, B_2), and there are edges between similar type pseudocenters, i.e. between (A_1, A_2) and (B_1, B_2). There are no edges between partitions (A_1, B_2) and (A_2, B_1), therefore this problem may appear more simple than the general 3D K-partite matching. However, it is still NP-Hard since the 3D K-partite matching problem is hard even for three partitions [16].

Here, we apply the following greedy method. First, we greedily select K-tuples of interactions in the descending order of S_{IN}. Notice, each selected K-tuple may intersect with at most K K-tuples from the optimal matching with a lower score.

When no interaction K-tuple can be created, then, separately for each PPI side, we greedily select pseudocenter K-tuples in the descending order of S_{PC}.

Complexity and Accuracy. Here we summarize the accuracy and complexity of the MAPPIS algorithm. In case the scoring functions S_{IN} and S_{PC} do not depend on inter atomic distances, i.e. consider only types of interactions and pseudocenters, the MAPPIS algorithm is an $(\beta, \gamma_{in} = 2K, \gamma_{pc} = K)$-approximation for any given β. Otherwise, γ_{in} and γ_{pc} depend on the accuracy of the superposition, i.e. β. Let $f(\beta)$ and $g(\beta)$ measure the maximal deviations of the scores S_{IN} and S_{PC} as a function of β. Then the algorithm approximation is $(\beta, \gamma_{in} = 2K \cdot f(\beta), \gamma_{pc} = K \cdot g(\beta))$. The time complexity depends mainly on the second combinatorial stage. Assume that the maximal depth of the filtering iterations is $K' \leq K$. Therefore, the time complexity is $O(n^{3K'} nK \ log(n)(\frac{\epsilon}{\beta})^6)$. In practice, the method quickly detects a high scoring solution and the exponential number of iterations is avoided (as the input structures are more similar the *bound* filter is more effective, thus $K' << K$). The practical running times are low as reported in Table 1.

Heuristic Improvement. Here we give a heuristic improvement that led to practically better running times without reduction of the final score. To define a 3D transformation for each PPI, instead of considering each triplet of points from $A_i \cup B_i$, we utilize the interface interaction information and consider only two pairs of interacting pseudo centers. Given two interactions from two PPIs, $(a_i^j, b_i^j) \in I_j$ and $(a_i^t, b_i^t) \in I_t$, $i = 1, 2$, we apply the least square fitting to compute a transformation T^*, that minimizes the RMSD between the pseudocenters: $\sqrt{(\sum_i |a_i^j - T^*(a_i^t)|^2 + \sum_i |b_i^j - T^*(b_i^t)|^2)\frac{1}{4}}$. This reduces the number of transformation to $\binom{|A_1|}{2} * \binom{|A_2|}{2}$, which is $O(n^4)$, instead of $O(n^6)$ as previously described. However, the approximation factors cannot be guaranteed.

4   Results

We have applied MAPPIS to several case studies. In all of the examples, we describe the details of a single solution with the highest score.

4.1   Small G Proteins: Their Regulators and Effectors

G proteins, which are also known as GTPases and GTP binding proteins, are a well studied group of GTP hydrolases involved in cell signaling [18, 19]. Their activity is regulated by three distinct families of regulatory proteins: (1) Guanine Dissociation Inhibitors (GDIs); (2) Guanine nucleotide Exchange Factors (GEFs) and (3) GTPase activating proteins (GAPs). In addition, G proteins regulate a large number of diverse proteins, known as downstream effectors. In the examples below, we apply MAPPIS to analyze the interfaces of G proteins with their effectors and regulators.

Interactions with GDIs. GDIs which interact only with the GDP-bound form of G proteins are responsible for the regulation and separation of the GTPases from the membrane into the cytoplasm. We used MAPPIS to align between 3 interfaces created by GDIs with G proteins of type Cdc42 and Rac (PDB: 1hh4, 1ds6, 1doa). These were recognized by MAPPIS to share 21 conserved interactions and the obtained alignment is correct and consistent with the results of both sequence and backbone alignments, as well as the study of Dvorsky et al. [19]. The advantage of MAPPIS in this case is the insight on the physico-chemical nature of the interactions created by the side-chain and backbone atoms of the proteins. Specifically, it provided the details of the interactions created by amino acids reported by Dvorsky et al. [19] to be important for the stability of the complex. In addition, it provides explanations for unfavorable substitutions of amino acids, such as of Alanine to Proline. For example, MAPPIS recognized that a substitution of the amino acids Ala331/P28 of GDIs (PDB: 1hh4/1ds6) preserves a hydrogen bond created by the backbone atoms.

Interactions with GEFs. GEF proteins accelerate GDP/GTP exchange as a response to the extracellular signal. In this study we align 5 interfaces of G proteins with GEFs from two different SCOP folds: (1) DBL homology domain (PDB: 1lb1, 1foe, 1kz7, 1ki1) and GEF domain of SopE toxin (1gzs) [19]. Figure 1(a) depicts the interactions shared by the interfaces. The residue of Thr37, which was reported by Dvorsky et al. [19] to be important for the interaction with Glu639 of GEF. MAPPIS indeed recognized such interactions created by the backbone atom of Thr37. Whereas the side chain of Thr37 was recognized to participate in hydrophobic aliphatic interactions with Leu777 of GEF (PDB:1lb1). The prominent residues Leu69 and Leu72 [19] were also recognized to participate in conserved hydrophobic aromatic interactions.

Interactions with GAPs. GAP proteins interact with the GTP bound state of G proteins, accelerating the rate of GTP hydrolysis. We applied MAPPIS to align 7 interfaces created by G proteins with GAPs from two different folds: (1) GTPase activation domain, type p50 RhoGAP (PDB: 1tx4, 1ow3, 1am4, 1grn, 2ngr) and (2) Four-helical up-and-down bundle (PDB: 1he1, 1g4u). While 10 interactions were recognized as conserved within the members of the first fold, only 3 of them are shared by all the 7 interfaces.

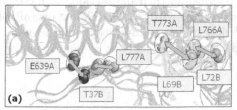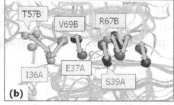

Fig. 1. The conserved interactions recognized by MAPPIS, represented as in Box 1. **(a)** Alignment of 5 PPIs of G proteins (blue) with GEF regulators from two different folds (green). **(b)** Interactions shared by 4 G proteins (blue) with effectors (green) of 3 different folds. Only 3 out of 7 conserved residues are labeled according to PDB:1c1y.

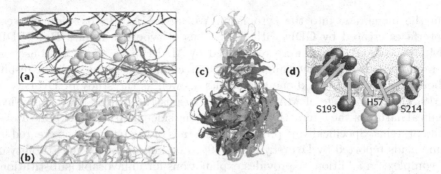

Fig. 2. Alignments of PPIS clusters. Shared interactions recognized by MAP-PIS represented as in Box 1. **(a)** Alignment of 5 PPIs from cluster #673(PDB: 1i9rAB,1jh5AB, 1d0gAB, 1a8mAB) [12]. The 4 shared aromatic interactions are represented by the pseudocenters. **(b)** Alignment of 6 PPIs from cluster #99 (PDB: 1l3bAD, 1l0oAB, 1b99AD, 1e7pAD, 1gttBC, 1iunAB). The PPIs are created by proteins of different overall folds but share 4 hydrophobic aliphatic interactions. **(c)** Alignment of 6 PPIs of Trypsin-like serine proteases (4sgb, 1ppf, 1acb, blue and red) and Subtilisin-like (1cse, 2sic, 1oyv, green and yellow). **(d)** The common interactions recognized by MAPPIS. The residues that are conserved in sequence in all the proteins are annotated according to PDB:4sgb.

Interactions with Effectors. Association of the G proteins with the effector proteins enables them to control a wide range of intracellular signaling pathways. We compared between the interfaces of 4 complexes of G proteins with effectors: (1) cH-p21 Ras with Phoshoinositide 3-kinase (PI3K) (PDB:1he8); (2-3) Rap1 with c-Raf1 RBD (PDBs: 1c1y,1gua); (4) CDC42 with PDZ domain (PDB:1nf3). In spite of the fact that the effectors of these complexes belong to three different folds and share almost no sequence similarity, all of the interfaces were recognized to share a pattern of 7 interactions. (see Figure 1(b)).

It must be noted that the interfaces in these examples are created by G proteins, which can be superimposed by multiple backbone alignment methods [20, 21]. However, these methods do not recognize the similarity of their physico-chemical properties and do not consider their interactions with the binding partners, which in many cases are proteins with totally different overall folds. For some of the examples presented below the superimposition and the matching problems can not be solved by standard protein backbone alignment methods.

4.2   PPI Clusters

We have applied our method to analyze the interactions shared by PPI clusters [12] created by iteratively applying a pairwise alignment method [11]. The new MAPPIS software with its ability to detect consensus binding organizations now allows to acquire additional insights on the interactions shared by all the members of the created clusters. The results that are automatically obtained

Table 1. Performance of MAPPIS

Case study	Num. of PPIs	Mean PPI size	Num. of interactions theor. : real : cons.	Run time (sec.)
G-proteins with GDIs	3	225	56 : 37 : 21	23
G-proteins with GEFs	5	124	59 : 39 : 4	54
G-proteins with GAPs	7	177	41 : 27 : 3	155
G-proteins with Effectors	4	133	40 : 30 : 7	22
Cluster 673 [12], TNF Family	5	165	49 : 33 : 4	48
Cluster 99 [12]	6	120	34 : 25 : 4	27
Serine Proteases	6	120	48 : 27 : 8	163

by MAPPIS are consistent with the manual biological inspections of Mintz et al. [12]. For example, when applied to cluster number 673 with 5 PPIs of members of the TNF (tumor necrosis factor) family, MAPPIS recognized an exceptional conservation of 4 aromatic interactions shared by all the interfaces (see Figure 2(a)). When applied to 6 PPIs of cluster number 99, MAPPIS revealed an pattern of 4 conserved hydrophobic interactions (see Figure 2(b)). All of the PPIs in this cluster are created by proteins of different folds, but the low-level function defined for most of them by GO is the Transferase activity [12]. An additional example is a cluster created by serine proteases, which are the most well studied example of functionally similar proteins with different overall folds: trypsin and subtilisin [4,7,10]. The MAPPIS solution is correct due to the correct alignment of the catalytic residues of these proteins. The advantage of MAPPIS is in the analysis of similarity of the created interactions (see Figure 2d).

4.3   Performance Evaluation

The general performance of MAPPIS is summarized in Table 1. The running times are measured on a standard PC, Intel(R) Pentium(R) IV 2.60GHz CPU with 2GB RAM. In each example, the table presents the average PPI size measured by the sum of pseudocenters of its two binding sites. In addition, we provide the mean number of interactions, represented by three values: (1) The

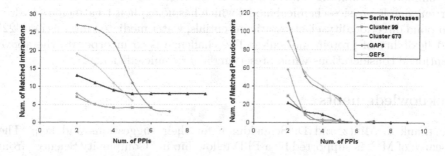

Fig. 3. Dependence of the number of matched interactions and matched non-interacting pseudocenters on the number of aligned PPIs

number of theoretical interactions defined by all pairs of complementary pseudo-centers that can potentially interact with each other; (2) The estimated number of "real" interactions after the limitation on the maximal number of interactions in which each pseudocenter can participate; (3) The number of conserved interactions shared by all of the aligned PPIs as detected by MAPPIS. Figure 3 presents the dependence of the common pattern of interactions on the number of aligned PPIs. In each example, the number of aligned PPIs ranges from two to the size of the dataset. The most interesting example is the alignment of PPIs of serine proteases, where we could enlarge the number of aligned PPIs (added PDBs:1csoEI, 1d6rAI, 1c9tAG). As can be seen, the common core converged to 8 common interactions, which indicates a strong conservation of the interaction pattern. Interestingly, the number of common pseudocenters converged to zero.

5   Summary and Conclusions

We presented a novel computational method, MAPPIS, for recognition of common physico-chemical properties and their interactions shared by a set of protein-protein interfaces (PPIs) without the need to assume similarity of sequential patterns or backbone patterns. We have shown its applications to several biological case studies. Our results are consistent with the available biological data. Computationally, the major advantages of our technique (e.g. over sub-graph isomorphism search strategies) are (1) polynomial time approximation algorithm for two PPIs, (2) for multiple PPI alignment, the practical convergence of the branch-and-bound stage to a high scoring solution is very quick and the exponential number of iterations is avoided. Practical running times range from seconds to several minutes.

Despite the guaranteed approximation to the given scoring function the method's approximation ratios are relatively high. The most problematic stage is the computation of the final matching, which gives a K-approximation. As a result there is a dependence on the pivot selection and on the order of the input PPIs. In our future research we intend to address the estimation of the biological significance of the recognized patterns. Specifically, we will explore what is the minimal number, combination, and type of interactions that are responsible for the similarity in binding and function. In addition, we consider to extend the current simple physico-chemical model which has no explicit treatment of hydrogen bond directionality, electrostatic potentials, water mediated interactions [22] and flexibility of protein molecules. The challenge is to incorporate the above mentioned considerations while preserving algorithmic efficiency.

Acknowledgments

We thank S. Mintz and D. Schneidman for their suggestions and help. The research of M.S. is supported by a PhD fellowship in "Complexity Science" from the Yeshaya Horowitz foundation. This research has been supported in part by the "Center of Excellence in Geometric Computing and its Applications" funded by the Israel Science Foundation. The research of H.J.W. is partially supported by the Hermann Minkowski-Minerva Center for Geometry at TAU. The research of R.N. has been funded in whole or in part with Federal funds from the NCI, NIH, under contract number NO1-CO-12400. The content of this publication does not necessarily reflect the view or policies of the Dep. of Health and Human Services, nor does mention of trade names, commercial products, or organization imply endorsement by the U.S. Government.

References

1. Lo Conte, L., Chothia, C., Janin, J.: The atomic structure of protein-protein recognition sites. J. Mol. Biol. **285** (1999) 2177–2198
2. Mintseris, J., Weng, Z.: Atomic contact vectors in protein-protein recognition. Proteins **53** (2003) 629–639
3. Falquet, L., Pagni, M., Bucher, P., Hulo, N., Sigrist, C., Hofmann, K., Bairoch, A.: The PROSITE database, its status in 2002. Nucleic Acids Res. **30** (2002) 235–238
4. Wallace, A.C., Laskowski, R.A., Thornton, J.M.: Derivation of 3D coordinate templates for searching structural databases: application to Ser-His-Asp catalytic triads in the serine proteinases and lipases. Protein Sci. **5** (1996) 1001–1013
5. Keskin, A., Tsai, C.H., Wolfson, H.J., Nussinov, R.: A new, structurally non-redundant, diverse dataset of protein-protein interfaces and its implications. Prot. Sci. **13(4)** (2004) 1043–55
6. Spriggs, R.V., Artymiuk, P.J., Willett, P.: Searching for patterns of amino acids in 3d protein structures. J. Chem. Inf. Comput. Sci. **43** (2003) 412–421
7. Schmitt, S., Kuhn, D., Klebe, G.: A new method to detect related function among proteins independent of sequence or fold homology. J. Mol. Biol. **323** (2002) 387–406
8. Kinoshita, K., Nakamura, H.: Identification of protein biochemical functions by similarity search using the molecular surface database eF-site. Protein Sci. **12** (2003) 1589–1595
9. Shulman-Peleg, A., Nussinov, R., Wolfson, H.J.: Recognition of functional sites in protein structures. J. Mol. Biol. **339(3)** (2004) 607–633
10. Russell, R.: Detection of protein three-dimensional side-chain patterns: new examples of convergent evolution. J. Mol. Biol. **279(5)** (1998) 1211–1227
11. Shulman-Peleg, A., Mintz, S., Nussinov, R., Wolfson, H.: Protein-protein interfaces: Recognition of similar spatial and chemical organizations. In Jonassen, I., Kim, J., eds.: Workshop on Algorithms in Bioinformatics. (2004) 194–205 LNCS, 3240.
12. Mintz, S., Shulman-Peleg, A., Wolfson, H.J., Nussinov, R.: Generation and analysis of a protein-protein interface dataset with similar chemical and spatial patterns of interactions. Proteins **in press** (2005) http://bioinfo3d.cs.tau.ac.il/Interfaces/.
13. Ambuhl, C., Chakraborty, S., Gartner, B.: Computing largest common point sets under approximate congruence. In: Proceedings of the 8th Annual European Symposium on Algorithms, Springer-Verlag (2000) 52–63
14. Heffernan, P.J., Schirra, S.: Approximate decision algorithms for point set congruence. Comput. Geom. Theory Appl. **4** (1994) 137–156
15. Akutsu, T., Halldorson, M.M.: On the approximation of largest common subtrees and largest common point sets. Theoretical Computer Science **233** (2000) 33–50

16. Shatsky, M., Shulman-Peleg, A., Nussinov, R., Wolfson, H.: Recognition of binding patterns common to a set of protein structures. In Miyano, S., ed.: RECOMB 2005, Cambridge MA. Volume 3500. LNCS (2005) 440–455
17. McDonald, I.K., Thornton, J.M.: Satisfying hydrogen bonding potential in proteins. J. Mol. Biol. **238** (1994) 777–793
18. Paduch, M., Jelen, F., Otlewski, J.: Structure of small G proteins and their regulators. Acta Biochim Pol. **48** (2001) 829–50
19. Dvorsky, R., Ahmadian, M.R.: Always look on the bright site of Rho: structural implications for a conserved intermolecular interface. EMBO Rep. **5** (2004) 1130–6
20. Shatsky, M., Nussinov, R., Wolfson, H.: MultiProt - a multiple protein structural alignment algorithm. In Guigo, R., Gusfield, D., eds.: Workshop on Algorithms in Bioinformatics. Volume 2452. Springer Verlag (2002) 235–250
21. Dror, O., Benyamini, H., Nussinov, R., Wolfson, H.: MASS: multiple structural alignment by secondary structures. Bioinformatics **19 Suppl. 1** (2003) i95–i104
22. Rodier, F., Bahadur, R., Chakrabarti, P., Janin, J.: Hydration of protein-protein interfaces. Proteins **60** (2005) 36–45
23. Connolly, M.L.: Measurement of protein surfaces shape by solid angles. J. Mol. Graph. **4** (1986) 3–6

Appendix. Physico-Chemical Score

Similarity between two superimposed interactions $i = (a, b)$ and $i' = (a', b')$ is measured by:

$$S(i, i') = S_{IN}(i) + S_{IN}(i') + S_{PC}(a, a') + S_{PC}(b, b')$$

$$S_{IN}(i) =$$
$$propen(i) \cdot \begin{cases} 0, & dist(i) > max\_dist(i) \\ (max\_dist(i) - dist(i))/(1 + charge\_comp(i)) & chem(i) = HB \\ (max\_dist(i) - dist(i))/(1 + shape\_comp(i)) & chem(i) = ALI \\ (max\_dist(i) - dist(i))/(1 + shape\_comp(i) + n_{PII}(i)) & chem(i) = PII \end{cases}$$

The similarity between two superimposed pseudocenters is defined by [16]:

$$S_{PC}(a, b) =$$

$$\begin{cases} 0, & dist(a, b) > max\_dist(a, b) \text{ or } chem(a) \neq chem(b) \\ 0, & shape(a, b) > 0.2 \text{ or } n_S(a, b) > 0.2 \\ (max\_dist(a, b) - dist(a, b))/(1 + charge(a, b)) & chem(a) = HB \\ (max\_dist(a, b) - dist(a, b))/(1 + shape(a, b) + n_{PII}(a, b)) & chem(a) = PII \\ (max\_dist(a, b) - dist(a, b) + v_{ALI}(a, b))/(2 + 20 * shape(a, b)) & chem(a) = ALI \end{cases}$$

- $dist(a, b)$ - the distance between a and b after the superimposition. $dist(i)$ - the distance between interacting pseudocenters a and b.
- $chem(a)$, $chem(i)$ - the physico-chemical property of the point a or interaction i. There are three types of properties: Hydrogen Bonding (HB), Aliphatic Hydrophobic (ALI) and Aromatic (PII).
- $max\_dist(a, b)$ - maximal allowed distance between a pair of pseudocenters, defined by $\epsilon =$3.0Å. $max\_dist(i)$ - the maximal distance allowed for the specific type of interaction. The default thresholds are $\gamma =$3.9Å for hydrogen bonds [17] and $\gamma =$8.0Å for hydrophobic aliphatic and aromatic interactions.

- $charge(a)$ - the partial atomic charge of the atom a, which can form hydrogen bonds. $charge(a,b) = |charge(a) - charge(b)|$ - measures the similarity of charges. $charge\_comp(i) = |charge(a) + charge(b)|$ - measures the complementarity of charges.
- $shape(a)$ - the average curvature of the surface region created by a. Calculated as an average of the solid angle shape functions [23] with spheres of radius 4,5,6 and 7Å . The sphere centers are located at projection point of a to the surface. $shape(a,b) = |shape(a) - shape(b)|$ - measures the similarity of shapes. $shape\_comp(i) = |1 - shape(a) - shape(b)|$ - measures the complementarity of shapes which sums to one.
- $n_S(a)$ - normal vector at projection point of a to the surface, $n_S(a,b) = n_S(a) \cdot n_S(b)$.
- $v_{ALI}(a,b)$ - the overlap of the hydrophobic group spheres of a and b, approximated by the difference between sum of radiuses and the distance between the centers.
- $n_{PII}(a)$ - for aromatic pseudocenters denotes the normal to the plane of the aromatic ring. $n_{PII}(i) = n_{PII}(a) \cdot n_{PII}(b)$ - represents the angle between two interacting aromatic ring.
- $propen(p)$ - the propensity of the physico-chemical property in the interface compared to the overall protein chain. The propensities of the pseudocenters were calculated by Mintz et al. [12]. $propen(i) = propen(a) \cdot propen(b)$.

Frequent Itemsets for Genomic Profiling

Jeannette M. de Graaf[1], Renée X. de Menezes[2,3],
Judith M. Boer[2], and Walter A. Kosters[1]

[1] Leiden Institute of Advanced Computer Science,
Universiteit Leiden, Leiden, The Netherlands
{graaf, kosters}@liacs.nl
[2] Center for Human and Clinical Genetics,
Leiden University Medical Center, Leiden, The Netherlands
{r.x.menezes, j.m.boer}@lumc.nl
[3] Laboratory of Pediatrics, Erasmus Medical Center,
Rotterdam, The Netherlands

Abstract. Frequent itemset mining is a promising approach to the
study of genomic profiling data. Here a dataset consists of real numbers describing the relative level in which a clone occurs in human DNA
for given patient samples. One can then mine, for example, for sets of
samples that share some common behavior on the clones, i.e., gains or
losses. Frequent itemsets show promising biological expressiveness, can
be computed efficiently, and are very flexible. Their visualization provides the biologist with useful information for the discovery of patterns.
Also it turns out that the use of (larger) frequent itemsets tends to filter
out noise.

1 Introduction

Frequent itemsets are often used in Data Mining research [11]; they can supplement the more traditional statistical approach [2]. The concept is simple, many
efficient algorithms are devised to detect different types of frequent itemsets,
and there is a rich literature describing associated topics. For instance, many researchers dealt with the problem of finding interesting sets, and the fuzzy logic
approach also gave a new impetus. The most well-known application is in the
area of market-basket analysis. In this case a frequent itemset is a set of products that is often purchased together. From such a set one can easily deduce
association rules of the form "if one buys X, one (often) buys Y too".

In this paper we apply the frequent itemset approach to explore copy number
changes in the genome. Chromosomal instability in tumors leads to DNA copy
number alterations with associated gain or loss of genes important in tumor
development [5]. Array-based comparative genomic hybridization (array CGH)
allows for high-throughput genome-wide screening of these DNA copy number
changes [1,7,9]. Typically, these experiments involve co-hybridization of a few
hundred fluorescently labeled patient DNA samples with normal reference DNA
onto microarrays containing several thousands of large-insert genomic clones

M.R. Berthold et al. (Eds.): CompLife 2005, LNBI 3695, pp. 104–116, 2005.

(relatively short pieces of DNA) such as bacterial artificial chromosomes (BACs). The resulting dataset is a database of clones, each consisting of a few hundred real numbers. Any such number describes the normalized log2-ratio of the number of clone copies found in a given patient sample compared with the reference DNA. When there is no copy number change in the patient DNA, the log2-ratio is expected to be 0 (no change). When the log2-ratio lies above a certain threshold or below another fixed threshold, the patient has a *gain* or a *loss*, respectively, for this clone. In principle, the boundaries between no change and change are very strict, allowing discretization of the data. However, factors such as tissue heterogeneity (i.e., a loss or a gain is present in only a subset of the cells) and the use of amplification procedures introduce more variation in measurements, making such boundaries less strict. And finally, we have the usual problems like measurement noise.

The database records can be viewed in (at least) two different ways. First, one can look at the clones as transactions, and view the samples as items; this is called here the *frequent sample sets model*. Note that this is the way in which the data is usually presented. Second, one can also see the samples as transactions, and the clones as items; this we call the *frequent clone sets model*. In this paper we treat both approaches, with emphasis on the first one. If we adhere to the first choice, we are interested in groups of samples, where the group elements share some common behavior; for the second choice, we try to find associations between clones. We shall provide many examples of the use of frequent itemsets in this biological setting.

For related work we refer to [8], where — among other things — minimal and related gain and loss zones are detected using frequent pattern mining. In [10] a method is discussed that deals with finding interesting association rules. The first step is to generate frequent itemsets. From these one can deduce a huge amount of association rules. The authors deal with a method to filter out, after all rules are discovered, the most interesting rules for biologists.

In the current paper we generate the frequent itemsets and extract useful information from those. We also show that frequent itemsets can be used to reduce the effect of noise. We mainly focus on visualizations, which are easily made and from which biologists can deduce information about certain relations between clones or patient samples. Our method is meant to be used as an exploratory tool, aiming at pattern discovery, that can be used in combination with other methods.

We shall not treat the database in any detail, but rather refer to the paper where it originates from [6]. In a few places we shall provide the necessary biological background, and we mention the biological consequences of the proposed methods. Anyway, there is a lot of data preparation involved, apart from some trivial data cleaning. In particular we mention the problem of distinguishing change from no change, as mentioned above, which is both of technical as well as biological nature.

The paper is organized in the following way. We first describe the method and illustrate it by using artificial data (Section 2 and Section 3), for both models mentioned earlier. In Section 4 we apply the techniques to the real life data from [6] and focus on the biological consequences of the proposed methods. We end with some conclusions and issues for further research.

2 Frequent Itemsets

Suppose we have a dataset $\mathcal{D}$ consisting of subsets (usually called *itemsets*) of a given finite set $\mathcal{I}$. The subsets have unique identifiers, so multiple occurrences of the same subset may appear. It is also possible to consider the dataset as a (time ordered) series, but this viewpoint is not taken here.

For any subset S of $\mathcal{I}$ we define its *support* as the number of elements in $\mathcal{D}$ that contain S. An itemset is called *frequent* if its support is larger than or equal to a pre-given support threshold *minsup*. If an itemset has k elements, it is called a k-itemset.

The first main problem in frequent itemset mining is to find all frequent itemsets for a given $\mathcal{D}$ and *minsup*. There exist many efficient implementations to tackle this problem. The fastest ones rely on so-called FP-trees and use the Apriori property [11]. For the experiments we used the implementation from [3].

In this paper we focus on data from array CGH studies. Here the original database consists of real numbers, but it is discretized to a database describing if a sample has a gain on a clone or not, or to a database showing if a sample has a loss on a clone or not. This is done because in CGH analysis one is often more interested in whether or not a patient has a gain (loss) at some clone, and not in the exact value. So the database consists of itemsets that are either sets of samples that have higher (lower) value than normal for a given clone, or sets of clones that have higher (lower) value than normal for a given sample. depending on whether we are more interested in the gains or the losses. In the first case, the clone is the identifier of the itemset, in the second case the sample is the identifier. One can think of the database as a two-dimensional array where rows correspond to clones and columns to samples (or the other way round in the second case). The transformed database contains only zeros and ones. If a clone occurs more (less) than normal for a given patient (its value being higher (lower) than some threshold), it is assigned a one on the corresponding array position, otherwise a zero.

3 Simulated Data

A dataset with similar structure to the one from array CGH studies was simulated as follows. A total of 150 samples, with 3200 observations per sample, are divided into three main groups of 50 samples each. Samples in each group are characterized by having in common a specific copy number effect in one of the chromosomes, as well as other effects in other chromosomes, as summarized in Table 1. The effect is assumed to hold for a given number of consecutive clones (shown between brackets) at the beginning of the affected chromosomes.

These three groups can be thought of as referring to patients with the same disease, but different genotypes. This is observed for example in many cancers, where various genotypic mechanisms can lead to the same result, as in the same kind of cancer. It is important to identify these different mechanisms since they often are associated with varying susceptibility levels to treatments and, as a consequence, varying chances of recovery. Sometimes these mechanisms share part of their structure, but differ in other parts.

Unaffected clone intensities are assumed to be independent of each other and to follow a normal distribution with mean 0 and standard deviation 1. Affected clone intensities are also assumed to be independent of each other and have a normal distribution with standard deviation 1, but their mean is taken as either 3 (if effect is a gain) or −3 (if effect is a loss).

Table 1. Summary of simulated effects (G = gain, L = loss)

Samples affected	Chromosomes								Gains/losses (total)
	1	3	7	10	11	13	18	20	
136–150		G(60)		G(40)			G(30)		130/0
121–135		G(60)				L(50)			60/50
101–120		G(60)							60/0
91–100									0/0
76–90	L(80)				L(60)				0/140
61–75	L(80)		G(50)						50/80
51–60	L(80)								0/80
36–50		G(60)					G(30)	G(20)	110/0
21–35		G(60)				L(50)			60/50
1–20		G(60)							60/0

In order to evaluate the effect of having more or less noise in the data, we have also simulated a dataset with the same structure and effects, where the standard deviation of the measurements was 0.6 instead of 1. This dataset is referred to as the *ideal dataset*: it corresponds to an "ideal" scenario, where there is very good separation between measurements with copy number and without. Of course, in such a case no special method has to be used to identify effects. In practice, however, it is more common to observe datasets with less perfect separation, as the first one. This dataset is called the *noisy dataset*.

3.1   The Frequent Sample Sets Model

We now regard the database as an ordered series of 3200 clones. Each record (i.e., clone, transaction) consists of 150 real numbers, corresponding to the samples (patients). As mentioned before we first transform the database into a database of zeros and ones after defining suitable thresholds for gains and losses.

In order to obtain insight in the data, and also to give a first (simple) application of frequent itemsets, in Figure 1 we show all frequent samples, i.e.,

1-itemsets. In the left hand side picture we have the ideal dataset, in the right hand side picture the noisy dataset. In a sense, these pictures give simple snapshots of the entire dataset: the picture on the left clearly reflects Table 1, while the dense regions of +'s and ×'s in the picture on the right also do so, but less convincing. The vertical lines denote the chromosome boundaries, with the chromosome numbers on top. We show the 1-itemsets for gains and the 1-itemsets for losses in one picture; gains (value > 2.0) have +'s, losses (value < -2.0) have ×'s. If a 1-itemset $\{i\}$ has at least $minsup = 30$ gains, those gains are plotted horizontally at y-level i, and similarly for the losses. For example, sample 80 has a series of ×'s for chromosomes 1 and 11, and single ×'s for clone 1306 (in chromosome 7) and clone 1623 (in chromosome 9). For the ideal dataset there are 115 frequent 1-itemsets for gains (meaning there are 115 patient samples having a gain on at least 30 clones), and 70 for losses — as expected. The noisy dataset has 150 frequent 1-itemsets for both, or equivalently: every 1-itemset is frequent here, so every patient has at least 30 gains and 30 losses!

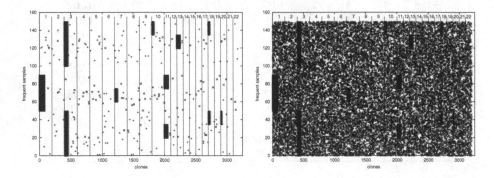

Fig. 1. Frequent samples (1-itemsets) for ideal (left) and noisy (right) dataset; gains (+) and losses (×)

The left hand side picture clearly shows the expected effects, the right hand side picture is much more diffuse. The supports on the left (the numbers of +/×'s in a single row) are smaller and the distribution is more crisp.

Note that these two pictures are the only ones that contain two types of itemsets in one image. In order to be frequent, an itemset should have at least some minimum number of gains or losses (but not together). In the sequel we also mention "combined gains and losses", which means that we add the numbers of gains and losses.

We now try to find larger sets of samples that share some common behavior, i.e., we look at k-itemsets with $k > 1$. We first examine gains; we let $minsup = 60$. In the plots from Figure 2 we depict the frequent 2-itemsets. Every horizontal series of +'s indicates the clones that have gains for *both* samples in the set. The frequent itemsets are depicted in the order in which they are generated by the algorithm from [3]; roughly speaking, larger supports occur for the higher numbered sample sets. Neighbouring sample sets usually have a non-empty intersection in this order (which is not the case if they are ordered by support). Again, the left hand side picture is for the ideal dataset. In this case we have 443 frequent 2-itemsets; the 2-itemset {138, 140} has the highest support: 122. This means that there are 122 clones on which sample 138 and sample 140 both have gains. Note that the gains series on chromosome 7 is not visible, since its length (50 clones) is smaller than *minsup* and samples 61–75 have no other gains. Therefore no combination of two samples from 61–75 (the only samples that have gains on chromosome 7) can reach the threshold 60. Furthermore, the samples 61–75 will not occur in any of the 443 frequent 2-itemsets.

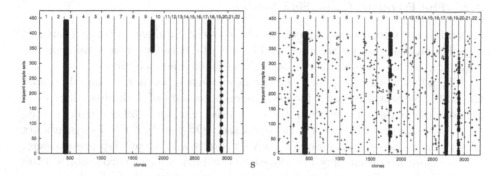

Fig. 2. Frequent 2-itemsets for ideal (left) and noisy (right) dataset; gains

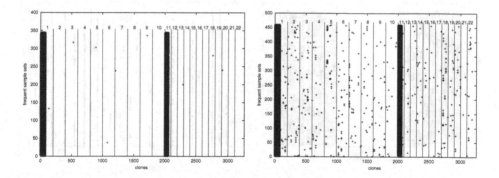

Fig. 3. Frequent sample sets (3-itemsets) for noisy dataset; losses; left: loss if value < -2.0, right: loss if value < -1.5

For the noisy dataset (Figure 2, right) there are 405 frequent 2-itemsets; the 2-itemset {139, 141} has the highest support: 105. Like in the ideal case we do not see the gains at chromosome 7 here either.

This example also reveals that a larger value of the size of the itemsets allows for better pictures, in particular for the noisy case. Patterns are much more visible now. In a next step one might decide to study chromosomes 3, 10,

18 and 20 in more detail, e.g., using the same method again on these specific chromosomes.

As a final picture we show the frequent 3-itemsets (at least having 80 common losses) for the noisy dataset, see Figure 3. In the left plot we have a loss if the dataset value is smaller than -2.0 (344 itemsets), in the right plot if the value is smaller than -1.5 (459 itemsets). This shows the dependence on the threshold defining gains/losses. Note that losses on chromosome 13 are not visible, because they only occur on 50 clones, and for samples which do not have losses elsewhere. It appears that the itemsets show a lot of overlap — a phenomenon that emerges even more for larger values of the itemset size.

3.2   The Frequent Clone Sets Model

As said in the introduction, we can also look at the database as being a series of samples. In that case we are interested in sets of clones that behave in a similar way, e.g., are all gains on at least some *minsup* common samples.

The picture below (Figure 4, left), which is just a "random" example, shows the 1136 frequent 7-itemsets (so each itemset consists of 7 clones) where each element has a value larger than 2.0 on at least *minsup* = 50 common (among the 7 elements) samples, for the noisy dataset. On the right the 7 elements are plotted for these sets. As observed above, there is a lot of overlap present here. Furthermore note that only the clones at the beginning of chromosome 3 are gains for at least 50 patient samples, which is consistent with Table 1.

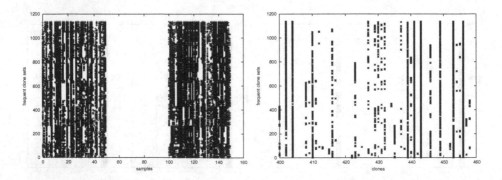

Fig. 4. Frequent clone sets (7-itemsets) for noisy dataset; gains; right: the set elements

4   Application to Colon Cancer Data

Nakao et al. [6] analyzed copy number changes in the genomes of 125 colorectal tumors using array CGH on microarrays containing 2463 BAC clones that covered the human genome at 1.5 Mb resolution. Their publicly available dataset contains normalized log2-ratios for 2124 clones (after filtering), located on chromosomes 1–22 and the X-chromosome (here referred to as 23). In this dataset any value larger than 0.225 is considered as a gain, any value smaller than −0.225 is considered as a loss. This threshold corresponds to values between 2 and 3 standard deviations from the mean. The total number of gains and losses varies between 2 and 1020 per sample.

The authors concluded that the majority of clones were infrequently gained or lost, with 95% of the changes occurring less than 35% of the time. However, high-frequency gains were detected on chromosomes 7p (35%), 7q (35%), 8q (42%), 11q (35%) and 20q (65%), and high-frequency losses were detected on 5q (35%), 8p (37%), 17p (46%), 18p (49%), 18q (60%), and 21q (35%). The distribution of alterations over the individual patients was not explored.

In Figure 5 we depict all 125 1-itemsets (combined gains and losses). In the left panel the samples are shown in the order in which they occur in the original dataset; in the right panel they are ordered with respect to their support. The 1-itemset {53} has the highest support: 1020.

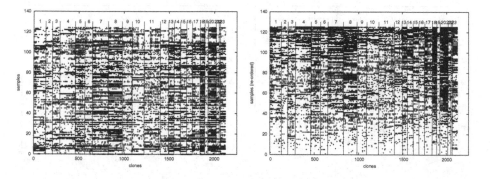

Fig. 5. All 1-itemsets, combined gains and losses; left: original order, right: ordered with respect to support

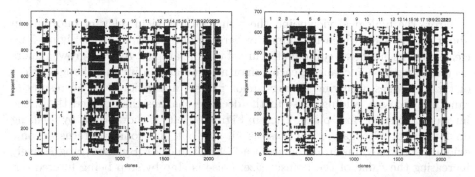

Fig. 6. Frequent sample sets; 2-itemsets; left: gains, right: losses

In Figure 6 we show the 985 frequent 2-itemsets for gains (left) and the 629 frequent 2-itemsets for losses (right), both for $minsup = 100$. Again a larger value of the itemset size gives rise to a clearer picture, showing e.g. common regions of gains on chromosomes 7, 8, 13, 20 and 23, which is consistent with the conclusions in [6]. However, the results for the synthetic noisy dataset are more outspoken, due to the random nature of this set.

This becomes even more apparent if we consider the 55 10-itemsets ($minsup$ = 100), see Figure 7. To the left we see the usual plot, showing a very small region in chromosome 11 having a gain, also detected in [6]. This region was not so clear from Figure 6, showing the importance of studying larger itemsets and thus filtering out more noise. To the right we plot for each set its 10 elements. This picture shows that the sets have quite a lot in common. It could have been worse: in the current situation there are no 11-itemsets; the 10-itemsets are all *maximal* (i.e., all their supersets are infrequent) and hence *closed* (i.e., all their supersets have lower support). The number of frequent itemsets depends on their size and on the support threshold $minsup$, as shown in Table 2. It is a challenging task to find combinations that give rise to interesting visualizations.

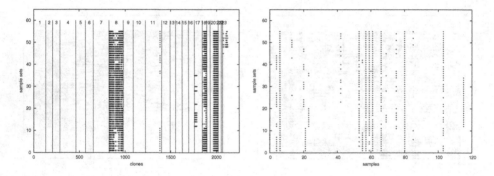

Fig. 7. Frequent sample sets (10-itemsets); combined gains and losses; right: the set elements

From the biological viewpoint, the visualizations yield relevant information about the dataset, which is commonly hard to obtain. First, from Figure 5 no clear patterns emerge, a common feature of tumor samples. As they grow, tumors accumulate genomic changes, each tumor accumulating different changes. Most of these changes occurring during tumor growth are believed to be results of random processes, adding noise to the decisive changes that turned the tissue into a tumor in the first place. Then in Figure 6 (left) it shows that, by focusing on 2-itemsets with gains, the noise is filtered out and some patterns become evident, such as gains in chromosomes 7, 8, 13 and 20. By then progressively increasing the value of the itemset size, noise is step-by-step being filtered out and only the most consistent patterns remain. Indeed, only 20 of the 125 samples (16%) in the dataset contribute to the 10-itemsets represented in Figure 7, but these have a consistent pattern of copy number changes in chromosomes 8, 18 and 20. Also chromosomes 11 and 17 show some activity. Changes in chromosome 23 (the X chromosome) are often ignored, as they mostly indicate that the sample and the control are of opposite genders, which is not of main interest.

Table 2. Number of frequent itemsets for different size and *minsup*: gains/losses/combined gains and losses

Size	*minsup* 80	*minsup* 90	*minsup* 100
1	90/84/97	87/79/96	84/76/95
2	1519/1002/3196	1236/800/2942	985/629/2743
3	6281/2222/37417	3675/1282/29634	2036/726/23228
4	10135/1618/179576	4001/647/112866	1601/285/71539
5	8090/621/425627	2147/213/210119	546/79/103318
6	3692/185/581939	556/52/220138	39/12/83637
7	972/30/507966	55/4/148282	0/0/43049
8	155/1/300636	2/0/65081	0/0/12865
9	9/0/117955	0/0/16428	0/0/1739
10	0/0/27494	0/0/1864	0/0/55
11	0/0/3048	0/0/43	0/0/0
12	0/0/79	0/0/0	0/0/0

We now look at the frequent clone sets model. Experiments showed that chromosome 20 was really dominant. Taking into account only clones 1–1800, finer patterns on other chromosomes can be discovered. As an example we show the 199 9-itemsets for gains (Figure 8), with *minsup* = 30. The right picture has the set elements (cf. Figure 4), all on chromosome 8. The four neighbouring clones near 900 are indeed of biological interest.

It is possible to use the frequent itemset approach for the discovery of particular phenomena. For example, there is exactly one 4-itemset, the set of samples {53, 59, 66, 80}, having 300 or more common gains and losses.

If one keeps track of the distance between consecutive common gains (and/or losses) one can order the frequent itemsets found. For example, for the 4-itemset mentioned before, 69% of the 313 common gains and losses are really consecutive; if one allows for at most one intermediate normal clone (a so-called *gap*), this percentage rises to 87%. In Figure 9 above we plot these last percentages for all

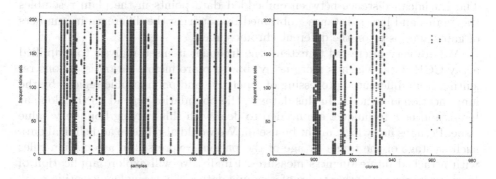

Fig. 8. Frequent clone sets (9-itemsets); gains; right: the set elements

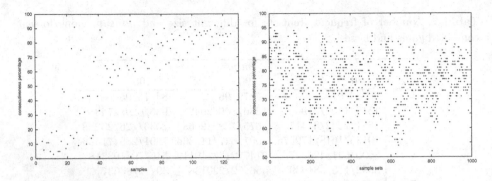

Fig. 9. Consecutiveness percentages; left: 1-itemsets, gains and losses; right: 2-itemsets, gains (sets on x-axis)

125 1-itemsets (gains and losses; ordered on the x-axis with respect to increasing support; left) and the 985 frequent 2-itemsets (gains; $minsup = 100$; right). Efficiently incorporating consecutiveness into frequent itemset mining seems non-trivial and is left for future work.

5 Conclusion and Further Research

We presented a method to discover patterns in array CGH datasets. We make use of frequent itemset mining in order to obtain combinations of samples or clones that share some common behavior. The method is flexible, fast (the generation of a picture usually takes a few seconds), capable of dealing with noise, and allows for different types of post-processing. In contrast with many other techniques the method is largely unsupervised, and allows for individual patient tracking.

Once given the frequent itemsets, one can use many different Data Mining techniques. It is for instance possible to use Self Organizing Maps (SOMs) and the like in order to obtain visualizations. In Figure 10 the 55 10-itemsets from Figure 7 are embedded in the unit square, using a push-and-pull network [4]. The Euclidean distance between embedded data points in the plain resembles the "gains and losses distance", obtained by squaring the difference in numbers of gains and losses on the different chromosomes.

We are very interested to extend the frequent itemset analysis to amplified array CGH data, which is more noisy due to reproducible ratio distortions resulting from differential processing of repetitive and polymorphic regions by the amplification enzyme [1]. In this dataset, the boundaries depend on the clone at hand, and new techniques are needed to deal with this varying boundary value issue. Perhaps fuzzy logic might be useful. We would also like to add clinical data such as stage of the tumor or age of the patient, expressed in association rules with attached interestingness measures. Finally, we will explore application of frequent itemsets to other types of genomic data, such as single nucleotide polymorphism genotyping data.

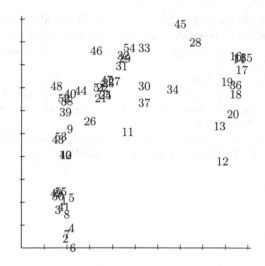

Fig. 10. Distance preserving embedding of the 55 10-itemsets from Figure 7

References

1. J. Cardoso, L. Molenaar, R.X. de Menezes, C. Rosenberg, H. Morreau, G. Möslein, R. Fodde and J.M. Boer, Genomic Profiling by DNA Amplification of Laser Capture Microdissected Tissues and Array CGH, Nucleic Acids Research 32 (2004) e146.1–146.13.
2. T. Hastie, R. Tibshirani and J. Friedman, The Elements of Statistical Learning, Springer, 2001.
3. W.A. Kosters and W. Pijls, Apriori: A Depth First Implementation, FIMI'03, Workshop on Frequent Itemset Mining Implementations 2003; CEUR Workshop Proceedings (online; B. Goethals and M.J. Zaki (eds.)).
4. W.A. Kosters and M.C. van Wezel, Competitive Neural Networks for Customer Choice Models, pp. 41–60 in: E-Commerce and Intelligent Models (J. Segovia, P.S. Szczepaniak and M. Niedzwiedzinski (eds.), Physica Verlag, Springer, 2002.
5. C. Lengauer, K. Kinzler and B. Vogelstein, Genetic Instabilities in Human Cancers. Nature 396 (1998) 643–649.
6. K. Nakao, K.R. Mehta, J. Fridlyand, D.H. Moore, A.N. Jain, A. Lafuente, J.W. Wiencke, J.P. Terdiman and F.M. Waldman, High-resolution Analysis of DNA Copy Number Alterations in Colorectal Cancer by Array-based Comparative Genomic Hybridization, Carcinogenesis 25 (2004) 1345–1357.
7. D. Pinkel, R. Segraves, D. Sudar, S. Clark, I. Poole, D. Kowbel, C. Collins, W.L. Kuo, C. Chen, Y. Zhai, S.H. Dairkee, B.M. Ljung, J.W. Gray and D.G. Albertson, High Resolution Analysis of DNA Copy Number Variation Using Comparative Genomic Hybridization to Microarrays, Nature Genetics 20 (1998) 207–211.
8. C. Rouveirol and F. Radvanyi, Local Pattern Discovery in Array-CGH Data, Proceedings Dagstuhl Workshop on Detecting Local Patterns, (J.F. Boulicaut, K. Morik and A. Siebes (eds.)), to appear in Lecture Notes in Artificial Intelligence, Springer, 2005.

9. S. Solinas-Toldo, S. Lampel, S. Stilgenbauer, L. Nickolenko, A. Benner, H. Dohner, T. Cremer and P. Lichter, Matrix-based Comparative Genomic Hybridization: Biochips to Screen for Genomic Imbalances, Genes Chromosomes Cancer 20 (1997) 399-407.
10. A. Tuzhilin and G. Adomavicius, Handling Very Large Numbers of Association Rules in the Analysis of Microarray Data, Proceedings of the Eighth ACM SIGKDD International Conference on Knowledge Discovery and Data Mining, 396–404, ACM Press, 2002.
11. C. Zhang and S. Zhang, Association Rule Mining, Models and Algorithms, Lecture Notes in Artificial Intelligence 2307, Springer, 2002.

Gene Selection Through Sensitivity Analysis of Support Vector Machines

Defeng Wang[1], Daniel S. Yeung[1], Eric C.C. Tsang[1], and Lin Shi[2]

[1] Department of Computing, The Hong Kong Polytechnic University, Hung Hom,
Kowloon, Hong Kong, China
{csdfwang, csdaniel, csetsang}@comp.polyu.edu.hk
[2] Department of Computer Science and Engineering, The Chinese University of Hong
Kong, Shatin, N.T., Hong Kong, China
lshi@cse.cuhk.edu.hk

Abstract. We present a novel approach to gene selection for microarry
data through the sensitivity analysis of support vector machines (SVMs).
A new measurement (sensitivity) is defined to quantify the saliencies
of individual features (genes) by analyzing the discriminative function
in SVMs. Our feature selection strategy is first to select the features
with higher sensitivities but meanwhile keep the remaining ones, and
then refine the selected subset by tentatively substituting some part with
fragments of the previously rejected features. The accuracy of our method
is validated experimentally on the benchmark microarray datasets.

1 Introduction

Gene selection is of fundamental and practical significance to research in biology
and medicine, especially in genetic diagnosis and drug discovery. Due to the new
advances in microarray technology, huge amounts of raw data are produced by
microarray devices, while researchers care about whether or not a small group
of genes is informative and sufficient such that the computation is reduced while
the accuracy is increased. Essentially, gene selection is a typical application of
the feature subset selection technique in machine learning.

The feature (gene) selection problem refers to the task of identifying and
selecting a representative subset of features (genes) as small as possible to represent a larger set of often mutually redundant features (genes) with different
associated measurement costs and/or risks. Feature selection methods can be
classified into the following three categories [1,2], based on the integration into
the learning method (see Fig. 1). In the *filter* approach, the selection of features
is independent of the learning algorithm: it filters out irrelevant attributes before induction occurs. *Wrapper* approach is to use an induction algorithm to
estimate the merit of the searched feature subset on the training data and to use
the estimated accuracy of the resulting classifier as its metric. The *Embedded*
approach embeds the selection within the induction algorithm.

As a popular supervised learning mechanism, support vector machines
(SVMs) have been widely used in pattern recognition, regression, image process-

M.R. Berthold et al. (Eds.): CompLife 2005, LNBI 3695, pp. 117–127, 2005.

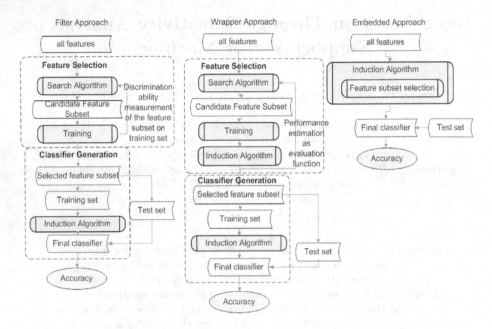

Fig. 1. Filter, wrapper and embedded feature selection approaches

ing, and bioinformatics, etc., and feature selection for SVMs has become an interesting topic with both theoretical and empirical significance. Weston et.al. [3] introduced a feature selection approach for SVMs by approximating the presence of each feature with a real number from 0 to 1, and optimizing the presence weights using the gradient descent method. However, the features are evaluated separately, and no correlation between them is considered. Guyon et al. [4] proposed SVM-RFE, a backward elimination method of gene selection using SVMs based on recursive feature elimination (RFE). This method has recently been extended by Rakotomamonjy [5] through an analysis two kinds of generalization error upper bounds. As SVM-RFE and its extension use a greedy strategy to perform backward elimination, they usually lead to suboptimal solutions.

In this paper, we introduce a new approach to feature selection for SVMs, feature selection via sensitivity analysis and refinement (FSSAR). In this framework, we first train the SVM and calculate the sensitivities of the features. We rank the features by their sensitivities and select the features with high sensitivities. Furthermore, the ranked list of features undergoes a refinement procedure and generates the final feature subset. Our method combines the simplicity of filter approach and the advantage of wrapper approach: we rank features after training SVMs for only once, which is more computationally efficient than RFE-based methods.

The paper is organized as follows. Section 2 provides the principles about SVMs. The framework and details about our FSSAR method are introduced in section 3. Section 4 presents the experimental results and comparisons. The last section gives the conclusion.

2   Support Vector Machines

Given ℓ training pairs $(\mathbf{x}_1, y_1), ..., (\mathbf{x}_\ell, y_\ell)$, where $\mathbf{x}_i \in \Re^d$ is an input vector labeled by $y_i \in +1, -1$ for $i = 1, ..., \ell$, support vector machines [6] search for a separating hyper-plane with largest margin, which is called an *optimal hyper-plane* $\mathbf{w}^T\mathbf{x} + b = 0$. This hyper-plane can classify an input pattern according to the following function $f(\mathbf{x}) = sgn(\mathbf{w}^T\mathbf{x} + b)$ where

$$sgn(k) = \begin{cases} +1, \text{ if } k \geq 0 \\ -1, \text{ if } k < 0 \end{cases}$$

In order to maximize the margin for linearly separable cases, we need to find the solution for the following quadratic programming problem

$$\min \frac{1}{2}\|\mathbf{w}\|^2 \tag{1}$$

$$s.t. \ \ y_i(\mathbf{w}^T\mathbf{x}_i + b) \geq 1, \forall i = 1, ..., \ell \tag{2}$$

In fact, there are many linearly non-separable problems in the real world. In order to solve these problems by linear SVMs, we have to modify the previous method by introducing non-negative slacking variables $\xi - i \geq 0$, $i = 1, ..., \ell$. The non-zero $\xi > 0$ are those training patterns that do not satisfy the constraints in Eq.(2). The optimal hyper-plane for this kind of problem could be found by solving the following quadratic programming problem

$$\min \frac{1}{2}\|\mathbf{w}\|^2 + C \sum_{i=1}^{\ell} \xi_i \tag{3}$$

$$s.t. \ \ y_i(\mathbf{w}^T\mathbf{x}_i + b) \geq 1 - \xi_i, \forall i = 1, ..., \ell \tag{4}$$

$$\xi_i \geq 0 \tag{5}$$

The problem is usually posed in its *Wolfe dual form* with respect to Lagrange multipliers $\alpha_i \in [0, C]$, $i = 1, ..., \ell$, which can be solved by standard quadratic optimization packages. The bias b can easily be calculated from any *margin vector* $\mathbf{x}_i$ satisfying $0 < \alpha_i < C$. The discriminative function is therefore given by

$$f(\mathbf{x}) = sgn(\mathbf{w}^T\mathbf{x} + b) = sgn(\sum_{i=1}^{\ell} \alpha_i y_i \mathbf{x}_i^T \mathbf{x} + b) \tag{6}$$

In a typical classification task, only a small number of the Lagrange multipliers α_i tend to be greater than zero. The respective training vectors are called *support vectors*, as $f(\mathbf{x})$ depends on them exclusively.

For some problems, improved classification can be achieved using nonlinear SVMs [6]. The basic idea of nonlinear SVMs is to map data vectors from the input space to a high-dimensional feature space using a nonlinear mapping Φ , and then proceed pattern classification using linear SVMs. However, the nonlinear mapping Φ is performed by employing kernel functions $K(\mathbf{x}_i, \mathbf{x})$, which obeys

Mercers conditions [6], to compute the inner products between support vectors $\Phi(\mathbf{x}_i)$ and the pattern vector $\Phi(\mathbf{x})$ in the feature space. For an unknown input pattern $\mathbf{x}$, we have the following discriminative function,

$$f(\mathbf{x}) = sgn(\sum_{i=1}^{\ell} \alpha_i y_i K(\mathbf{x}_i^T \mathbf{x}) + b) \tag{7}$$

Theorem 1. *If the ℓ training samples belonging to sphere of radius R are linearly separable with the margin M ($M = frac1\|\mathbf{w}\|$), then the expectation of the error probability has the bound [6],*

$$EP_{err} \leq (1/\ell) \cdot E\{R^2 \|\mathbf{w}\|^2\} \tag{8}$$

where the expectation is taken over all training sets of size ℓ.

This theorem justifies that the performance of SVMs depends not only on the margin M, but also the radius R, which is controlled by the mapping function Φ and can be calculated via solving a quadratic programming problem [7].

3 Feature Selection for SVMs Using FSSAR

3.1 Sensitivity Analysis of Discriminative Functions

Intuitively, the same perturbation in different features will cause the discriminative hyper-plane deviate to various extents. Greater deviation in the hyper-plane corresponds to feature with stronger impact. Here, we present an example of linearly non-separable classification problem to illustrate this idea.

In a 2-D (x^1 and x^2) linearly non-separable case, we use a linear SVM as the classifier (see Fig. 2). Fig. 2(b) and Fig. 2(c) show the discriminative hyper-planes before and after feature x^1 and x^2 feature are perturbed by $\triangle$ respectively. According to the scaling and translation invariant properties of SVMs, the margin width does not change when features are disturbed, however, the position of discriminative hyper-plane does change accordingly. Assume the displacements of the hyper-plane, due to the perturbations of x^1 and x^2, are d_1 and d_2 respectively, so that:

$$d_1 = \triangle \times \left| \frac{\partial f(\mathbf{x})}{\partial x^1} \right| \Big/ \sqrt{\left(\frac{\partial f(\mathbf{x})}{\partial x^1} \right)^2 + \left(\frac{\partial f(\mathbf{x})}{\partial x^2} \right)^2} \tag{9}$$

$$d_2 = \triangle \times \left| \frac{\partial f(\mathbf{x})}{\partial x^2} \right| \Big/ \sqrt{\left(\frac{\partial f(\mathbf{x})}{\partial x^1} \right)^2 + \left(\frac{\partial f(\mathbf{x})}{\partial x^2} \right)^2} \tag{10}$$

where $\mathbf{x} = [x^1, x^2]^T$ and $f(\mathbf{x}) = \mathbf{x} \cdot \mathbf{w} + b$. $f(\mathbf{x}) = 0$ is the discriminative hyper-plane.

Eq. (9) and (10) indicate that the displacement of the hyper-plane is only related to the partial derivative of the discriminative function, if the same perturbation is given to different features. Accordingly, we give our definition of sensitivity of each feature.

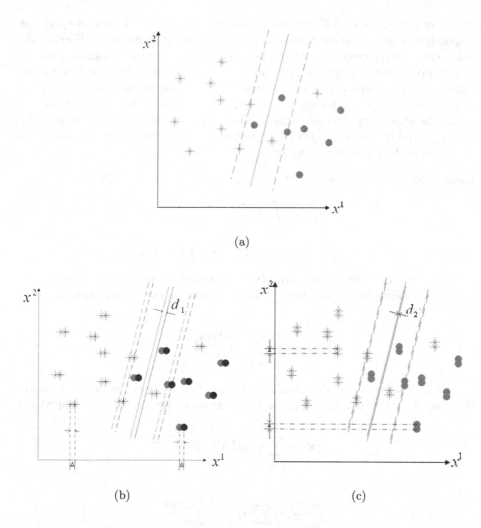

Fig. 2. An example illustrating the impacts of feature perturbations on the decision boundary (a) A linearly non-separable dataset and the decision boundary generated by SVMs (b) with perturbation $\triangle$ in x^1 (c) with perturbation $\triangle$ in x^2

Definition 1. *The sensitivity of feature x^i for a linear SVM is*

$$S(i) = \left| \frac{\partial f(\mathbf{x})}{\partial x^i} \right| \tag{11}$$

where $f(\mathbf{x}) = \sum_{j=1}^{N} \alpha_j y_j x_j^T \mathbf{x} + b$, and $\frac{\partial f(\mathbf{x})}{\partial x^i} = \sum_{j=1}^{N} \alpha_j y_j x_j^i$.

Non-linear support vector machine maps the originally lineally non-separable data points to a high-dimensional kernel space, where the patterns become linearly separable. We are aiming at reducing the dimension of input space instead of that of kernel space. A linear discriminative boundary in the kernel space corresponds to a non-linear discriminative boundary in the input space. We consider the average dis-placement of $f(\mathbf{x})$, due to the feature perturbation, in a narrow region around the discriminative boundary as the sensitivity of the feature, because the partial derivatives on the points far away from the discriminative plane contribute little to the deviation of discriminative plane. To formalize this, we define the sensitivity $S(i)$ of feature x^i as the integration of the absolute value of the partial derivative of $f(\mathbf{x})$ with respect to input feature x^i within region N_ε (assume the *pdf* of $\mathbf{x}$ is $p(\mathbf{x})$):

Definition 2. *The sensitivity of feature x^i for a nonlinear SVM is*

$$S(i) = \int_{N_\varepsilon} \left| \frac{\partial f(\mathbf{x})}{\partial x^i} \right| p(\mathbf{x}) d\mathbf{x} \qquad (12)$$

where $N_\varepsilon = \{\mathbf{x} | -\varepsilon < f(\mathbf{x}) < \varepsilon\}$, in which ε is a small number.

Since it is difficult to estimate $p(\mathbf{x})$, to approximate the above sensitivity, we only consider the support vectors on or within the two margin bounds $|f(\mathbf{x})| = \pm 1$:

$$S(i) = \sum_{-1 \leq f(\mathbf{x}_j) \leq 1} \left| \frac{\partial f(\mathbf{x})}{\partial x^i} \right|_{\mathbf{x} = \mathbf{x}_j} \qquad (13)$$

Without loss of generalization, we suppose the Langrange multipliers remain the same if the perturbation of given feature is small. Hence, for nonlinear SVMs,

$$f(\mathbf{x}) = \sum_{j=1}^{N} \alpha_j y_j K(\mathbf{x}_j, \mathbf{x}) + b,$$

and

$$\frac{\partial f(\mathbf{x})}{\partial x^i} = \sum_{j=1}^{N} \alpha_j y_j \frac{\partial K(\mathbf{x}_j, \mathbf{x})}{\partial x^i}$$

The kernel derivatives for the commonly used kernel functions can be found in Table 1.

3.2   Feature Selection via Sensitivity Analysis and Refinement (FSSAR)

Sensitivity measurement is able to reflect the significance of a feature on the classification result. Intuitively, features with high sensitivities are preferred. However, selecting features only based on their sensitivity rankings may not lead to a subset of features that is most informative, due to the possible dependencies among them. Note that the quality of the selected feature subset can be measured by the R^2W^2 error bounds [6,7] in using these features to train SVM. The motivation of our FSSAR algorithm is to select the features that can

Table 1. Typical kernel functions and their derivatives

Kernel Type	Kernel Function $K(\mathbf{x}_j, \mathbf{x})$	Kernel Derivative $\frac{K(\mathbf{x}_j, \mathbf{x})}{x^j}$
RBF kernel	$exp\left(\frac{-\|\mathbf{x}_j - \mathbf{x}\|^2}{2\delta^2}\right)$	$\frac{1}{\delta^2}(x_j^i - x^i)K(\mathbf{x}_j, \mathbf{x})$
Polynomial kernel	$((\mathbf{x}_j \cdot \mathbf{x}) + 1)^d$	$d\,((\mathbf{x}_j \cdot \mathbf{x}) + 1)^{d-1} \cdot x_j^i$
Sigmoid kernel	$tanh\,(\gamma \mathbf{x}_j \cdot \mathbf{x} + \theta)$	$\frac{\gamma}{cosh^2(\gamma \mathbf{x}_j \cdot \mathbf{x})} \cdot x_j^i$

generate the minimum $R^2 W^2$ error bounds. And our strategy is first to select the features with higher sensitivities but at the same time keep the remaining ones, and then refine the selected subset by tentatively substituting some part with fragments of the previously rejected features. Obviously, the $R^2 W^2$ error bounds of the selected feature subset in training SVM monotonously decrease in this algorithm.

Given a desired number of features m, our FSSAR algorithm selects the m features according to two steps. The process is illustrated in Fig. 3.

The first step is for initial partition. We train SVMs with all the features and get a ranking of their sensitivities. Based on this ranking, the features are divided into two lists, i.e. the selected list $M(|M| = m)$ and the remaining list $N(|N| = d - m)$. Then we split list N into n fragments, each containing δ elements.

The second step is refinement. We tentatively and separately substitute the last δ features in M with each fragment in N, and record the $R^2 W^2$ error bounds $E(M^i)$ after each substitution, where M^i is the list M with the last δ elements substituted by the ith fragment of N. Compare $\min\left(E(M^i)\right)$ with $E(M)$ to see if any one of these substitutions reduces the $R^2 W^2$ error bound. If $\min\left(E(M^i)\right) < E(M)$, we update M with M^* that corresponds to $\min\left(E(M^i)\right)$, train SVMs with the updated M, and re-rank features in M by their sensitivities. Then the next round of refinement starts. If we can not find any improvement in

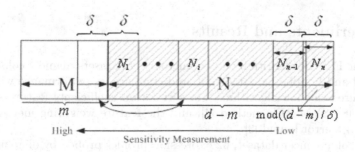

Fig. 3. The process of FSSAR

terms of R^2W^2 error bound after these substitutions, the algorithm terminates and returns the current best feature subset M. The pseudocode of FSSAR is presented as follows.

FSSAR Algorithm.
Input: whole set of features $S(|S| = d)$, desired number of features m, fragment size δ
Output: selected feature subset M
Step1 (Initial partition):
    Train SVM with S and rank S by sensitivities;
    Split S into two sorted lists: selected feature list M and remaining list N, where $|M| = m$, $|N| = d - m$;
    Divide N into n fragments each with size δ,

$$N_i = \begin{cases} \{N[(i-1) \times \delta + 1], N[(i-1) \times \delta + 2], \dots, N[i \times \delta]\}, & \text{for } i = 1 \sim n-1 \\ \{N[d - m - \delta + 1], \dots, N[d-m]\}, & \text{for } i = n \end{cases}$$

    Calculate the R^2W^2 error bound $E(M)$;
Step2 (Refinement):
    Exchange the last δ elements in M and N_i and get M^i,

$$M^i = \{M(1), M(2), \dots, M(m - \delta)\} \cup N_i$$

$$N_i = \{M(m - \delta + 1), M(m - \delta + 2), \dots, M(m)\}$$

    Train SVM with M^i, and calculate R^2W^2 error bound $E(M^i)$;
    Let $E(M^*) = \min \left(E(M^i) \right)$ $(i = 1, 2, \dots, n)$;
    If $E(M^*) < E(M)$
        $M = M^*; E(M) = E(M^*)$;
        Train SVM with M, and sort the features in M by their sensitivity values;
        Go to the beginning of **step2**;
Else
    stop and return the selected feature subset M.

---

4   Experiments and Results

We test the FSSAR algorithm on two microarray datasets, namely colon cancer dataset [8] and lymphoma dataset [9], and compare its performance with other three feature selection algorithms, i.e. correlation coefficients feature selection algorithm [8], SVM-RFE method [4], and the feature weighting method based on the R^2W^2 error bound [3].

In the colon cancer dataset, 62 expression profiles probed by oligonucleotide arrays contain 40 tumor and 22 control profiles that must be discriminated based upon the expression of 2000 genes. We randomly split the profiles into a training set of 50 samples and a testing set of 12 samples. And the proportions of tumor and control profiles are similar in both training and testing sets.

In the lymphoma dataset, the goal is to separate cancerous and control tissues in a large B-Cell lymphoma problem. The dataset contains 96 expression profiles concerning 4026 genes, where 62 cancerous samples are in any of the classes DLCL, FL and CLL, and the remaining 34 are labeled as control. The dataset was randomly split into a training set of size 60 and a testing set of size 36 with similar proportions of cancerous and control profiles.

We run each of the four algorithms independently for 100 times and use the statistical results to reflect their overall performances. In each run of a specific algorithm, we use the training and testing sets generated by independently randomly splitting the raw datasets. We perform our FSSAR algorithm and other three algorithms to select 20, 50, 100, 250, and 1000 features from the above two datasets. The fragment size δ in FSSAR is set to $0.1 \times m$, and the regularization parameter C for the linear SVM is set to 1000. Test accuracies are listed in Table 2. From the results, one can find that FSSAR generally achieves higher accuracies with lower standard deviations than other three methods.

Table 2. Means and standard deviations of the test accuracies of correlation coefficient (CC) feature selection algorithm, feature weighting method based on the R^2W^2 bound, SVM-RFE method, and FSSAR

Dataset	# of genes	$CC(\%)$	$R^2W^2(\%)$	SVM-RFE(%)	FSSAR(%)
	20	78.2 ± 10	79.3 ± 14	81.8 ± 9	84.1 ± 5
	50	77.8 ± 11	81.6 ± 11	83.7 ± 9	86.9 ± 3
Colon Cancer	100	79.4 ± 11	83.2 ± 10	84.2 ± 10	86.3 ± 4
	250	82.3 ± 9	84.1 ± 9	83.1 ± 9	86.5 ± 5
	1000	83.3 ± 9	82.8 ± 8	83.4 ± 9	85.8 ± 4
	20	78.6 ± 10	87.5 ± 7	91.7 ± 4	95.2 ± 3
	50	86.6 ± 6	91.2 ± 5	93.1 ± 5	95.7 ± 2
Lymphoma	100	90.5 ± 5	92.1 ± 4	93.5 ± 5	94.8 ± 2
	250	92.4 ± 5	92.4 ± 4	92.9 ± 4	94.5 ± 3
	1000	93.0 ± 5	91.9 ± 4	92.2 ± 4	94.3 ± 3

To test the performance of the proposed FSSAR algorithm in nonlinear problems, we synthesized an XOR problem with 52 features akin to [3]. Only the first two features are relevant to the toy problem. The probabilities of classes $y = +1$ and $y = -1$ are equal. If $y = -1$, the first two features x_1, x_2 are drawn from $N\left((-0.8, -3)^T, I\right)$, or $N\left((0.8, 3)^T, I\right)$ from with equal probability. If $y = +1$, x_1, x_2 are drawn with equal probability from $N\left((3, -3)^T, I\right)$ or $N\left((-3, 3)^T, I\right)$. The remaining features are randomly drawn from $N(0, 20)$.

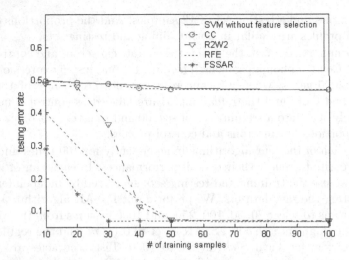

Fig. 4. Comparison of feature selection methods on a synthetic nonlinear problem with many irrelevant features

For the four feature selection methods in comparison, we selected the best two features. We chose a RBF kernel of width $\delta = 0.1$ and parameter $C = 100$ for the nonlinear SVM. Fig. 4 shows the performance of SVMs without feature selection, the correlation coefficients feature selection method, the feature weighting method based on the R^2W^2 error bound, SVM-RFE method and our FSSAR algorithm. The performance is measured in terms of the averaged test errors on the 400 independent test points over 30 runs for each training set. Due to the relatively small number of given features, the fragment size δ in FSSAR is fixed to 1. One can find that our FSSAR algorithm achieves lower testing error rate than other four methods on training sets with 10 to 100 samples.

We consider the number of SVM training as the measurement of the time complexity, because SVM training is the most time-consuming part in this algorithm. FSSAR trains SVM for only once in the initialization part, and for $h \cdot (n + 1)$ times in the refinement part, where h is the number of refinements before FSSAR converges, and n is the number of fragments in the remaining list N. Generally, we have $n \ll d$. Actually, in our experiments, h is less than 10. Therefore, the total running time of FSSAR is moderate.

5   Conclusion

A new feature selection method for SVMs via sensitivity analysis and refinement is proposed in this paper. The sensitivity values can reflect the significances of the features, but only considering sensitivities would be too greedy. This potential weakness is compensated by refining the selected features using the previously rejected features. To evaluate the quality of current subset of features, we use the theoretical bounds on the generalization error for SVMs, which is computationally attractive.

The FSSAR algorithm is compared with other three feature selection algorithms, i.e. correlation coefficients feature selection algorithm, SVM-RFE method, and the feature weighting method based on the R^2W^2 error bound. The experiments are performed on two microarray datasets, i.e. colon cancer and lymphoma, as well as a synthetic dataset. From these experiments, one can conclude that once given the desired number of features the FSSAR algorithm results in the feature subset with competitive quality.

Acknowledgment

This work was supported by the Hong Kong Research Grant Council under Grants G-YD87 and G-T891.

References

1. Guyon, I., Elisseeff, A.: An introduction to variable and feature selection. Journal of Machine Learning Research **3** (2003) 1157–1182
2. Blum, A., Langley, P.: Selection of relevant features and examples in machine learning. Artificial Intelligence **97** (1997) 245–271
3. Weston, J., Mukherjee, S., Chapelle, O., Pontil, M., Poggio, T., Vapnik, V.: Feature selection for svms. In: Advances in Neural Information Processing Systems. Volume 13. (2001) 668–764
4. Guyon, I., Weston, J., Barnhill, S., Vapnik, V.: Gene selection for cancer classification using support vector machines. Machine Learning **46** (2002) 389–422
5. Rakotomamonjy, A.: Variable selection using svm-based criteria. Journal of Machine Learning Research **3** (2003) 1357–1370
6. Vapnik, V.: The Nature of Statistical Learning Theory. Springer-Verlag, New York (1999)
7. Chapelle, O., Vapnik, V.: Choosing multiple parameters for support vector machines. Machine Learning **46** (2002) 131–159
8. Alon, U., Barkai, N., Notterman, D., Gish, K., Ybarra, S., Mack, D., Levine, A.: Broad patterns of gene expression revealed by clustering analysis of tumor and normal colon cancer tissues probed by oligonucleotide arrays. Cell Biology **96** (1999) 6745–6750
9. Alizadeh, A., Eisen, M., et al.: Distinct types of diffuse large b-cell lymphoma identified by gene expression profiling. Nature **403** (2000) 503–511

The Breakpoint Graph in Ciliates[*]

R. Brijder[1], H.J. Hoogeboom[1], and G. Rozenberg[1,2]

[1] Leiden Institute of Advanced Computer Science, Leiden University,
Niels Bohrweg 1, 2333 CA Leiden, The Netherlands
rbrijder@liacs.nl
[2] Department of Computer Science, University of Colorado,
Boulder, CO 80309-0347, USA

Abstract. The gene assembly process in ciliates (single-cell organisms) is interesting from both the biological and computational point of view. This paper studies the computational nature of the gene assembly process. Motivated by the breakpoint graph known from another branch of DNA transformation research, we introduce the reduction graph as a tool for the study of this process, and illustrate its usefulness by proving a number of properties of gene assembly.

1 Introduction

Ciliates are single-cell organisms that have two functionally different nuclei, one called micronucleus and the other called macronucleus. At some stage in sexual reproduction a micronucleus is transformed into a macronucleus in a process called *gene assembly*. This is the most involved DNA processing in living organisms known today. The reason that gene assembly is so involved is that the genome of the micronucleus is dramatically different from the genome of the macronucleus — this difference is particularly pronounced in the stichotrichs group of ciliates which we consider in this paper. The investigation of gene assembly [4] turns out to be very exciting from both biological and computational points of view.

Another branch of DNA transformation research is *sorting by reversal*, see, e.g., [6] and [7]. Two different species can have several contiguous segments on their genome that are very similar, although there relative order (and orientation) may differ on both genomes. In the theory of sorting by reversal one tries to determine the number of operations needed to reorder such a series of genomic 'blocks' from one species into that of another. The theory is still being refined [1]. An essential tool is the *breakpoint graph* (or reality and desire diagram) which is used to capture both the present situation, the genome of the first species, and the desired situation, the genome of the second species.

Motivated by the breakpoint graph, we introduce the *reduction graph* into the theory of gene assembly. The intuition of 'reality and desire' remains in place,

---

[*] This research was supported by the Netherlands Organization for Scientific Research (NWO) project 635.100.006 "VIEWS".

M.R. Berthold et al. (Eds.): CompLife 2005, LNBI 3695, pp. 128–139, 2005.

but the technical details are different. Instead of one operation, the reversal, we have three operations as described in [9]. Furthermore, these operations are irreversible and can only be applied on special positions in the string, called *pointers*. Also, instead of two different species, we deal with two different nuclei — the present situation is a gene in its micronuclear form, and the desired situation is the gene in its macronuclear form. Surprisingly, where the breakpoint graph in the theory of sorting by reversal is mostly useful to determine the number of needed operations, the reduction graph has different uses in the theory of gene assembly, providing valuable insights into the gene assembly process.

For example, the reduction graph allows for a direct characterization of the *intermediate* strings that may be constructed during the transformation of a given gene from its micronuclear form to its macronuclear form. Also, it allows one to determine the number of loop recombination operations (see Figure 1 below) needed in this transformation. These results may be experimentally verified to allow for a validation of the model. Due to space constraints, the proofs of the theorems are omitted.

2   Background: Gene Assembly in Ciliates

In ciliates, genes can occur in two forms: in their micronuclear and in their macronuclear form. A gene in its micronuclear form consists of relevant genetic segments called MDSs (macronuclear destined sequences) separated by IESs (internally eliminated sequences). During sexual reproduction, a micronucleus is converted into a macronucleus. The genes are converted to their macronuclear form by excising the IESs and by splicing, permutating, and possibly inverting the MDSs. This process is referred to as *gene assembly* [4].

Above is an example gene in its micronuclear form with five MDSs $M_1, \ldots, M_5$, where $\bar{M}_i$ denotes the inverse of M_i (M_i rotated by 180 degrees). This form can be described by the string $\bar{M}_3 M_4 M_2 \bar{M}_1 M_5$. After gene assembly the string $M_1 M_2 \cdots M_5$ is obtained (without the intermediate IESs).

In general, the structure of MDS M_i for $1 < i < \kappa$, M_1 and M_κ can be depicted as:

respectively. The symbols p_i and $\bar{p}_i$ for $2 \leq i \leq \kappa$ represent single stranded DNA sequences, and u_i for $2 \leq i \leq \kappa$ represent double stranded DNA sequences. DNA sequence $\bar{p}_i$ is the Watson-Crick complement of DNA sequence p_i. Each double strand $\dfrac{p_i}{\bar{p}_i}$ is called a *pointer*, and is considered part of both M_{i-1} and M_i. In the macronucleus, the MDSs $M_1, M_2, \ldots, M_\kappa$ are spliced together by "gluing" each M_j with M_{j+1} on $\dfrac{p_{j+1}}{\bar{p}_{j+1}}$ (for $1 \le j < \kappa$). This gluing is irreversible (in the gene assembly process) and thus after gluing M_j with M_{j+1}, the occurrences of $\dfrac{p_{j+1}}{\bar{p}_{j+1}}$ in M_j and M_{j+1} are not considered pointers anymore. Consequently, the molecular operations can be seen as operations that remove pointers. This is an important property of gene assembly and appears explicitly in the formal models of the gene assembly process. The gene assembly process is accomplished through the following three molecular operations.

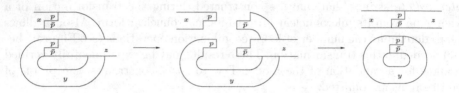

Fig. 1. The loop recombination operation

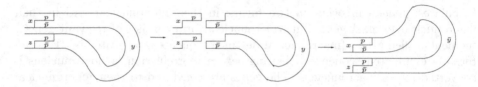

Fig. 2. The hairpin recombination operation

Loop recombination. This operation is applicable to a gene pattern which has two identical pointers separated by a single IES y. An application results in the excitation from the genome of a circular molecule consisting of IES y only (Figure 1).

Hairpin recombination. The operation is applicable to a gene pattern containing a pair of pointers in which one pointer is an inversion of the other. An application results in the inversion of the sequence of the genome that is between the mentioned pair of pointers (Figure 2).

Double-loop recombination. The operation is applicable to a gene pattern containing two pairs of identical pointers for which the sequence between the first pair of pointers overlaps with the sequence between the second pair of pointers. An application results in the interchanging of the sequence between the first two (of the four) pointers and the sequence between the last two pointers in the gene pattern (Figure 3).

For a given gene in its micronuclear form, a sequence of these molecular operations is *successful* if it transforms the pattern into its macronuclear form.

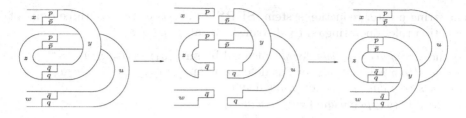

Fig. 3. The double-loop recombination operation

3   The String Pointer Reduction System

Motivated by the molecular operations discussed in Section 2, three equivalent types of formal systems were considered in [3] and [4]. They are believed to be sound models of the gene assembly process. In this paper we consider the string pointer reduction system, which we will recall now.

We define the alphabet $\Delta = \{2, 3, \ldots, \kappa\}$. For $D \subseteq \Delta$, we define $\bar{D} = \{\bar{a} \mid a \in D\}$ and $\Pi_D = D \cup \bar{D}$; also $\Pi = \Pi_\Delta$. We will use the the alphabet Π to formally denote the pointers — the intuition is that the pointer $\dfrac{p_i}{\bar{p}_i}$ will be denoted by either i or $\bar{i}$. Accordingly, elements of Π will also be called *pointers*.

We use the "bar operator" to move from Δ to $\bar{\Delta}$ and back from $\bar{\Delta}$ to Δ. Hence, for $p \in \Pi$, $\bar{\bar{p}} = p$. For a string $u = x_1 x_2 \cdots x_n$ with $x_i \in \Pi$, the *inverse* of u is the string $\bar{u} = \bar{x}_n \bar{x}_{n-1} \cdots \bar{x}_1$. For $p \in \Pi$, we define $\mathbf{p} = p$ if $p \in \Delta$ and $\mathbf{p} = \bar{p}$ if $p \in \bar{\Delta}$, i.e., $\mathbf{p}$ is the "unbarred" variant of p. The *domain* of a string $v \in \Pi^*$ is $dom(v) = \{p \in \Delta \mid p \text{ or } \bar{p} \text{ occurs in } v\}$. A *legal string* is a string $u \in \Pi^*$ such that for each $p \in \Pi$ that occurs in u, u contains exactly two occurrences from $\{p, \bar{p}\}$. For a pointer p and a legal string u, if both p and $\bar{p}$ occur in u then we say that both p and $\bar{p}$ are *positive* in u; if on the other hand only p or only $\bar{p}$ occur in u, then both p and $\bar{p}$ are *negative* in u. So, every pointer occurring in a legal string is either positive or negative in it.

For each gene in its micronuclear form, we associate a legal string through the homomorphism π_κ defined by:

$$\pi_\kappa(M_1) = 2, \quad \pi_\kappa(M_\kappa) = \kappa, \quad \pi_\kappa(M_i) = i(i+1) \quad \text{for } 1 < i < \kappa,$$

and $\pi_\kappa(\bar{M}_j) = \overline{\pi_\kappa(M_j)}$ for $1 \leq j \leq \kappa$.

Example 1. The gene in its micronuclear form in Section 2 described by $\bar{M}_3 M_4 M_2 \bar{M}_1 M_5$ corresponds to legal string $u = \overline{43}452 3\bar{2}5$. Pointers 2, 3 and 4 (and their inverses) are positive in u, and pointers 5 and $\bar{5}$ are negative in u.

Example 2. The gene (in its micronuclear form) that encodes the actin protein in the stichotrich *Sterkiella nova* is described by $\delta = M_3 M_4 M_6 M_5 M_7 M_9 \bar{M}_2 M_1 M_8$ (see [8], [2], and [4]). The associated legal string is $\pi_9(\delta) = 3445675678 9\bar{3}2289$.

The string pointer reduction system (**SPRS** for short) consists of three types of reduction rules operating on legal strings. For all $p, q \in \Pi$:

- the *string negative rule* for p is defined by $\mathbf{snr}_p(u_1 p p u_2) = u_1 u_2$.
- the *string positive rule* for p is defined by $\mathbf{spr}_p(u_1 p u_2 \bar{p} u_3) = u_1 \bar{u}_2 u_3$.
- the *string double rule* for p, q is defined by
  $\mathbf{sdr}_{p,q}(u_1 p u_2 q u_3 p u_4 q u_5) = u_1 u_4 u_3 u_2 u_5$.

where $u_1, u_2, \ldots, u_5$ are arbitrary strings over Π. We define $Snr = \{\mathbf{snr}_p \mid p \in \Pi\}$, $Spr = \{\mathbf{spr}_p \mid p \in \Pi\}$ and $Sdr = \{\mathbf{sdr}_{p,q} \mid p, q \in \Pi\}$ to be the sets containing all the reduction rules of a specific type.

Note that each of these rules is defined only on legal strings that satisfy the given form. For example, $\mathbf{snr}_2$ is not defined on legal string 2323. It is important to realize that for every non-empty legal string there is at least one reduction rule applicable. Indeed, every legal string for which no string positive rule and no string double rule is applicable must have only negative pointers and no overlapping pointers and thus a string negative rule is applicable.

The string negative (positive, double, resp.) rule corresponds to the loop (hairpin, double-loop, resp.) recombination operation. Note that the fact (pointed out at the end of Section 2) that the molecular operations remove pointers is explicit in **SPRS**. Each of the three molecular operations from Section 2 has its own (biological) complexity. Thus, in order to characterize the complexity of the gene assembly process in various strands of ciliates, we may restrict ourselves to subsets of the rules, and consider rules from, e.g., $\{Snr, Spr\}$, i.e., without the string double rules.

Definition 1. *Let u and v be legal strings and $S \subseteq \{Snr, Spr, Sdr\}$. A composition φ of reduction rules from S is called an $(S\text{-})$reduction of u, if φ is applicable to (defined on) u. A successful reduction φ of u is a reduction of u such that $\varphi(u) = \lambda$ (λ denotes the empty string). We then also say that φ is successful for u. We say that u is reducible to v in S if there is a S-reduction φ of u such that $\varphi(u) = v$. We simply say that u is reducible to v if u is reducible to v in $\{Snr, Spr, Sdr\}$. We say that u is successful in S if u is reducible to λ in S.*

Because (as pointed out already) for every non-empty legal string there is at least one reduction rule applicable, we easily obtain Theorem 9.1 in [4] which states that every legal string is successful in $\{Snr, Spr, Sdr\}$.

Example 3. Let $S = \{Snr, Spr\}$, $u = 3245\bar{4}5\bar{3}\bar{2}$, and $v = \bar{5}45\bar{4}$. Then u is reducible to v in S, because $(\mathbf{snr}_3 \, \mathbf{spr}_2)(u) = v$. Applying $\varphi = \mathbf{spr}_{\bar{5}} \, \mathbf{spr}_4 \, \mathbf{snr}_{\bar{2}} \, \mathbf{spr}_3$ to u yields λ, thus φ is successful for u. On the other hand, $u = 3232$ is not reducible to any v in S, because none of the rules in Snr and none of the rules in Spr is applicable for this u.

4   Reduction Graphs

We are ready now to define the main notion of this paper: the reduction graph. The reduction graph is a two-sorted graph and it is defined for a legal string u

and a subset $D \subseteq dom(u)$. A two-sorted graph is a graph $G = (V, E_1, E_2, f, l, s, t)$ that has two separate sets of edges E_1 and E_2, two special vertices s and t, every vertex (except s and t) is labelled through labelling function f, and every edge is labelled through labelling function l.

Isomorphism between two-sorted graphs is defined in the usual way. Two-sorted graphs $G = (V, E_1, E_2, f, l, s, t)$ and $G' = (V', E_1', E_2', f', l', s', t')$ are *isomorphic*, denoted by $G \approx G'$, if there is a bijection $\alpha : V \to V'$ such that $\alpha(s) = s'$, $\alpha(t) = t'$, $f(v) = f'(\alpha(v))$ for all $v \in V$, $l((x, y)) = l'((\alpha(x), \alpha(y)))$, and $(x, y) \in E_i$ iff $(\alpha(x), \alpha(y)) \in E_i'$, for all $x, y \in V$ and $i \in \{1, 2\}$.

	2		3		$\bar{2}$		$\bar{4}$		3		4	

Fig. 4. Part of a genome with three pointer pairs corresponding to the same gene

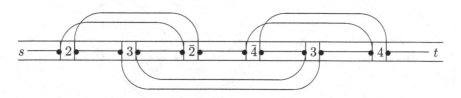

Fig. 5. The reduction graph corresponding to the underlying genome

The intuition behind the reduction graph is as follows. Figure 4 depicts a part of a genome with three pointer pairs corresponding to the same gene g. The reduction graph (of g with respect to these pointer pairs) introduces two vertices for each of these pointers and two special vertices s and t representing the ends. It connects adjacent pointers through *reality edges* and connects pointers corresponding to the same pointer pair through *desire edges* in a way that reflects how the parts will be glued after a molecular operation is applied on that pointer (recall Figures 1, 2, and 3). The resulting reduction graph is depicted in Figure 5. Thus, every reality edge corresponds to a certain DNA segment. If such a DNA segment contains other pointers of g, then these pointers form the label of that reality edge. The formal definition is given below.

Definition 2. *Let $D \subseteq \Delta$ and let u be a legal string, such that $u = \delta_0 p_1 \delta_1 p_2 \ldots p_n \delta_n$ where $\delta_0, \ldots, \delta_n \in \Pi_D^*$ and $p_1, \ldots, p_n \in \Pi_{dom(u) \backslash D}$. The reduction graph of u with respect to D is a two-sorted graph*

$$G_{u,D} = (V, E_1, E_2, f, l, s, t)$$

where

$$V = \{I_1, I_2, \ldots, I_n\} \ \cup \ \{I_1', I_2', \ldots, I_n'\} \ \cup \ \{s, t\} \ \textit{are the vertices,}$$

$E_1 = E_{1,r} \cup E_{1,l}$ *are the reality edges, where* $E_{1,r} = \{e_0, e_1, \ldots, e_n\}$ *with*

$$e_i = (I'_i, I_{i+1}) \text{ for } 1 \le i \le n-1, e_0 = (s, I_1), e_n = (I'_n, t),$$

$E_{1,l} = \{\bar{e} \mid e \in E_{1,r}\}, \text{ where for } e = (x, y), \text{ we define } \bar{e} = (y, x),$

$E_2 = \{(I'_i, I_j), (I_i, I'_j) \mid i, j \in \{1, 2, \ldots, n\} \text{ with } i \ne j \text{ and } p_i = p_j\} \cup$
$\{(I_i, I_j), (I'_i, I'_j) \mid i, j \in \{1, 2, \ldots, n\} \text{ and } p_i = \bar{p}_j\}$ *are the desire edges,*

$f(I_i) = f(I'_i) = \mathbf{p}_i$ *for* $1 \le i \le n$ *is the vertex labelling function, and*

$l(e_i) = \delta_i, l(\bar{e}_i) = \bar{\delta}_i$ *for* $0 \le i \le n$ *and* $l(e) = \lambda$ *for* $e \in E_2$ *is the edge labelling function.*

When $D = \emptyset$, we simply refer to $G_{u,D}$ as the *reduction graph of* u. The following example should make the notion of reduction graph more clear.

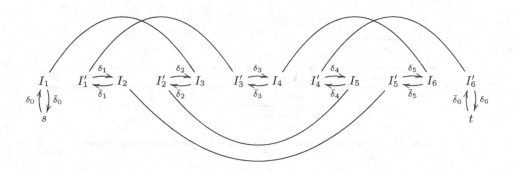

Fig. 6. The reduction graph $G_{u,D}$ as defined in Example 4 (the vertex labels are omitted)

Example 4. Let $u = 5236882\bar{5}437\bar{7}46$ be a legal string and $D = \{5, 6, 7, 8\} \subseteq dom(u)$. Thus, $\{2, 3, 4\} = dom(u) \backslash D$, and $u = \delta_0 \, 2 \, \delta_1 \, 3 \, \delta_2 \, \bar{2} \, \delta_3 \, \bar{4} \, \delta_4 \, 3 \, \delta_5 \, 4 \, \delta_6$ with $\delta_0 = 5$, $\delta_1 = \lambda$, $\delta_2 = 688$, $\delta_3 = 5$, $\delta_4 = \lambda$, $\delta_5 = 7\bar{7}$ and $\delta_6 = 6$. Notice that $\delta_1, \delta_2, \ldots, \delta_6 \in \Pi_D^*$. This example corresponds to the situation in Figure 4.

The reduction graph $G_{u,D}$ of u with respect to D is given in Figure 6. Note that for every desire edge e, we represent both e and $\bar{e}$ by a single undirected edge. The graph is drawn in a form that closely relates to the linear ordering of u. The desire edges that cross correspond to positive pointers, and the desire edges that do not cross correspond to negative pointers.

Since the exact identity of the vertices in a reduction graph is not essential for the problems considered in this paper, in order to simplify the pictorial notation of reduction graphs we will replace the vertices (except for s and t) by their labels. Figure 7 gives $G_{u,D}$ in this way. In this figure we have reordered the vertices, making it transparent that $G_{u,D}$ has one linear and one cyclic connected component.

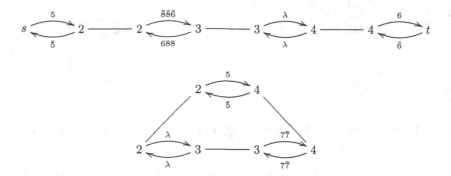

Fig. 7. The reduction graph of Figure 6 in a simplified form

Note that a reduction graph is an undirected graph in the sense that if $e \in E_1$ ($e \in E_2$, resp.) then also $\bar{e} \in E_1$ ($\bar{e} \in E_2$, resp.). If we think of the reduction graph as an undirected graph by considering edges (x, y) and (y, x) as one undirected edge $\{x, y\}$, then every vertex (except for s and t) is connected to exactly two (undirected) edges. Thus the reduction graph has exactly one connected component that has a linear structure with s and t as endpoints and possibly one or more connected components that have a cyclic structure (called *cyclic components*).

A *walk* in a two-sorted graph G is a string $\pi = e_1 e_2 \cdots e_n$ over $E = E_1 \cup E_2$ with $n \geq 1$ such that there are vertices $x_1, \ldots, x_{n+1}$ such that $e_i = (x_i, x_{i+1})$ for $1 \leq i \leq n$. We say that π is *alternating* if moreover for all $1 \leq i < n$, e_i and e_{i+1} are not both in E_1 and not both in E_2. The *label of* π is the string $l(\pi) = l(e_1)l(e_2) \cdots l(e_n)$. If a two-sorted graph G has a unique alternating walk from s to t, then the label this walk is called the *reduct of G*, denoted by $red(G)$. Thus, the reduct exists for a reduction graph $G_{u,D}$ of a legal string u with respect to a $D \subseteq dom(u)$. It is then also called the *reduct of u to D*, and denoted by $red(u, D)$. A direct consequence of the definition of reduction graph is that $red(u, dom(u)) = u$ for each legal string u. It is also clear that if two-sorted graphs G_1 and G_2 are isomorphic, then $red(G_1) = red(G_2)$.

Example 5. If we take u and D from Example 4, then $red(u, D) = 5\bar{8}\bar{8}66$, which is easy to see in Figure 7.

We now define functions on reduction graphs, called *reduction functions*, that simulate the effect (up to isomorphism) of each of the three string pointer reduction rules on a reduction graph. For a vertex label p, the *p-reduction function*, denoted by rf_p, merges edges that form an alternating walk 'over' vertices labelled by p, concatenating their labels, and removes all vertices labelled by p.

Example 6. If we take $G_{u,D}$ from Example 4, then $rf_2(G_{u,D})$ is given in Figure 8. Consider, e.g., the edges (s, I_1), (I_1, I_3), and (I_3, I_2') of $G_{u,D}$ labelled by 5, λ, and $\bar{8}\bar{8}6$, respectively. Since they form a walk 'over' vertices labelled by 2, in $rf_2(G_{u,D})$ they are replaced by a single edge (s, I_2') labelled by $5\bar{8}\bar{8}6$.

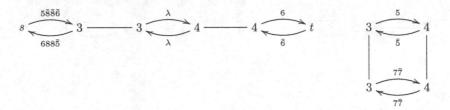

Fig. 8. The graph obtained when applying rf_2 to the reduction graph of Figure 7

It is easy to see that reduction functions do not affect the reduct: for each reduction graph $G_{u,D}$ and all $p \in dom(u)\backslash D$, $red(G_{u,D}) = red(rf_p(G_{u,D}))$. Reduction functions satisfy also another important property. The theorem states that the reduction functions simulate the effect (up to isomorphism) of each of the three string pointer reduction rules on a reduction graph.

Theorem 1. *Let u be a legal string, $D \subset dom(u)$ and $p, q \in \Pi$ with $\mathbf{p}, \mathbf{q} \in dom(u)\backslash D$. Let $G_{u,D}$ be the reduction graph of u with respect to D.*

- *If $\mathbf{snr}_p$ is applicable to u, then $rf_{\mathbf{p}}(G_{u,D}) \approx G_{\mathbf{snr}_p(u),D}$.*
- *If $\mathbf{spr}_p$ is applicable to u, then $rf_{\mathbf{p}}(G_{u,D}) \approx G_{\mathbf{spr}_p(u),D}$.*
- *If $\mathbf{sdr}_{p,q}$ is applicable to u, then $rf_{\mathbf{q}}(rf_{\mathbf{p}}(G_{u,D})) \approx G_{\mathbf{sdr}_{p,q}(u),D}$.*

Example 7. If we again take $u = 52368825437\bar{7}46$ and $D = \{5, 6, 7, 8\} \subseteq dom(u)$ from Example 4, then $\mathbf{spr}_2(u) = 5\bar{8}\bar{8}6\bar{3}5437\bar{7}46$. By constructing the reduction graph $G_{\mathbf{spr}_2(u),D}$ of $\mathbf{spr}_2(u)$ with respect to D, one may verify that $G_{\mathbf{spr}_2(u),D}$ is isomorphic to $rf_2(G_{u,D})$. Thus, omitting the identity of the vertices of $G_{\mathbf{spr}_2(u),D}$ as usual, we again obtain the graph of Figure 8 (depicting $rf_2(G_{u,D})$).

Let **oper** be one of the SPRS rules, applicable to u and using pointers *not* in D, as in the above theorem. By the theorem, we can construct the reduction graph $G_{\mathbf{oper}(u),D}$ of **oper**(u) with respect to D from $G_{u,D}$ by applying one or two reduction functions. As these do not change the reduct, we have $red(u, D) = red(\mathbf{oper}(u), D)$. By applying this argument iteratively, we see that if legal string u is reducible to legal string v, then $red(u, dom(v)) = red(v, dom(v)) = v$. We will strengthen this statement in the next section.

5   Properties of the Reduction Graph

In the section we show that a number of interesting properties can be proven using the reduction graph. From a biological point of view, Theorem 2 characterizes which gene patterns can occur *during* the transformation of a given gene from its micronuclear form to its macronuclear form using only a specific subset of the three types of recombination operations. Theorem 3 allows one to determine the number of loop recombination operations that are necessary in this transformation.

We first define the notion of *pointer removal operations*. For a subset $D \subseteq \Delta$, the D-removal operation, denoted by rem_D, is the homomorphism defined by $rem_D(a) = \lambda$ if $a \in D \cup \bar{D}$ and $rem_D(a) = a$ if $a \notin D \cup \bar{D}$.

Example 8. Let $u = 3245\bar{4}5\bar{3}2$ and $D = \{4, 5\}$. Then $rem_D(u) = 32\bar{3}2$. Note that $2, 3 \notin D$. Note also that $\varphi = \mathbf{snr}_3 \, \mathbf{spr}_2$ is applicable to both u and $rem_D(u)$, but for $rem_D(u)$, φ is also successful.

The next theorem gives a characterization of reducibility given a chosen set of reduction rules $S \subseteq \{Snr, Spr, Sdr\}$.

Theorem 2. *Let u and v be legal strings, $D = dom(v) \subseteq dom(u)$ and $S \subseteq \{Snr, Spr, Sdr\}$. Then u is reducible to v in S iff $rem_D(u)$ is successful in S and $red(u, D) = v$.*

The following corollary follows directly from the previous theorem and the fact that every legal string is successful in $\{Snr, Spr, Sdr\}$. The corollary implies that it takes only linear time $O(|u|)$ to determine whether or not u is reducible to v.

Corollary 1. *Let u and v be legal strings and $D = dom(v) \subseteq dom(u)$. Then u is reducible to v iff $red(u, D) = v$.*

Example 9. Consider again legal string u and $D = \{5, 6, 7, 8\}$ from Example 4. By Example 5, $red(u, D) = 58\bar{8}\bar{6}6$. By the previous corollary, there is no reduction φ of u that removes exactly the pointers from D, because there is no legal string v such that $v = red(u, D)$ and $dom(v) = D$.

It turns out that the cyclic components in the 'full' reduction graph $G_{u,\emptyset}$ of a legal string u reveals important properties of u. For example, if $\mathbf{snr}_p$ is applicable to u, then $u = u_1 p p u_2$ for some strings u_1 and u_2. Therefore, $G_{u,\emptyset}$ contains the following cyclic component.

Now, $G_{\mathbf{snr}_p(u),\emptyset} \approx rf_{\mathbf{p}}(G_{u,\emptyset})$, thus this (and only this) cyclic component is removed by $rf_{\mathbf{p}}$. It turns out that in the **spr** and **sdr** cases, no cyclic components are removed, and consequently the following theorem holds.

Theorem 3. *Let N be the number of cyclic components in the reduction graph of a legal string u. Then every successful reduction of u has exactly N string negative rules.*

Example 10. Let $u = 23\bar{2}434$ be a legal string. The reduction graph of u is depicted in Figure 7, but with all reality edges labelled by λ. By Theorem 3 it follows that every reduction of u has exactly one string negative rule. It can be shown that there are exactly four successful reductions of u. These are $\mathbf{snr}_2 \, \mathbf{spr}_3 \, \mathbf{spr}_{\bar{4}}$, $\mathbf{snr}_3 \, \mathbf{spr}_2 \, \mathbf{spr}_{\bar{4}}$, $\mathbf{snr}_3 \, \mathbf{spr}_{\bar{4}} \, \mathbf{spr}_2$ and $\mathbf{snr}_4 \, \mathbf{spr}_3 \, \mathbf{spr}_2$. Notice that each of these reductions has exactly one string negative rule.

Example 11. By constructing the reduction graph $G_{u,\emptyset}$ of legal string $u = 3445675678\overline{93}2289$ corresponding to the actin gene of *Sterkiella nova*, it follows that $G_{u,\emptyset}$ has two cyclic components. Thus exactly two string negative rules are present in every successful reduction φ of u. One such reduction is $\varphi = \mathbf{spr}_3 \ \mathbf{sdr}_{8,9} \ \mathbf{snr}_7 \ \mathbf{sdr}_{5,6} \ \mathbf{spr}_2 \ \mathbf{snr}_4$. Therefore, the transformation of the actin gene from its micronuclear form to its macronuclear form requires exactly two loop recombination operations (Figure 1).

The previous theorem implies that it takes only linear time $O(|u|)$ to determine how many string negative rules are needed to successfully reduce legal string u. The theorem also allows for a generalization of a number of results in [5] (and Chapter 13 in [4]) concerning the characterization of successfulness. For example, since every legal string u is successful in $\{Snr, Spr, Sdr\}$, the following holds.

Corollary 2. *Let u be a legal string. Then u is successful in $\{Spr, Sdr\}$ iff the reduction graph of u has no cyclic component.*

It is easy to see that for legal string u and $D \subseteq dom(u)$, $G_{rem_D(u),\emptyset}$ is isomorphic to $G_{u,D}$ modulo the labels of the edges. Then, by Theorem 2 and Corollary 2, we have the following corollary. In this result it is particularly apparent that both the linear component *and* the cyclic components of reduction graphs reveal crucial properties concerning reducibility.

Corollary 3. *Let u and v be legal strings with $dom(v) \subseteq dom(u)$, and $G_{u,D}$ be the reduction graph of u with respect to $D = dom(v)$. Then u is reducible to v in $\{Spr, Sdr\}$ iff $G_{u,D}$ has no cyclic components and $red(G_{u,D}) = v$.*

6   Conclusion

This paper introduces the concept of breakpoint graph (or reality and desire diagram) into gene assembly models, through the notion of reduction graph. The reduction graph provides valuable insight into the gene assembly process, because for example it allows one to characterize which gene patterns can occur during the transformation of a given gene from its micronuclear form to its macronuclear form. Formally, in the string pointer reduction system we characterize whether legal string u is reducible to legal string v for a given set of reduction rule types. The characterization is independent from the chosen subset of the three types of string pointer rules, and it allows us to determine whether a legal string u is reducible to a legal string v in linear time $O(|u|)$. In addition, it allows one to determine the number of loop recombination operations that are necessary in the transformation of a given gene from its micronuclear form to its macronuclear form.

Acknowledgment

The authors are indebted to two anonymous referees for their helpful comments.

References

1. A. Bergeron, J. Mixtacki, J. Stoye, On Sorting by Translocations, *The 9th Annual Int. Conf. on Research in Comp. Molecular Biology (RECOMB 2005)*.
2. A.R.O. Cavalcanti, Ciliates IES MDS Database, website: http://oxytricha.princeton.edu/dimorphism/.
3. A. Ehrenfeucht, T. Harju, I. Petre, D.M. Prescott, G. Rozenberg, String and graph reduction systems for gene assembly in ciliates. *Mathematical Structures in Computer Science* **12** (2002) 113–134.
4. A. Ehrenfeucht, T. Harju, I. Petre, D.M. Prescott, G. Rozenberg, *Computation in Living Cells: Gene Assembly in Ciliates*, Springer 2004.
5. A. Ehrenfeucht, T. Harju, I. Petre, G. Rozenberg, Characterizing the micronuclear gene patterns in ciliates. *Theory of Computing Systems* **35** (2002) 501–519.
6. J. Meidanis, J.C. Setubal, *Introduction to Computational Molecular Biology*. PWS Publishing Company 1997.
7. P.A. Pevzner, *Computational Molecular Biology: An Algorithmic Approach*, MIT Press 2000.
8. D.M. Prescott, M. DuBois, Internal eliminated segments (IESs) of Oxytrichidae. *Journal of Eukariotic Microbiology* **43** (1996) 432–441.
9. D.M. Prescott, A. Ehrenfeucht, G. Rozenberg, Molecular operations for DNA processing in hypotrichous ciliates. *European Journal of Protistology* **37** (2001) 241–260.

ProSpect: An R Package for Analyzing SELDI Measurements Identifying Protein Biomarkers

Andreas Quandt[1], Alexander Ploner[1], Chuen Seng Tan[1],
Janne Lehtiö[2], and Yudi Pawitan[1]

[1] Karolinska Institute, Department for Medical Epidemiology and Biostatistics,
Stockholm SE-17177, Sweden
Andreas.Quandt@meb.ki.se
[2] Cancer Centrum Karolinska, Karolinska vägen 23,
171 64 Solna, Sweden

Abstract. Protein expression profiling is a multidisciplinary research field which promises success for early cancer detection and monitoring of this widespread disease. The surface enhanced laser desorption and ionization (SELDI) is a mass spectrometry method and one of two widely used techniques for protein biomarker discovery in cancer research. There are several algorithms for signal detection in mass spectra but they are known to have poor specificity and sensitivity. Scientists have to review the analyzed mass spectra manually which is time consuming and error prone. Therefore, algorithms with improved specificity are urgently needed. We aimed to develop a peak detection method with much better specificity than the standard methods.

The proposed peak algorithm is divided into three steps: (1) data import and preparation, (2) signal detection by using an Analysis of Variance (ANOVA) and the required F-statistics, and (3) classification of the computed peak cluster as significant based on the false discovery rate (FDR) specified by the user.

The proposed method offers a significantly reduced preprocessing time of SELDI spectra, especially for large studies.

The developed algorithms are implemented in R and available as open source packages *ProSpect*, *rsmooth*, and *ProSpectGUI*. The software implementation aims a high error tolerance and an easy handling for user which are unfamiliar with the statistical software R. Furthermore, the modular software design allows the simple extension and adaptation of the available code basis in the further development of the software.

1 Introduction

Since gene expression analysis [11,15] and protein expression profiling [17] have been introduced to cancer research, it seems possible to clarify the development of this widespread disease. Scientists are using these methods to look for patterns within the given data which are characteristic for case or control. These patterns, called biomarkers, help to detect cancer in early stages and to monitor

M.R. Berthold et al. (Eds.): CompLife 2005, LNBI 3695, pp. 140–150, 2005.

the development of cancer [10]. Especially, protein biomarkers promise success for it because cancer is increasingly beginning recognized as a proteomic disease [4,19].

The modern cancer research is a multidisciplinary field where (A) molecular biologists/physicians (experiment design and data analysis), (B) computational scientists (data handling and data processing), and (C) statisticians (data processing and data analysis) work together to analyze large scale datasets, produced by genomic and proteomic techniques, to get reasonable results.

1.1 Techniques for Proteomic Biomarker Detection

Today, proteomic biomarker detection means the analysis of proteins and peptides within a range of ca. 3 - 120 kDa. Currently, there is none technique which is optimized for analyzing the whole range. Per default, the two-dimensional gel electrophoresis (2-DE) [7,12] is used for analyzing molecules over 30 kDa and the surface-enhanced laser desorption and ionization (SELDI) [6,5] is the preferred method for analyzing the range below. In general, SELDI provides a better solution than 2-DE and the possibility for automation. SELDI has been used successfully to find biomarkers for ovarian cancer, prostate cancer and to the most common method for proteomic characterization of breast cancer [19]. Especially, the possibility to analyze body fluids is a great advantage of this method and makes it perfect to find the ideal biomarker [14].

1.2 Peak Detection in SELDI Mass Spectra

The ProteinChip technology developed by Ciphergen Biosystems (Fremont, CA, USA) is based on SELDI and frequently used for detecting protein biomarkers. The platform provides high throughput capabilities for mass screens with reproducible results and the possibility to analyze tissue, cell lines, and different body fluids like blood serum and urine [8,18,22].

In the analysis of mass spectra, it is a real challenge to determine whether the peaks are due to real proteins or measurement noise. Existing peak detection algorithms are known to have poor specificity [3] and sensitivity. They tend to pick false peaks or fail to detect true peaks [6] in following situations: (1) baseline level drift, (2) misspecification of the reference peak shape, and (3) low signal to noise ratio of the peak of interest.

Scientists visually inspect the mass spectra and try to ascertain that the most interesting peaks are replicated over several spectra. This approach offers only limited reproducibility, is time consuming and error prone. Hence, algorithms which improve the automated process and the specificity of the results are urgently needed.

Currently available programs apply their peak detection algorithms to each spectrum separately. This single spectrum approach is problematic to realize, especially when the noise is of high frequency. *Tan et al.* developed a new peak detection algorithm [20] which mimics the multi-spectral visual validation. The algorithm improves the characterization of the background noise and avoids or at least minimizes the visual inspection by scientists. The statistical analysis is based on a one-way ANOVA to study the variability of spectra. Therefore, the spectra are cut into windows with a fixed number of paired observations from each spectrum. From this framework, the F-statistic is used to identify regions where the variance between spectra is significantly greater than the variability within spectra.

2   Methods

The peak detection algorithm by *Tan et al.* [20] can be separated in three major parts: (1) Data preparation, (2) calculation of the F-statistics, and (3) detection of significant peaks (figure 1). Most nucleotide biomarkers are detected in a range from 3 kDa to 10 kDa but generally, scientists analyze their spectra up to 150kDa by default. Hence, the algorithm has not only to provide a high specificity and sensitivity, it is also important to finish the calculations in acceptable time.

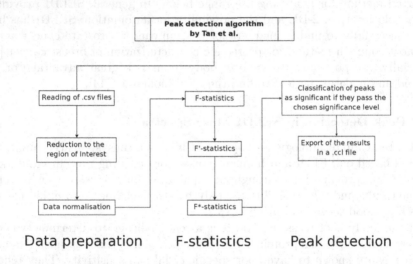

Fig. 1. Flow chart of the peak detection algorithm by *Tan et al.*

2.1   Data Preparation

Ciphergen's software provides the export of experimental data as comma separated values (CSV). The .csv files have to contain pairs of data points which are separated into m/z and intensity values. In the next step, the data are stored within an array, a multidimensional matrix. The first dimension is used to order the data by their time of flight and the second dimension to order them by spectrum.

In the next step, the data get reduced to the region of interest. That means the region where the user is interested in to look for potential biomarkers. The reduction step is based on the alignment of all m/z values across the spectra. The m/z values of each spectrum are located within the chosen range but the corresponding TOF values are different. In addition, the amount of values within each column differ in the most cases and the data array would be irregular and not allowed by definition. Hence, the shifted columns and the different number of values within each column have to be corrected to keep the array as a regular rectangular block. There are two different types of reduction procedures: The first possibility is to cut outer points and create a rectangular matrix of minimal common values. The opposite is to add missing values and get a rectangular matrix of maximal common points. Both procedures correct the shifting across the spectra but not the shifting of values within the columns.

After the reduction step, the data are normalized. Often, it is more helpful to compare the relative differences between data instead of their absolute differences. Normalizing the data means to transfer absolute differences which can be measured into the relative differences. A widely used methodology is the loess correction. This method calculates a smoothing curve corresponding to the minimum of sum of squares of all data pairs of a spectrum and has to be repeated for each spectrum separately. Usually the loess correction is based on all available data points. Within the range from 3 kDa to 10 kDa, there are approximately 50,000 data points. That means, normalizing the data by using loess is very intensive in terms of run time and computer memory (RAM). We were looking for different possibilities to reduce the run time and the use of RAM. One possibility is to use different levels of subsampling. For example, a subsampling of 10 collects each 10th data pair to calculate the smoothing curve. Smoothing curves, based on subsampling are almost identical to the original smoothing curve with less run time and use of memory. But even with this modification, computing the baseline correction was still too slow. Therefore, another was implemented. We developed a new smoothing method, named robust smoother, which is based on maximum likelihood [13] and available as a separate R package *rsmooth*. The robust smoother was written in Fortran and needs significantly less run time and computer memory than the loess method.

2.2 Calculation of the F-Statistics

Afterwards, F-statistics are calculated for a series of one-way ANOVAs for sequential non-overlapping windows of data which cut the spectra into equal-sized bins along the m/z axis. The window size can be varied via the argument *winSize* which specifies the number of data points collected in a window. The ANOVA with the spectrum as factor is calculated for the data points within each window. The F-statistics describing the effect of the spectra compared to measurement noise have to be modified and corrected in several steps (figure 1) before the theoretical F-distribution (figure 2) can be used to calculate p-values.

In the first step, the denominator of the F statistic (MSE) is smoothed to increase the statistical power of the ANOVA [20]:

$$F' = \frac{MSR}{MSE'} \tag{1}$$

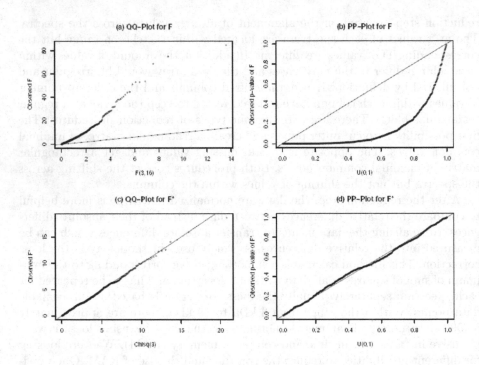

Fig. 2. QQ-plot for the corrected modified F-statistics (F*)

The smoothed MSE value (MSE') is based on a summary function of the neighboring MSE values. Examples for these summary functions are mean and median, which can specified via the argument *mean*. The second tuning parameter to influence the calculation of MSE' is *B.mse*, the range of neighboring MSE values. *B.mse* has to be chosen with some care: in our experience, peaks in the range from 1 kDa to 10 kDa are rather narrow in the sense that they cover a comparatiively small range of m/z values; additionaly, this area contains a lot of noisy peaks, and it is more difficult to separate significant peaks from the noise. Therefore it is generally recommended to smooth the MSE-values stronger by choosing a higher *B.mse*. The opposite is the case in the area over 10 kDa. There, peaks are wider and easier to classify. Hence, *B.mse* should be smaller.

After smoothing the MSE, the F'-statistics still have to be corrected. Figure 2 shows that the observed F'-statistics still do not follow the theoretical F-distribution. It is obvious though that the observed F'- statistics differs from the theoretical distribution only by a constant factor c. Therefore, the modified F-statistics (F') is further modified to yield F* (formula 2). Similar to F', the F*-statistics is calculated by a smoothing step. The user has to choose a mean value and a B-value (B.Fprime) again. Figure 2 shows the observed F* are following the theoretical distribution.

$$F^* = c * F'$$ (2)

$$F_{Theoretical} \sim F^*$$ (3)

In the next step, the significant peaks are detected. The peak detection is based on the F*-statistics of the data set, the specified significance criterion, and the chosen significance level. We suggest the use the false discovery rate (FDR) proposed by *Benjamini and Hochberg* [2] to detect significant peaks: detection is generally based on multiple testing, and peaks are classified as significant if the null hypothesis of no difference between the spectra for their cluster is rejected. Compared to p- values, the FDR provides an inherent multiplicity correction, which provides a more realistic significance criterion. By specifying a cut- off for the FDR, researchers can decide how many false positive peaks they are willing to accept. For analyzing large data sets the problem of a high rate of false positives is well known, and a strong motivation for using the FDR as significance criterion [9,21]. *Reiner et al.* used the FDR for identifying differentially expressed genes in microarray apllications, and suggested that controlling the probability of at least one false rejection among many hypothesis appears to be over-conservative and will result in reduced experimental efficiency due to unnecessary loss of power" [16]. Instead of controlling the chance of any false positive result (like e.g. via a Bonferroni correction), the FDR controls the expected proportion of false positives among significant results. A FDR threshold is determined from the observed p-value distribution, and hence is adaptive to the amount of signal in the data. *Benjamini and Hochberg* defined the FDR in the following formula

$$FDR = E(V/R; R > 0) \tag{4}$$

where V is the number of false positives and R is the number of rejected hypotheses [2].

Depending on the chosen window size, it is possible that a several neighboring cluster detect the same peak as significant. In this case, the peak labeling can be improved by merging the clusters before the results are exported in a .ccl file. The .ccl file format is based on the extended Markup Language and is used by Ciphergen's software to save and to re-import data files which were analyzed by the peak detection algorithm implemented in Ciphergen's software.

2.3   Implementation in R

R [1] is based on the statistical language S and was developed by *R.Ihaka and R. Gentleman* in the middle of the 1990s. The advantages of high modularity and the chance for fast developing and adapting of code by using additional code packages were the main reasons for the implementation of the algorithm in R. The use of key functions improves the usability of *ProSpect* and gives the user full control over the parameter set at the same time. That is important for users like molecular biologists and physicians who are mostly unfamiliar with the statistical software R and also involved in the analysis of the data sets. Key functions like *summaryPeaks()* and *plotPeaks()* help the user to analyze the data at different steps of the peak detection algorithm and to visualize them by using the Portable document file format (PDF).

Our collaborators noticed a lack of usability during the first practical tests of *ProSpect*. The most criticized problems were (1) spelling errors and the handling

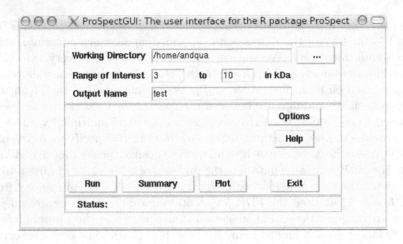

Fig. 3. The start-up window of ProSpectGUI

of case sensitive arguments and values, and (2) a lack of knowledge for most of the statistical parameter and their useful range. These problems are mainly due to the command line interface of R. The logical solution to improve the usability is to develop a Graphical User Interface (GUI), called *ProSpectGUI*, which is also available as a R package.

The start-up window of *ProSpectGUI* is shown in figure 3 and contains three frames which are visible by a borderline realized as relief. Frames are containers used to place and to collect widgets for example by different tasks.That is useful for keeping the GUI design flexible and to add and replace groups of widgets easily. Before a run can be started, the user has to specify a minimal set of arguments. Afterwards, the event buttons which are connected to one of the key functions of *ProSpect*can be used. If a run is started, the status line at the bottom of the start-up window show a message that the calculation is still running. Advanced users get access to more control over the process by pushing the Options button. Figure 4 shows the Options window that opens afterwards and offers different tuning parameters for each of the three key functions (1)*findPeaks()*, (2)*plotPeaks()*, and (3)*summaryPeaks()*. The user can optimize the offered parameter set and save it. Later, the same set can be used to analyze other experimental data in similar runs.

3   Results

Two different datasets were used to validate our procedure, after applying the standard pre-processing steps in Ciphergen's software [20]:

(1) Spike-in data. Bovine insulin at approximately 5733 Da was spiked into human blood serum, at seven levels of dilution: one control with only serum and six standard spike-ins with 10,6,4,2,1, and 0.4 μl of standard solution in serum. Each dilution was performed in independent duplicates, resulting 14

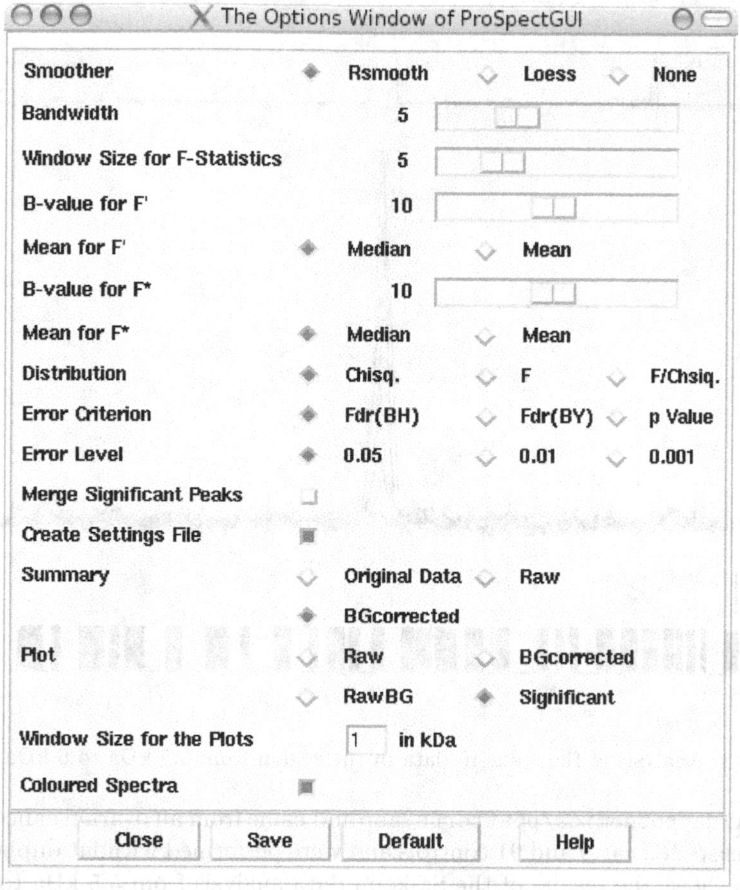

Fig. 4. The options window of ProSpectGUI serving the full set of tuning parameter for the R package ProSpect

spectra which were analyzed by low laser intensity in the region from 5 to 6.5 kDa, with 3765 points per spectrum.

(2) Lung cell line data (H69). In this study we analyzed four spectra of a lung cell line resistant to chemotherapy versus four spectra of sensitive lung cell line by low laser intensity. The data were restricted to a range from 3 to 10 kDa, with 4295 points per spectrum.

Furthermore, we used two larger datasets to demonstrate the routine application of the *ProSpect* software bundle to more realistic clinical studies. The first dataset was a drug sensitivity study with 30 spectra (independent duplicates of 8 patients sensitive to a drug and 7 patients resistant to it) performed at four different chip surfaces. All together, 128 spectra with low laser intensity were analyzed (for one chip only 28 spectra were available, due to quality control reasons). The second dataset is publicly available from http://home.ccr.cancer.

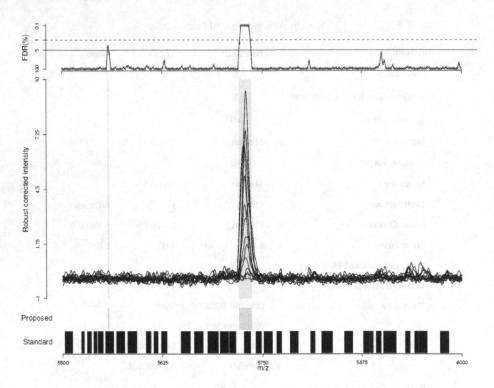

Fig. 5. Analysis of the spike-in data in the region from 5.5 kDa to 6 kDa [20]

gov/ncifdaproteomics/ppatterns.asp and came from an ovarian cancer study
that consists 162 cases and 91 controls and were performed without duplications.
Figure 5 shows the results of the Spike-in data analysis from 5.5 kDa to 6 kDa.
The graph on top visualizes the computed FDR values and shows the location
of potential biomarkers within the overlaid spectra. The bottom row of Figure 5
indicates the location of peaks identified by *ProSpect* as significant on an FDR
cut-off of 5% and the peak locations identified by Ciphergen's software as standard method. In the region from 5.5 kDa to 6 kDa, the standard method detected
19 cluster (48 windows) and the proposed method implemented in *ProSpect* only
detected 2 cluster (7 windows).

After a manual review of these spectra, the result was *ProSpect* detected 1
false positive region compared with 30 from Ciphergen Software. For data (2)
the result was similar, the implemented peak detection algorithm in *ProSpect*
shows with 80% sensitivity a lower false discovery rate (8%) than the standard
method (30%).

4   Conclusions

ProSpect is modular programmed and provides a simple mechanism to extend
available code or to replace parts of it in the further development of the software.

The new developed robust smoother (R package *rsmooth*) allows the usage of *ProSpect* for high throughput analysis screening of SELDI mass spectra to detect peak locations which differ between the analyzed spectra. The concept of key functions improves the usability for *ProSpect* itself and was used to developed the GUI add-on *ProSpectGUI* which helps user who are unfamiliar with R's command line interface.

The bundle of the R packages *ProSpect*, *ProSpectGUI*, and *rsmooth* was successful tested and detects potential biomarkers in SELDI mass spectra better than the standard software from Ciphergen.

5   Outlook

The results for comparing *ProSpect* and Ciphergen's software show not only that the peak detection of *ProSpect* is better. It becomes also visible that the peak quantification of *ProSpect* is not better compared to the standard software.

To label detected peaks on their real top is one of the challenges in Bioinformatics/Biostatistics. An improved peak labeling would afford the automatic search for posttranslational modifications and provide a better comparison of different experiments. Hence, a new algorithm will be developed and implemented in *ProSpect* which addresses this problem.

Furthermore, it will be possible to use clinical information to group the spectra within the created plots. This feature will help to look for reasons why the detected signals are different between the analyzed spectra.

References

1. http://www.r-project.org.
2. Y. Benjamini, Y. & Hochberg. Controlling the false discovery rate: a practical and powerful approach to multiple testing. *J. R. Stat. Soc*, B 57:289–300, 1995.
3. Kevin R Coombes, Herbert A Fritsche, Charlotte Clarke, Jeng-Neng Chen, Keith A Baggerly, Jeffrey S Morris, Lian-Chun Xiao, Mien-Chie Hung, and Henry M Kuerer. Quality control and peak finding for proteomics data collected from nipple aspirate fluid by surface-enhanced laser desorption and ionization. *Clin Chem*, 49(10):1615–23, Oct 2003.
4. Ruth Etzioni, Nicole Urban, Scott Ramsey, Martin McIntosh, Stephen Schwartz, Brian Reid, Jerald Radich, Garnet Anderson, and Leland Hartwell. The case for early detection. *Nat Rev Cancer*, 3(4):243–52, Apr 2003.
5. Eric Fung, Deb Diamond, Anja Hviid Simonsesn, and Scot R Weinberger. The use of SELDI ProteinChip array technology in renal disease research. *Methods Mol Med*, 86:295–312, 2003.
6. Eric T Fung and Cynthia Enderwick. ProteinChip clinical proteomics: computational challenges and solutions. *Biotechniques*, Suppl:34–8, 40–1, Mar 2002.
7. J Klose. Protein mapping by combined isoelectric focusing and electrophoresis of mouse tissues. A novel approach to testing for induced point mutations in mammals. *Humangenetik*, 26(3):231–43, 1975.

8. Lihua Li, Hong Tang, Zuobao Wu, Jianli Gong, Michael Gruidl, Jun Zou, Melvyn Tockman, and Robert A Clark. Data mining techniques for cancer detection using serum proteomic profiling. *Artif Intell Med*, 32(2):71–83, Oct 2004.

9. JG Liao, Yong Lin, Zachariah E Selvanayagam, and Weichung Joe Shih. A mixture model for estimating the local false discovery rate in DNA microarray analysis. *Bioinformatics*, 20(16):2694–701, Nov 2004.

10. JS MacNeil. Better biomarkers for the diagnostics labyrinth. *Genome Technol.*, pages 24–33, 2004.

11. Romain Molist, Michelle Gerbault-Seureau, Jerzy Klijanienko, Philippe Vielh, and Bernard Dutrillaux. Potential rapid assessment of breast cancer prognosis using induced chromosome condensation performed on cytological specimens. *Lab Invest*, 84(4):433–9, Apr 2004.

12. PH O'Farrell. High resolution two-dimensional electrophoresis of proteins. *J Biol Chem*, 250(10):4007–21, May 1975.

13. Yudi Pawitan. *In All Likelihood: Statistical Modelling and Interence Using Likelihood.* Oxford University Press, 2001.

14. M Sullivan Pepe, R Etzioni, Z Feng, JD Potter, ML Thompson, M Thornquist, M Winget, and Y Yasui. Phases of biomarker development for early detection of cancer. *J Natl Cancer Inst*, 93(14):1054–61, Jul 2001.

15. CM Perou, T Sørlie, MB Eisen, M van de Rijn, SS Jeffrey, CA Rees, JR Pollack, DT Ross, H Johnsen, LA Akslen, O Fluge, A Pergamenschikov, C Williams, SX Zhu, PE Lønning, AL Børresen-Dale, PO Brown, and D Botstein. Molecular portraits of human breast tumours. *Nature*, 406(6797):747–52, Aug 2000.

16. Anat Reiner, Daniel Yekutieli, and Yoav Benjamini. Identifying differentially expressed genes using false discovery rate controlling procedures. *Bioinformatics*, 19(3):368–75, Feb 2003.

17. K.A. Resing and N.G. Ahn. Proteomics strategies for protein identification. *FEBS Lett*, 579(4):885–9, 2005.

18. M. Shiwa, Y. Nishimura, R. Wakatabe, A. Fukawa, H. Arikuni, H. Ota, Y. Kato, and T. Yamori. Rapid discovery and identification of a tissue-specific tumor biomarker from 39 human cancer cell lines using the SELDI ProteinChip platform. *Biochem Biophys Res Commun*, 309(1):18–25, 2003.

19. R.I. Somiari, S. Somiari, S. Russell, and C.D. Shriver. Proteomics of breast carcinoma. *J Chromatogr B Analyt Technol Biomed Life Sci*, 815(1-2):215–25, 2005.

20. Chuen-Seng Tan, Alexander Ploner, Andreas Quandt, Janne Lehtiö, and Yudi Pawitan. Signal detection of seldi measurements for identifying protein biomarkers. Biostatistics (submitted Feb. 2005).

21. DB Weatherly, JA Atwood, TA Minning, C Cavola, RL Tarleton, and R Orlando. A heuristic method for assigning a false discovery rate for protein identifications from mascot database search results. *Mol Cell Proteomics*, Feb 2005.

22. Z. Xiao, D. Prieto, T.P. Conrads, T.D. Veenstra, and H.J. Issaq. Proteomic patterns: their potential for disease diagnosis. *Mol Cell Endocrinol*, 230(1-2):95–106, 2005.

Algorithms for the Automated Absolute Quantification of Diagnostic Markers in Complex Proteomics Samples

Clemens Gröpl[1], Eva Lange[1], Knut Reinert[1], Oliver Kohlbacher[2], Marc Sturm[2], Christian G. Huber[3], Bettina M. Mayr[3], and Christoph L. Klein[4]

[1] Free University Berlin, Algorithmic Bioinformatics, D-14195 Berlin, Germany
[2] Eberhard Karls University Tübingen, Simulation of Biological Systems, D-72076 Tübingen, Germany
[3] Saarland University, Instrumental Analysis and Bioanalysis, D-66123 Saarbrücken, Germany
[4] European Commission - Joint Research Centre - Institute for Health and Consumer Protection, EVCAM, - TP 580 - I-21020 Ispra(VA), Italy

Abstract. HPLC-ESI-MS is rapidly becoming an established standard method for shotgun proteomics. Currently, its major drawbacks are two-fold: quantification is mostly limited to relative quantification and the large amount of data produced by every individual experiment can make manual analysis quite difficult. Here we present a new, combined experimental and algorithmic approach to absolutely quantify proteins from samples with unprecedented precision. We apply the method to the analysis of myoglobin in human blood serum, which is an important diagnostic marker for myocardial infarction. Our approach was able to determine the absolute amount of myoglobin in a serum sample through a series of standard addition experiments with a relative error of 2.5%. Compared to a manual analysis of the same dataset we could improve the precision and conduct it in a fraction of the time needed for the manual analysis. We anticipate that our automatic quantitation method will facilitate further absolute or relative quantitation of even more complex peptide samples. The algorithm was developed using our publically available software framework OpenMS (www.openms.de).

1 Introduction

The accurate and reliable quantification of proteins and peptides in complex biological samples has numerous applications ranging from the determination of diagnostic markers in blood and the discovery of these markers to the identification of potential drug targets. HPLC-MS-based shotgun proteomics is rapidly becoming the method of choice for this type of analysis. Currently, the huge amount of data being produced and difficulties with absolute quantification of individual peptides are the major problems with this method.

In this work, we propose an HPLC-MS-based approach for the absolute quantification of myoglobin in human blood serum and demonstrate the viability of this approach using reference material developed by the European Commission Joint Research Centre. Myoglobin is a low-molecular weight (17 kDa) protein present in the cytosol of cardiac and skeletal muscle. Due to these characteristics, myoglobin appears in blood

M.R. Berthold et al. (Eds.): CompLife 2005, LNBI 3695, pp. 151–162, 2005.

after tissue injury earlier than other biomarkers, such as creatine kinase MB isoenzyme (CK-MB) and cardiac troponins [1]. It is of pivotal importance in clinical diagnosis as early biomarker of myocardial necrosis. Serum myoglobin has been used in routine practice since the development of automated non-isotopic immunoassays [2]. Currently, the National Academy of Clinical Biochemistry [3], the International Federation of Clinical Chemistry and Laboratory Medicine (IFCC) [4], and the American College of Emergency Physicians [5] have recommended use of myoglobin as early marker of myocardial necrosis. Unfortunately, results from different analytical procedures for myoglobin determination have significant biases as a result of the lack of assay standardization. Results from National External Quality Assurance Schemes showed a bias of over 100% for serum myoglobin [6, 7]. Standardization of any measurand requires a reference measurement system, including a reference measurement procedure and (primary and secondary) reference materials (RM) [8]. The joint HPLC-MS/bioinformatics approach has been used to develop a reference method that can be used to standardize myoglobin assays [9, 10] and subsequently to reduce the bias observed between commercial myoglobin assays, to standardize and harmonize measurement results, and to improve quality of diagnostic services.

In the experimental part of this work, myoglobin was separated from the highly abundant serum proteins by means of strong anion-exchange chromatography. Subsequently, the myoglobin-fraction was trypsinized and the peptides were analyzed by capillary reversed-phase high-performance liquid chromatography-electrospray ionization mass spectrometry (RP-HPLC-ESI-MS) using an ion-trap mass spectrometer operated in full-scan mode. In order to avoid quantification errors by artifacts in the sample preparation we added a constant amount of horse myoglobin to each sample in the additive series. We chose horse myoglobin as internal standard, since the tryptic horse peptides corresponding to their human counterparts elute roughly at the same time and are sufficiently different from the human peptides, such that corresponding peptides have different mass. To achieve an absolute quantification, known amounts of human myoglobin were added to aliquots of the sample. Each of the samples was measured in four replicates.

The raw data acquired by the instrument was analysed automatically using a newly developed algorithm that detects and quantifies all ions belonging to peptides in the sample. The cornerstone of the algorithm is formed by a two-dimensional model of the peptide isotope pattern and its elution profile. This model is then applied to accurately and automatically integrate the raw data into (relative) intensities proportional to the amount of the peptide. Using standard statistical tools, we can then determine the true concentration of myoglobin in our samples.

Our results indicate that the algorithms outperforms manual analysis of the same data set in terms of accuracy. It allows an accurate determination of myoglobin in serum from a set of HPLC-MS raw data sets without manual intervention. The relative errors observed were as low as 2.5% and thus below the errors observed in manual analysis of the same data set. Moreover, these results could be obtained in a fraction of the time required for the manual analysis.

Besides its use in the reference method for myoglobin quantitation, we anticipate that our automatic method is generic enough such that it will facilitate quantitative analysis of even more complex proteomics data without labeling techniques (see for example [11]) and thus allow for other types of analyses and high-throughput applications such as detecting diagnostic markers, or the analysis of time series. The algorithm was implemented within our publically available software framework OpenMS (www.openms.de).

The outline of the paper is as follows. In Section 2 we describe the overall experimental setup and algorithmic techniques applied to the data. Section 3 gives detailed results of the manual and automatic analysis of these data. We conclude with a brief discussion of the method, its advantages and limitations in Section 4.

2   Methods

In the following two subsections we describe the experimental protocol to produce the data and the algorithmic approach taken to conduct the quantification and analysis in an automated fashion.

2.1   Sample Preparation and Data Generation

We give a brief summary of sample preparation and data generation (more details and optimizations of experimental conditions will be described elsewhere). Briefly, myoglobin-depleted human serum (blank reference serum, from the European Commission - Joint Research Centre - Institute of Reference Materials and Measurements, Geel, Belgium, IRMM) was spiked with 0.40-0.50 ng/μl human myoglobin (from IRMM). This concentration represented the target value to be quantitated. To this spiked serum sample, 0.50 ng/μl horse myoglobin (Sigma, St. Louis, MO) were added as internal standard. For the additive series, known amounts of human myoglobin standard were added to the serum sample, resulting in concentrations of added myoglobin standard between 0.24 and 3.3 ng/μl. Usually, 6-7 standard additions were performed. The myoglobin fraction was isolated from 20 μl human serum by means of strong anion-exchange chromatography upon collection of the eluate between 4.2 and 4.8 min eluting from a ProPac SAX-10 column (250×4.0 mm i.d. with 50×4.0 mm i.d. precolumn, Dionex, Sunnyvale, CA). The column was operated with a gradient of 0-50 mmol/l sodium chloride in 10 mmol/l TRIS-HCl, pH 8.5, in 10 min, followed by a 4 min isocratic hold at 50 mmol/l sodium chloride and a finally a gradient of 50-500 mmol/l sodium chloride in 2 min at a volumetric flow rate of 1.0 ml/min. After evaporation of part of the solvent in a vacuum concentrator, the myoglobin fraction was adjusted to a defined weight of 100.0 mg using an analytical balance. The proteins in the myoglobin fraction where digested for two hours at 37 °C with trypsin (sequencing grade, from Promega, Madison, WI) using Rapigest^{TM} (Wates, Milford, MA) as denaturant following standard digestion protocols. The digested fractions were transferred to glass vials and analyzed by reversed-phase high-performance liquid chromatography-electrospray ionization mass spectrometry (RP-HPLC-ESI-MS). Desalting and separation of the peptides was accomplished with a 60×0.20 mm i.d. monolithic capillary column (home-made, commercially available from LC-Packings, Amsterdam, NL) and 6 min isocratic elution with 0.050% trifluoroacetic acid in water, followed by a 15 min gradient of 0-40%

acetonitrile in 0.050% trifluoroacetic acid at a volumetric flow rate of $2.4\,\mu l$/min. The eluting peptides were detected in a quadrupole ion trap mass spectrometer (Esquire HCT from Bruker, Bremen, Germany) equipped with an electrospray ion source in full scan mode (m/z 500-1500). Each measurement consisted of ca. 1830 scans. The scans were roughly evenly spaced over the whole retention time window with an average of 0.9 scans per second. The sampling accuracy in mass-to-charge dimension was 0.2 Th. The instrument software was configured to store the measurement data in its most un-processed form available (described below). The raw data was converted to flat files of size ca. 300 MB each using Bruker's CDAL library. With the upcoming mzData stan-dard data format for peak list information [12, 13], this step should become much easier in the near future. Quantitation of the myoglobin peptides in the serum sample was then conducted as decribed in the next section.

2.2   Feature Finding

By the term *feature finding* we refer to the process of transforming a file of raw data as acquired by the mass spectrometer into a list of features. Here a *feature* is defined as the two-dimensional integration with respect to retention time (rt) and mass-over-charge (m/z) of the eluting signal belonging to a single charge variant of a peptide. Its main attributes are average *mass-to-charge ratio*, centroid *retention time*, *intensity*, and a *quality* value.

In our study, the raw data set exported from the instrument consisted of *profile spec-tra*, but no baseline removal or noise filtering had been performed. In particular, no *peak picking* had taken place (where peak picking denotes the process of transforming a pro-file spectrum to a stick spectrum by grouping the raw data points into one-dimensional "peaks", which have a list of attributes similar to those of features). Features are com-monly generated from raw data by forming groups with respect to one dimension after the other, thereby reducing the dimensionality one by one. However better results can be achieved using a genuinely two-dimensional approach.

Theoretical model of features.   Each of the chemical elements contributing to the sum formula of a peptide has a number of different isotopes occuring in nature with certain abundancies [14]. The mass differences between these isotopes can be approximated by multiples of 1.000495 Da up to the imprecision of the instrument. Given these param-eters, and the empirical formula of a peptide, one can then compute its the theoretical stick spectrum. In our study, such an *isotope pattern* has 3-6 detectable masses. Since the lightest isotopes are by far most abundant for the elements C, H, N, O, and S, it is common to use the corresponding stick as a reference point, called *monoisotopic s peak*.

If the peaks for consecutive isotope variants are clearly separated in the profile spec-tra, they can be picked individually and combined to isotope patterns afterwards. How-ever, in our raw data set, having a sampling accuracy of 0.2 Th, this is the case only for charge 1. Already for charge 2 the profiles of peaks overlap to such an extent that such a two-step approach is not feasible. Moreover, as the mass and charge increases, the whole isotope pattern at a given fixed value of m/z becomes more and more bell-shaped and eventually converges to a normal distribution. In our case, neither extreme is a good

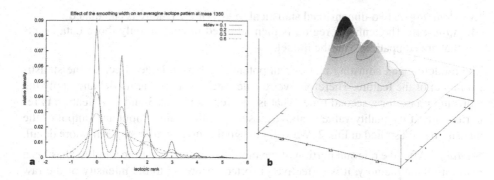

Fig. 1. (a) Effect of the smoothing width on the theoretical isotope distribution of a peptide of mass 1350 Da. Increasing the smoothing width can emulate the effect of low instrument resolution. **(b)** A two-dimensional model for a feature of charge two and mass 1350 *Da*.

approximation. Instead, we model the m/z profile of the raw data points belonging to a single isotope pattern by a mixture of normal distributions, as shown in Fig. 1 (left).

One of our design goals was that the algorithm should not rely on information about specific peptides given in advance. Therefore the empirical formula of a peptide of a given mass is approximated using so-called *averagines*, that is, average atomic compositions taken from large protein databases. For example, an averagine of mass 1350 contains "59.827" C atoms, "4.997" N atoms etc. We calculated the isotopic distributions of the tryptic myoglobin peptides and found that they are well approximated by averagines (see Fig. 3 in the Appendix). If necessary, an even better approximation could be used that takes into account that the peptides are digested by a specific protease (in our case trypsin), which results in a bias of the amino acids at the end of a peptide. The theoretical m/z distribution is then obtained by convoluting the sticks of the theoretical isotope pattern with a normal distribution to simulate the measurement inaccuracy. The left part of Fig. 1 shows the effect of the smoothing width on an averagine isotope distribution at mass 1350 Da.

The signal of a single charge variant of a peptide extends over a certain interval of retention time. As a model for the retention profile, we currently use a normal distribution with variable width. More sophisticated models that incorporate fronting and tailing effects that are observed especially for high intensity peaks are known (see e.g. [15, 16]). These shall be investigated in subsequent work.

It is natural to assume that isotope pattern and elution profile are independent from each other. Consequently, our theoretical model for features is a product of a model for the m/z domain and a model for the retention time domain. An example of a two-dimensional feature model is shown in the right part of Fig. 1.

Algorithm. The algorithm for feature finding consists of four main phases:

1. *Seeding.* Data points with high signal intensity are chosen as starting points of the feature detection.
2. *Extension.* The region around each seed is conservatively extended to include all potential data points belonging to the feature.

3. *Modeling.* A two-dimensional statistical model of the feature is calculated.
4. *Adjusting.* The tentative region is then adjusted to contain only those data points that are compatible with the model.

The modeling and adjusting phases can potentially have a large effect on the statistical model of the feature. Therefore we re-calculate the statistical model and apply the adjusting phase for a second time. That is, we repeat phases 3 and 4. A feature is reported only if its quality value is above a user-specified value. Input and output of the algorithm is illustrated in Fig. 2. We will now go through the four stages in more detail.

Seeding. After the relevant portion of the input file (a retention time window) has been read into main memory, it is (effectively) sorted according to the intensity of the raw data points. In a greedy fashion we consider the most intense data point as a so-called *seed* for the formation of a feature. This is motivated by the fact that the most intense data points are very likely to belong to a feature. A seed is considered for the next phase (extension) only if it is not already contained in a feature. We stop when the seed intensity falls below a threshold. (The actual implementation does not sort the raw data physically, but uses a priority queue instead, from which the seeds are extracted in order of intensity. This way the low-intensity data points need not be sorted.)

Extension. Given a seed, we conservatively determine a *region* around it that very likely contains all data points of the feature. The region grows in all directions simultaneously, preferring the strongest raw data points near the *boundary*.

Initially, the region is empty and the boundary set only contains the seed. In each step, a data point in the boundary is selected and moved into the region. Then the boundary is updated by exploring the neighborhood of the selected data point. The selected data point is chosen based on a *priority* value, and the boundary set is implemented as a priority queue (This should not to be confused with the priority queue used for seeding). The priority of a data point is never decreased by an update of the boundary. If the updated priority of a neighboring data point exceeds a certain threshold, it is moved into the boundary. The seed extension stops when the intensity of all data points in the boundary falls below a certain threshold.

The priority values of raw data points are not identical to their intensities. Their purpose is to control the growth of the feature, such that a number of constraints are met: The boundary should be a relatively 'thin' layer around the region. It should be resistant to noise in the data and allow for 'missing' raw data points. Data points close to the region should be preferred. We compute the priority values as follows: When a data point is extracted from the priority queue, we explore a cross-like neighborhood around it in four directions ("m/z up", "m/z down", "rt up", "rt down"). The priority is calculated by multiplying the intensity of the data point with a certain function of the distance from the extracted point. Currently we use triangular shapes that go to zero at distance 2.0 s in rt and 0.5 Th in m/z.

The criteria controlling the growth of the boundary and the stopping of the seed extension are adapted during the seed extension process based on the information gathered so far. This is done as follows:

1. We compute an intensity threhold for stopping the extension phase. The The threshold is a fixed percentage of the 5 th largest inensity (we do not choose the largest for robustness reasons).

2. We maintain a running average of the data point positions, weighted by their intensities. The neighborhood of a boundary point is not further explored if it is too distant from the centroid of the feature. This is important to avoid collecting low intensity data points (baseline) when the seed has a relatively low intensity.

Modeling.   Given a region, we fit a two-dimensional statistical *model* to it. The point intensity of the two-dimensional model is the product of two one-dimensional models, one explaining the isotope pattern and one explaining the elution profile. The raw data points are considered empirical samples from this distribution.

The fit in m/z dimension examines different distributions implied by charge states in a range provided by the user (currently 1 to 3). For each charge, we try a number of smoothing widths of the averagine isotope pattern (currently 0.15, 0.2, 0.25, 0.3, and 0.35 Th). The correct charge state is likely to provide the best fit to the data points. In addition we also fit a normal distribution using maximum likelihood estimators. As a measure of confidence in the charge prediction we report on the distance to the fit with the second best charge hypothesis. The fit in retention time dimension uses a maximum likelihood normal approximation.

The quality of fit of the data against a model is measured using the squared correlation

$$\frac{(\sum_x f(x)g(x))^2}{\sum_x f(x)^2 \sum_x g(x)^2} \, ,$$

where f = observed, g = model, x = data point position. Other methods like the χ^2-test have already been implemented in OpenMS and can be used if desired.

Adjusting.   At this stage of the algorithm, we have a region of data points and a statistical model for it. But the region is very likely to contain data points not belonging to the feature. To discard those, and keep only those data points which are consistent with the statistical model, we re-assemble the data points contained in the feature similar to the extension phase using a modified priority that takes the model into account. Using a model is the main difference of this phase compared to the extension phase.

To combine the theoretical and observed intensities, we use the geometric mean of the observed intensity of a data point and its prediction by the model as the priority for extension. This is based on the following considerations: Since the normal distribution decays exponentially at its tails, data points not explained by the model are effectively cut off. Moreover the geometric mean compensates for inaccuracies when the intensity of the data points decays faster than predicted. Of course many other strategies for adjusting can be considered and should be tested in the future.

3   Results

We present results from a series of 32 RP-HPLC-ESI-MS measurements performed as described in Section 2.1 (four replicates of eight different spiked concentrations). The quantification was performed using the eleventh tryptic peptide of human myoglobin, HGATVLTALGGILK, here denoted *T11hu*, with and without the tenth tryptic peptide

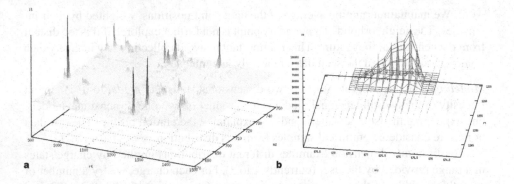

Fig. 2. (a) The raw data map drawn as a 3D picture. Each sample resulted in of these two-dimensional data sets. (b) Feature finding isolates individual peptides out of this map. The figure shows one of the peptides used for quantification: the raw data is shown as red sticks, the optimal model describing the feature is plotted on top in blue.

of horse myoglobin, HGTVVLTALGGILK, denoted *T10ho*, as an internal standard. These two peptides are sufficiently similar to behave similarly in terms if ionization and still can be separated easily in both RT and m/z dimension.

To assess the quality of the automated analysis, we also report the results of a manual expert analysis of the same data set that was performed earlier by one of the authors [17]. Manual quantification was performed using the Bruker instrument software and Microsoft Excel. The peak areas were calculated from extracted ion chromatograms with an isolation width of ± 0.5 Da after smoothing with a gauss filter.

Automated analysis was performed using the features found by the algorithm described in Section 2.2 without further manual intervention. We provided approximate masses and approximate retention times of the peptides used for quantification and restricted the feature finding to a large window of the raw data (RT = 900–1600 sec, m/z = 600–1000 Th.) to speed up the process. The algorithm then identified features in the 32 data sets, integrated the feature areas and performed the statistical analysis detailed in the following table:

Method	OpenMS	Manual
Computed concentration [ng/μl]	**0.474**	**0.382**
Lower bound of 95% interval [ng/μl]	0.408	0.315
Upper bound of 95% interval [ng/μl]	0.545	0.454
True value [ng/μl]	*0.463*	*0.463*
Relative deviation from true value [%]	**+2.46**	**−17.42**
Lower bound of 95% interval [%]	−11.82	−32.04
Upper bound of 95% interval [%]	+17.62	−1.84

Both manual and automated analysis were able to estimate the true concentration of myoglobin in the serum sample with very high precision. While manual analysis of these large data sets amounts to half a day of work, automated analysis of the datasets could be performed in less than 2 hours on a 2.6 GHz Pentium IV machine with 1 GB of RAM running Linux. The regression results are shown in Fig. 4 in the Appendix.

The results of several additional independent studies for myoglobin quantification all yielded relative quantification errors below 8% (data not shown). Automated analysis of the data sets yielded comparable or better results in these experiments.

4 Conclusion

Analyzing complex shotgun proteomics data is still a major challenge; the size of the data, its complex nature, and the lack of established algorithms to reduce these data to their essentials are all restricting the use of this powerful technique to rather trivial experimental setups.

We present an algorithm for the automated data reduction for quantification purposes based on a statistical modeling of peptide isotope patterns and elution profiles. The algorithm is robust and handles large data efficiently. In contrast to the tiresome manual analysis of large datasets, this automated technique allows the analysis of a large number of samples, thus enabling more complex experiments. As a result, we can even use this technique to absolutely quantify individual proteins from the serum sample through standard addition techniques with extremely high accuracy (2.5% relative error). In most routine applications, this level of accuracy might not be viable or necessary, but the automated analysis clearly saves valuable time over manual approaches and even results in improved accuracy of the analysis.

The present approach can be used for direct quantification of a target peptide or a number of target peptides in a complex biological matrix, such as human blood serum with simultanous identification. It is clearly demonstrated that the method can be used for quantitative determination of human myoglobin in serum and therefore is a suitable candidate for serving as a reference method. It can be used for value assignment of a candidate CRM as under investigation by IRMM. By using the present method in combination with a matrix-based CRM, in vitro diagnostics (IVD) manufacturers could demonstrate traceability of their working methods and kits as used in clinical chemistry, fulfill the legal requirements, and further improve quality of products and services by harmonization and standardization.

While it is clear that the proposed technique is (experimentally) too involved and costly for most routine applications, it clearly is a viable technique for high accuracy applications, for example as reference methods in standardization. The algorithms proposed here nevertheless are not limited to this application and can be used in a wide range of other proteomics experiments, e.g. relative differential proteomics for the identification of diagnostic markers or drug targets. The ability to analyze data on a larger scale will enable a wider range of experimental setups encompassing a larger number of samples and repeats, something that is currently not viable due to the limits of manual analysis. The method can also be distributed trivially on a compute farm allowing for extremely rapid analysis of data.

The algorithms proposed are clearly first steps only. More advanced statistical models accounting for asymmetric peak shapes, strongly overlapping features, or low signal-to-noise ratios are not yet accounted for in its current state. Extensions addressing these issues are currently being implemented and will hopefully yield even better performance in future versions of the software.

References

1. Panteghini, M., Pagani, F., Bonetti, G.: The sensitivity of cardiac markers: an evidence-based approach. Clin. Chem. Lab. Med. **37** (1999) 1097–1106
2. Wu, A.H., Laios, I., Green, S., Gornet, T.G., Wong, S.S., Parmley, L., Tonnesen, A.S., Plaisier, B., Orlando, R.: Immunoassays for serum and urine myoglobin: myoglobin clearance assessed as a risk factor for acute renal failure. Clin Chem **40** (1994) 796–802
3. Wu, A., Apple, F., Gibler, W., Jesse, R., Warshaw, M., Valdes, J.R.: National academy of clinical biochemistry standards of laboratory practice: recommendations for use of cardiac markers in coronary artery diseases. Clinical Chemistry **45** (1999) 110–121
4. M., P., Apple, F., Christenson, R., Dati, F., Mair, J., Wu, A.: Use of biochemical markers in acute coronary syndromes. Clin. Chem. Lab. Med. **37** (1999) 687–693
5. Fesmire, F., Campbell, M., Decker, W., Howell, J., Kline, J.: Clinical policy: critical issues in the evaluation and management of adult patients presenting with suspected acute myocardial infarction or unstable angina. Ann. Emerg. Med. **35** (2000) 521–544
6. College of American Pathologists: Cardiac markers survey (2003) Northfield, IL.
7. Panteghini, M.: Recent approaches to the standardization of cardiac markers. Scand. J. Clin. Lab. Invest. **61** (2001) 95–102
8. Panteghini, M. In: Standardization of cardiac markers. Totowa (2003) 213–229
9. Dati, F., Linsinger, T., Apple, F., Christenson, R., Mair, J., Ravkilde, J., et al.: IFCC project for standardization of myoglobin immunoassays. Clin. Chem. Lab. Med. **40** (2002) S311
10. Dati, F., Panteghini, M., Apple, F., Christenson, R., Mair, J., Wu, A.: Proposals from the IFCC committee on standardization of markers of cardiac damage (C-SMCD): strategies and concepts on standardization of cardiac marker assays. Scand. J. Clin. Lab. Invest. **230** (1999) 113–123
11. Silva, J., Denny, R., Dorschel, C., Gorenstein, M., Kass, I., Li, G., McKenna, L., Nold, M., Richardson, K., Young, P., Geromanos, S.: Quantitative rpoteomic analysis by accurate mass retention time pairs. Anal. Chem. **77** (2005) 2187–2200
12. Orchard, S., Hermjakob, H., Apweiler, R.: The proteomics standards initiative. Proteomics **3** (2003) 1374–1376
13. HUPO Proteomics Standards Initiative. http://psidev.sourceforge.net/ (2005)
14. de Hoffmann, E., Charette, J., Stroobant, V.: Mass Spectrometry. 2nd edn. John Wiley and Sons (2001)
15. Marco, V.B.D., Bombi, G.G.: Mathematical functions for the representation of chromatographic peaks. Journal of Chromatography A **931** (2001) 1–30
16. Pai, S.C.: Temporally convoluted gaussian equations for chromatographic peaks. Journal of Chromatography A **1028** (2004) 89–103
17. Mayr, B.M.: Die Kopplung der Flüssigchromatographie mit der Elektrospray-Ionisations-Massenspektrometrie als Werkzeug für die Genomanalyse und die Quantitative Proteomforschung. PhD thesis, Universität des Saarlandes (2005)

A   Appendix

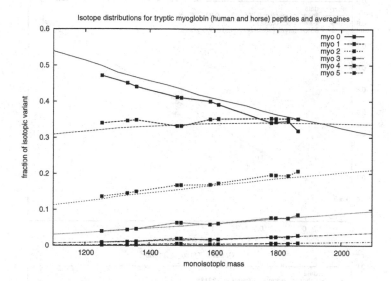

Fig. 3. The comparision shows that the isotope distributions for tryptic myoglobin (human and horse) peptides are well approximated by averagines

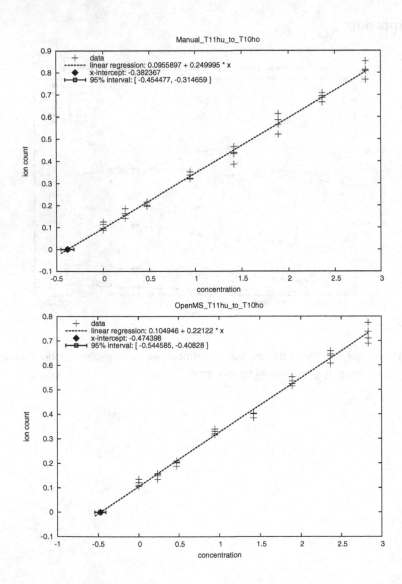

Fig. 4. Regression results for manual (top) and automated (bottom) analysis of myoglobin in serum samples. The automated analysis yields smaller standard deviations between replicates of the same sample and tighter error bounds on the absolute concentration computed.

Detection of Protein Assemblies in Crystals

Evgeny Krissinel and Kim Henrick

European Bioinformatics Institute, Hinxton, Cambridge CB10 1SD, UK
keb@ebi.ac.uk
http://www.ebi.ac.uk/msd-srv/prot_int/cgi-bin/piserver

Abstract. The paper describes a new approach to the prediction of probable biological units from protein structures obtained by means of protein crystallography. The method first employs graph-theoretical technique in order to find all possible assemblies in crystal. In second step, found assemblies are analysed for chemical stability and only stable oligomers are left as a potential solution. We also discuss theoretical models for the assessment of protein affinity and entropy loss on complex formation, used in stability analysis.

1   Introduction

Considerable part of protein functionality in biological systems is associated with ability of proteins to bind each other and form stable complexes (assemblies, or biological units). Data on multimeric state of protein complexes and spatial arrangement of their subunits may often provide a deeper insight into the functioning of machinery of life and role of particular proteins in it.

Experimental means for the identification of spatial structure of protein complexes are limited. Because of their relatively large size, protein assemblies are not a good object for NMR studies. Some proteins may exist in dynamic equilibrium between different multimeric states, which also complicates NMR analysis. Electron microscopy is suitable for studying large complexes, but it yields rather low-resolution structures. About 80% of entries in Protein Data Bank [1] represent structures solved by means of X-ray diffraction on protein crystals. In these experiments, crystal structure is identified in the form of atomic coordinates in the asymmetric unit (ASU), unit cell geometry and space symmetry group. However, protein crystallography does not identify true protein associations among all protein contacts in a crystal. At the same time, it is reasonable to expect that protein assemblies do not dissociate during the crystallisation process and therefore protein crystals should contain assemblies as subunits.

Identification of protein assemblies in crystals is, in general, a non-trivial task. The asymmetric unit may be chosen in many different ways, and it does not necessarily coincide with the biological unit. An asymmetric unit may be made from more than one assembly, or a few ASUs may be required to make an assembly, or assembly may be made from several incomplete ASUs. A further complication arises if one assumes that a few different complexes may co-exist in dynamic equilibrium, then crystal may be made from more than one assembly.

M.R. Berthold et al. (Eds.): CompLife 2005, LNBI 3695, pp. 163–174, 2005.

Two approaches to the problem have been proposed so far [2,3]. Both of them are based on the scoring of individual protein interfaces (identified as crystal contacts between monomeric chains) in order to conclude about their biological relevance. PQS server at EBI-MSD [2] scores interfaces mostly on the basis of interface area, with a point system for the hydrophobic effect of complexation, hydrogen bonds, salt bridges and disulphide bonds. The assemblies are built up by the progressive addition of monomeric chains that are bonded by high-scored interfaces. PITA software [3] uses a sophisticated statistical potential to score the interfaces [4] and looks for the solution by iterative bipartitioning of the largest possible assembly in crystal until the minimum-cut interface score exceeds a predefined threshold.

In this paper, we propose another approach based on the consistent enu-meration of all assemblies that are possible in a given crystal, with subsequent analysis for chemical stability. The analysis is based on the evaluation of free energy of complex dissociation, which includes the free energy of binding and the entropy change term. As found, the new approach predicts protein assemblies with a higher success rate than its predecessors.

2   Graph-Theoretical Detection of Protein Assemblies in Crystals

We now want to find all different assemblies in crystal that are allowed by sym-metry considerations and content of ASU. We do not assume that crystal is necessarily made from identical assemblies, so that we are looking to find all possible *sets* of different assemblies that fill all the crystal space in a systematic manner. One can note that each such set is unambiguously identified by the inner-assembly interfaces. We will refer to such interfaces as engaged. Then the search may be formulated as enumeration of all possible interface engagements that obey the following rules:

1. Due to crystal symmetry, if an interface of a particular type (that is, between given monomeric chains in a particular relative position) is engaged, all other interfaces of the same type in crystal are also engaged.
2. An interface cannot be engaged if doing so results in assembly that contains identical chains in parallel orientations.

Rule 2 originates from the consideration that if an assembly contains two molecules in parallel orientation, then due to translation symmetry in crystal this assembly must have infinite size. As a consequence of this rule, assembly size cannot exceed the size of unit cell.

The described task may be efficiently addressed by a backtracking scheme, a procedure commonly used in graph matching algorithms [5]. Imagine crystal as a graph where monomeric chains represent vertices, and interfaces between the chains represent edges. The vertices may be calculated by applying all symmetry operations of the required space symmetry group to the chains in ASU, and translating the obtained unit cell according to the cell dimensions and geometry.

The obtained graph is periodic in three dimensions, with period equal to the size of unit cell in the respective dimension. The periodicity allows one to imitate calculations for an infinite crystal on a single unit cell by applying a periodic shift to the inter-cell edges.

1. Calculate periodic graph representing the crystal
2. List all unique interfaces as I_k
3. Make empty sets of engaged and tested interfaces $\{I\} := \emptyset$, $\{T\} := \emptyset$
4. **call** $Backtrack(\{I\}, \{T\})$
5. **stop**

procedure $Backtrack$ (interface sets $\{I\}$, $\{T\}$)
B.1   **copy** $\{T\}$ to $\{T_1\}$
B.2   **for** all interfaces I_k not found in $\{I\}$ and $\{T_1\}$ **do**
B.3   **copy** $\{I\}$ to $\{I_1\}$
B.4   **add** I_k to $\{I_1\}$ and $\{T_1\}$ (engage interface I_k)
B.5   **do**
B.6   Identify assemblies formed by interfaces in $\{I_1\}$
B.7   Identify induced interfaces and add them to $\{I_1\}$ and $\{T_1\}$
B.8   **until** no interfaces are induced
B.9   **if** no assembly contains identical parallel chains **then**
B.10   output set of assemblies as possible solution
B.11   **if** more stable assemblies may be found **then**
B.12   **call** $Backtrack(\{I_1\}, \{T_1\})$
B.13   **endif**
B.14 **done**

Fig. 1. The assembly enumeration algorithm, see text for details

The assembly enumeration algorithm is schematically depicted in Fig. 1. It represents a recursive backtracking scheme, which explores all unique combinations of engaged interfaces $\{I\}$. Each such combination corresponds to a set of assemblies which is noted for further analysis of chemical stability. Set $\{T\}$ and its local copies $\{T_1\}$ are used in order to avoid redundant combinations of the interfaces. In steps B.5-8 the algorithm looks for "induced" interfaces and engages them. "Induced" interface is identified as one that appears to be internal to assembly formed by previously engaged interfaces. For example, engaging interfaces $A_1 : A_2$ and $A_2 : A_3$ in trimer (A_1, A_2, A_3) induces interface $A_1 : A_3$.

It may be shown that the total number of unique interface combinations is $N_I!$, where N_I is the total number of unique interfaces. Factorial complexity becomes prohibitive for many PDB entries where $N_I \geq 10$. Therefore, in step B.11 of the algorithm, we terminate those branches of the recursion tree which definitely do not lead to stable assemblies. This technique is borrowed from graph-matching algorithms [5]. The termination condition is derived from the chemical stability analysis and will be described in Section 4.

3   Analysis of Chemical Stability

Most of assemblies, emerging from the graph-theoretical search, represent unstable structures, which dissociate in dilute solutions. In what follows, we consider assembly as unstable if equilibrtium constant of dissociation is greater than 1. Then protein complex $(A_1, A_2 \ldots A_n)$ dissociates into subunits A_i (any subunit may be a multimer) if the free energy change upon dissociation ΔG_{diss} is negative:

$$\Delta G_{diss} = -\Delta G_{int} - T\Delta S < 0 \tag{1}$$

where ΔG_{int} represents free energy of binding of subunits A_i and ΔS is the rigid-body entropy change upon dissociation. Consider terms of Eq. (1) in more detail.

3.1   Free Energy of Protein Binding

The binding energy ΔG_{int} is calculated as a free energy of interface formation between subunits A_i. There are many factors that contribute into protein association energy [6,7,8,9,10,11,12,13,14], but it is widely acknowledged that major contributions are due to the interaction of protein surface with the solvent and formation of hydrogen bonds and salt bridges across the interfaces:

$$\Delta G_{int} = \Delta G_s(A_1, A_2 \ldots A_n) - \sum_{i=1}^{n} \Delta G_s(A_i) - E_{hb}N_{hb} - E_{sb}N_{sb} \tag{2}$$

In Eq. (2), $\Delta G_s(A)$ stands for the solvation free energy of folding. It may be approximated as [11,16]

$$\Delta G_s(A) = \sum_k \Delta\sigma_k(a_k - a_k^r) \tag{3}$$

where summation is done for all atoms in structure A, a_k stands for the atom's solvent-accessible surface area, $\Delta\sigma_k$ and a_k^r are atomic solvation parameters and surface area in reference state, respectively. $\Delta\sigma_k$ and a_k^r depend on the atom type and charge state in residue. Eq. (2) takes into account that atom charge state may change with changing a_k due to interface formation.

Eq. (2) measures the effect of each of N_{hb} hydrogen bonds and N_{sb} salt bridges between all the subunits A_i by average free energy contributions E_{hb} and E_{sb}, respectively. The strength of a hydrogen bond is estimated to be between 2 and 10 kcal/mol [17]. However, upon disengaging an interface, all potential hydrogen bonding partners become satisfied by hydrogen bonds to water. The only effect that remains here is the decreasing entropy of solvent due to the loss of mobility by bound molecules. Estimations show a contribution of about $E_{hb} \approx 0.6 - 1.5$ kcal/mol per bond [18,19]. Experimental data on the stabilisation effect of salt bridges are limited. Known studies suggest that free energy contribution of a salt bridge is very close to that of a hydrogen bond $E_{sb} \approx 0.9 - 1.25$ kcal/mol [20,21].

3.2   Entropy of Protein Complex Formation

Entropy contribution into the free energy of complex dissociation ΔG_{diss} (cf. Eq. (1)) originates from the change of the vibrational mode pattern and regain of rotational and translational degrees of freedom by subunits A_i upon dissociation. Entropy of subunit A may be represented as

$$S(A) = S_{rb}(A) + S_{vib}(A) + S_{surf}(A) \tag{4}$$

where $S_{rb} = S_{trans} + S_{rot}$ stands for the rigid-body (translational and rotational) entropy term, S_{vib} – entropy of internal vibrational modes and S_{surf} – entropy of surface atoms with fractional degrees of freedom.

There are no rigorous theoretical models for the rigid-body entropy of sizeable objects in liquids. Translational entropy contribution S_{trans} may be approximated by the Sackur-Tetrode equation, which was originally derived for the case of small molecules in gas phase [22,23,24]

$$S_{trans}(A) = R \log \left[\left(\frac{2\pi m(A)kT}{h^2} \right)^{3/2} \left(v e^{5/2} \right) \right] \tag{5}$$

where $m(A)$ is molecular weight and v is the volume open to a molecule. Eq. (5) was found to be a reasonable approximation in liquid phase, too, after corresponding adjustment of the value of v [25].

Rotational rigid-body entropy term can be estimated as [23,24]

$$S_{rot}(A) = R \log \left[\frac{\sqrt{\pi}}{\sigma(A)} \left(\frac{8\pi^2 kTe}{h^2} \right)^{3/2} \sqrt{J_1(A)J_2(A)J_3(A)} \right] \tag{6}$$

where J_1, J_2 and J_3 are the principle moments of inertia and σ is the symmetry number. This expression seems to be a good approximation in liquids, where rotational entropies were found to differ by only 2% from gas phase values [26].

Vibrational entropy may be estimated as a sum of S_{vib} for all frequences in the molecule's vibration spectra [26]

$$S_{vib} = \sum_k \left[R \frac{hc\nu_k}{kT} \left(\exp\left(\frac{hc\nu_k}{kT} \right) - 1 \right)^{-1} - R \log \left(1 - \exp\left(-\frac{hc\nu_k}{kT} \right) \right) \right] \tag{7}$$

where ν_k is kth frequency. Calculation of vibration spectra for protein structures is a computationally hard procedure. As was shown in Ref. [26], usually the value of TS_{vib} is less than 0.5 kcal/mol at normal temperatures, and one can expect that its change at dissociation $T\Delta S_{vib}$ would be much less than that. We therefore neglect vibrational entropy in our model.

The last entropy contribution in Eq. (4), $S_{surf}(A)$, is associated with the mobility of surface (side-chain) atoms. In first approximation, this term may be considered as proportional to the surface area of structure A:

$$S_{surf}(A) = F \sum_k a_k = F W_S(A) \tag{8}$$

where $W_S(A)$ is solvent-accessible surface area of subunit A.

Eqs. (4-8) allow one to estimate a subunit's entropy in solution as

$$S(A) \approx C + \frac{3}{2}R\log\left(m(A)\right) + \frac{1}{2}R\log\left(\frac{J_1(A)J_2(A)J_3(A)}{\sigma(A)}\right) + FW_S(A) \quad (9)$$

This expression contains two empirical parameters: surface entropy factor F, introduced in Eq. (8), and constant entropy term C, which depends on the poorly defined volume v (cf. Eq. (5)). Authors of Ref. [26] estimate uncertainty in S_{trans} as 20-40% of the estimate given by Eq. (5), however state that the expression for S_{rot} (Eq. (6)) is rather precise. We therefore introduce in Eq. (9) the empiric parameter C in attempt to compensate the uncertainty in the definition of v and possibly to account, in first approximation, for other entropy terms, such as conformational entropy, for which no feasible model can be proposed.

Using Eq. (9), entropy change upon complex dissociation in Eq. (1) may be estimated as

$$\begin{aligned}\Delta S &= \sum_{i=1}^{n} S(A_i) \; - \; S(A_1, A_2 \dots A_n) \\ &= (n-1)C \; + \; \frac{3}{2}R\log\left(\frac{\prod_i m(A_i)}{\sum_i m(A_i)}\right) \; + \; FW_I(A_1, A_2 \dots A_n) \\ &\quad + \; \frac{1}{2}R\log\left(\frac{\prod_{i;k} J_k(A_i)\sigma(A_1, A2 \dots A_n)}{\prod_k J_k(A_1, A_2 \dots A_n)\prod_i \sigma(A_i)}\right)\end{aligned} \quad (10)$$

where $W_I(A_1, A_2 \dots A_n)$ is buried surface area of subunits A_i in the complex.

3.3   Dissociation Pattern

Eqs. (1-3,10) allow one to estimate stability of a protein assembly if its dissociation pattern, or set of subunits $\{A_i\}$, is known. For the purpose of our study it is enough to find at least one dissociation pattern for which $\Delta G_{diss} < 0$ in order to detect instability and to remove the assembly from further consideration.

In order to be a potential dissociation pattern, set of subunits $\{A_i\}$ should satisfy the following conditions:

1. All multi-chain subunits must represent connected stable assemblies.
2. From symmetry considerations, identical interfaces can not be internal to a subuint and separate two subunits in the same dissociation pattern.

Dissociation patterns may be found using a backtracking scheme similar to that shown in Fig. 1. Represent assembly as a graph in which vertices and edges correspond to monomeric chains and interfaces between them, respectively. Then starting point for the algoritm in Fig. 1 would be a non-empty set of all interfaces $\{I\}$ found in assembly (step 3), loop B.2 runs over all interfaces found in $\{I\}$ and not found in $\{T\}$, in steps B.4 and B.7 the algorithm disengages interface I_k and any interfaces induced by that, steps B.9-B.11 are replaced for the calculation of ΔG_{diss} and stability analysis of the subunits calculated in step B.6. Each subunit is analysed for stability by a recursive application of the bactracking scheme to the subunit. The recursion should terminate once a negative ΔG_{diss} is encountered or all subunits contain only monomeric chains.

Dissociation pattern of stable assemblies may be of a potential interest, too. In general, a protein complex may dissociate in a few different ways, the most efficient of which would be the one with lowest ΔG_{diss}. Dissociation pattern with lowest ΔG_{diss} may be easily identified by the backtracking scheme described above, because it enumerates all possible dissociation patterns for *stable* complexes.

4   Implementation

As described above, our procedure is based on the exchaustive enumeration of all potential assemblies in crystal and their dissociation patterns, using recursive backtracking schemes. Backtracking algorithms are known to be NP-complete and therefore they may be computationally untractable unless a proper termination condition is employed.

Suppose that algorithm in Fig. 1 has generated a set of assemblies that all appear to be unstable, so that $\Delta G_{int}^r + T\Delta S^r > 0$, where index r stands for the recursion level. Then entropy of dissociation on the next level of recursion ΔS^{r+1} should be not less than ΔS^r because any dissociation pattern on level $r + 1$ results in the same or larger number of stable subunits than that on level r, while the assembly size only increases with increasing recursion level (cf. Eq. (10)). Maximum energy of binding on level $r + 1$ cannot be lower than $\Delta G_{int}^r + \sum_k \Delta G_{int}(I_k)$ where summation is done for all hydrophobic interfaces that still may be engaged, i.e. those with $\Delta G_{int}(I_k) < 0$ and not found in the interface sets $\{I_1\}$ and $\{T_1\}$ (cf. Fig. 1; $\Delta G_{int}(I_k)$ is calculated using Eq. (2) for $n = 2$). Therefore the termination condition is

$$\Delta G_{int}^r + \Delta S^r + \sum_{I_k \notin \{I_1\}, \{T_1\}} \min\left(\Delta G_{int}(I_k), 0\right) \geq 0 \tag{11}$$

where all quantities are calculated for the volume of one unit cell. Despite a very general nature of this estimate, we found that it works very efficiently, especially if interfaces in the backtracking scheme are ordered by increasing $\Delta G_{int}(I_k)$.

In our implementation, we define interface as protein surface area which becomes inaccessible to solvent upon bringing two chains into contact. For the surface area calculations, a method similar to that used in program AREAIMOL of the CCP4 Program Suite [27] was employed. Recipies for the calculation of hydrogen bonds and salt bridges are found in Refs. [6,15].

Parameters E_{hb}, E_{sb} (cf. Eq. (2)) and C, F (10) were chosen by a fitting procedure using a benchmark set of 218 structures published in Ref. [3]. Since only multimeric states are known for the benchmark structures, we assumed that correct oligomers are the ones of the required multimeric state and lowest

Table 1. Empirical parameters entering Eqs. (2,10), obtained through the fitting of multimeric states found in the benchmark set of 218 PDB entries from Ref. [3]

E_{hb}, kcal/mol	E_{sb}, kcal/mol	C, kcal/mol	F, kcal/(mol*Å^2)
0.51	0.21	11.7	$0.57 \cdot 10^{-3}$

ΔG_{diss} (1). Then the parameters were fitted such as to satisfy the following system of inequalities for as many structures as possible:

$$\begin{cases} \Delta G_{diss} > 0 \text{ for correct oligomers} \\ \Delta G_{diss} \leq 0 \text{ for all other multimeric states not lower than the correct one} \end{cases} \tag{12}$$

The described algorithm is implemented as a web-server available at URL given in the title. The server provides pre-calculated data for all PDB entries solved by means of X-ray crystallography, and allows to upload PDB and mmCIF coordinate files for interactive processing. Calculation time depends drastically on the number of different interfaces in crystal, however most of entries are solved in a few-minute time. The server also provides a detail annotation of interfaces and structures, visualisation of assemblies and database search tools.

5   Results and Discussion

The resulting values of empirical parameters, used in Eq. (10), are listed in Table 1. As seen from the Table, energy effect of hydrogen bonds and salt bridges appears to be somewhat smaller than the estimates given in the above discussion, but well within a reasonable range. Given that significant interfaces normally have 10-20 and more hydrogen bonds, their contribution to the free energy of binding G_{int} appears to be comparable with that of hydrophobic interactions. Entropy contribution from the frozen motion of surface atoms in interfaces, F, is quite small, just over 0.5 kcal/mol per 10^3Å^2 of interface area. Most of entropy change at complex formation comes from the constant entropy term, C, followed by the mass- and moment of inertia- dependent terms (cf. Eq. (10)). Mathematically, the system of inequalities (12) appears slightly underfit, which means that the used benchmark set may be insufficient for the calibration purposes, and the results may still be improved if a larger data set is used.

Table 2 presents the assembly classification results obtained for the benchmark set of 218 PDB entries [3], used for the calibration of empirical parameters. Each row of the Table corresponds to one of 5 oligomeric classes present in the benchmark set, and columns give the classification counts obtained for that class. As seen from the Table, we have obtained a nearly uniform success rate across different oligomeric classes, with the lowest rate of 87% for tetramers. Tetramers have also been found as the least predictable oligomeric class in Ref. [3], with considerably larger differences between the classes. The overall success rate is 90%, which is higher than the one reported in Ref. [3] (84%). On comparison,

Table 2. Assembly classification obtained for the benchmark set of 218 PDB entries from Ref. [3]. The rows give counts of multimeric states obtained for assemblies annotated as monomeric, dimeric, trimeric, tetrameric and hexameric in the benchmark set. Counts represented as $N + M$ stand for N homomers and M heteromers obtained, otherwise only homomers are listed.

	1mer	2mer	3mer	4mer	6mer	Other	Sum	Correct
1mer	50	4	0	1	0	0	55	91%
2mer	6	68+11	0	2+1	0	0	76+12	90%
3mer	1	0	22	0	1	0	24	92%
4mer	2	3	0	27+6	0	0	32+6	87%
6mer	0	0	0	1	10+2	0	11+2	92%
						Total:	198+20	90%

the PQS server at EBI-MSD gives 78% of correct answers, however this figure is less indicative because PQS was not optimised for the used benchmark set.

A detail study of misclassified cases shows a typical misestimate of ΔG_{diss} (Eq. (1)) within ±5 kcal/mol. This value could be taken as a precision limit for the models proposed in Section 3 if multimeric states in the benchmark set are trusted. There is, however, one example of misclassification that is far beyond any reasonable precision range for the method. PDB entry 1qex contains two identical chains, which should form a homo-trimer [3]. Our procedure, as well as PQS [2], suggests that it is actually a homo-hexamer shown in Fig. 2. Calculation results indicate that the most favourable dissociation pathway for this assembly is through a detachement in the isthmus between the two identical trimers with $\Delta G_{diss} \approx 90$ kcal/mol. Such high value of the dissociation barrier implies that the structure could well be hexameric.

The example of 1qex may indicate that not all multimeric states given in the used benchmark set are correct. A probable source of errors may be that only one oligomer from a few of them in chemical equilibrium is reliably detected in experiment. However, we tend to explain most of misclassifications by neglecting the specific experimental conditions, such as concentration, pH, tem-

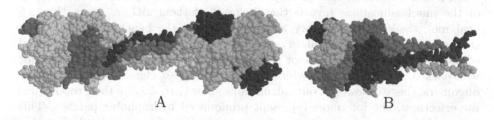

Fig. 2. Homo-hexamer found for PDB entry 1qex (A), and homo-trimer (B) which should be the correct multimeric state according to data in Ref. [3], see discussion in the text. The images were obtained using the Rasmol software [28].

Table 3. Assembly classification obtained for the new entries deposited into PDB through EBI-MSD deposition site. The reference classification has been done in MSD by manual curation. See Table 2 for used notations.

	1mer	2mer	3mer	4mer	5mer	6mer	8mer	10mer	12mer	Sum	Correct
1mer	131	11	0	4	0	2	2	0	0	150	87%
2mer	12+6	88+12	0+1	4	0	1	0+2	0	0	105+21	79%
3mer	1	0+2	6+2	0	0	0+1	0	0	0	7+5	66%
4mer	1+1	5+2	0	25+5	0	0	1+2	0	0	32+10	71%
5mer	0+1	0	0	0	2+1	0	0	0	0	2+2	75%
6mer	0+1	2+1	0	0	0	13+2	0	0	0	15+4	79%
8mer	0	1	0	0	0	0	0+2	0	0	1+2	67%
10mer	0	0	0	0	0	0	0	2	0	2	100%
12mer	2	0	0	0	0	0	0	0	5+1	7+1	75%
									Total:	321+45	81%

perature and presence of other agents, in our models. A thorough account of all affecting factors is difficult and if done then requires a quite detail description of experimental conditions from a user.

Most structures are deposited into PDB without experimental evidence of their oligomeric states. The benchmark set of 218 PDB entries published in Ref. [3] contains all structures with oligomeric states that are currently known to us as experimentally verified. Biological unit assignments in PDB is based mainly on the curators' scientific experience. Table 3 compares automatic assembly classification, obtained by us, with manual curation results for 366 new entries deposited recently into PDB at the EBI-MSD deposition site. As seen from the Table, most (75%) of the depositions were classified as monomers and dimers, which is reproduced at 87% and 79% success rate, respectively. Success rate for other oligomeric classes varies from 66% to 100%, however these figures are less indicative because of too few structures present. Overall, 81% of automatic an manual classifications agree with each other.

The most frequent misclassifications in Table 3 are dimers instead of tetramers, then monomers instead of dimers and vice versa. These are special cases when a larger assembly may or may not be divided in two parts. A detail study of the misclassifications reveals that in most of them ΔG_{diss} lies within ± 5 kcal/mol, the same uncertainty as that found for the benchmark set. A few strongest exceptions to this observation are shown in Table 4. Visual inspection of these assemblies reveals a poor packing quality of their interfaces (except for well-packed 1y6x and 1y7p), which fact could suggest classification into lower oligomeric classes. However, our calculations show that, despite their topological imperfectness, the interfaces represent pronounced hydrophobic patches. This means that the interfaces may be stronger than visually appears, which makes higher oligomeric states possible. A definite answer as to what the oligomeric state actually is in these cases, as well as in cases with low $|\Delta G_{diss}|$, may be given only by experimental study.

Table 4. The strongest misclassifications in Table 3. See text for details

PDB entry	1y6x	1ywk	1v7y	1wq5	2bh8	1y7p	1ylf
Assigned state	1mer	1mer	1mer	1mer	4mer	2mer	1mer
Calculated state	4mer	6mer	2mer	2mer	8mer	6mer	6mer
ΔG_{diss}, kcal/mol	16.5	9.9	9.0	15.3	9.2	36.1	16.2

6   Conclusion

We have described here a novel method for the calculation of biological units from protein crystallography data. In difference of its predecessors, our method is based on the stability analysis of all assemblies allowed by crystal symmetry and geometry of unit cell. We estimate the free energy of dissociation using theoretical models for free energy of protein binding and rigid-body entropy of protein assemblies. This approach allows us not only to predict the multimeric states and 3D arrangements of monomeric units with 80-85% accuracy, but also to guess on the probable dissociation patterns of assemblies.

The described procedure is implemented as a web-server available at URL given in the title of this paper. The server provides a detail summary of all crystal contacts and monomeric chains, list of probable protein assemblies, as well as searching for alike interfaces in the PDB archive.

Although our models neglect specific conditions, such as concentration and pH, which may affect formation of assemblies, predictive power of the method appears to be sufficiently high. Further studies are needed to improve the theoretical models of protein affinity and entropy change upon assembly formation.

Acknowledgement

E.K. is supported by the research grant No. 721/B19544 from the Biotechnology and Biological Sciences Research Council (BBSRC) UK. The authors thank Mr. A. Hussain for his work on the comparison of assigned and calculated oligomeric states using the described software.

References

1. Berman, H.M., Westbrook, J., Feng, Z., Gilliland, G., Bhat, T.N., Weissig, H., Shindyalov, I.N. and Bourne, P.E.: The Protein Data Bank. Nucleic Acids Res. **28** (2000) 235–242.
2. Henrick, K. and Thornton, J.: PQS: a protein quaternary structure file server. Trends in Biochem.l Sci. **23** (1998) 358–361.
3. Ponstingl, H., Kabir, T. and Thornton, J.: Automatic inference of protein quaternary structure from crystals. J. Appl. Cryst. **36** (2003) 1116–1122.
4. Ponstingl, H., Henrick, K., and Thornton, J.: Discriminating between homodimeric and monomeric proteins in the crystalline state. Proteins **41** (2000) 47–57.
5. Krissinel, E. and Henrick, K.: Common subgraph isomorphism detection by backtracking search. Softw. Pract. Exper. **34** (2004) 591–607.

6. Baker, E.N. and Hubbard, R.E.: Hydrogen bonding in globular proteins. Prog. Biophys. Molec. Biol **44** (1984) 97–179.
7. Janin, J., Miller, S., and Chothia, C.: Surface, subunit interfaces and interior of oligomeric proteins. J. Mol. Biol. **204** (1988) 155–164.
8. Argos, P.: An investigation of protein subunit and domain interfaces. Protein Eng. **2** (1988) 101–113.
9. Miller, S.: The structure of interfaces between subunits ofdimeric and tetrameric proteins. Protein Eng. **3** (1989) 77–83.
10. Janin J. and Chothia, C.: The structure of protein-protein recognition sites. J. Biol. Chem. **265** (1990) 16027-16030.
11. Horton, N. and Lewis, M.: Calculation of the free energy of association for protein complexes. Protein Sci. **1** (1992) 169–181.
12. Janin, J. and Rodier, F.: Protein-protein interaction at crystal contacts. Proteins: Struc. Func. Genet. **23** (1995) 580–587.
13. Jones, S. and Thornton, J.M.: Protein-Protein interactions: a review of protein dimer structures. Prog. Biophys. Molec. Biol. **63** (1995) 31–65.
14. Jones, S. and Thornton, J.M.: Principles of protein-protein interactions. Proc. Natl. Acad. Sci. USA **93** (1996) 13–20.
15. Xu, D., Tsai, C.-J. and Nussinov, R.: Hydrogen bonds and salt bridges across protein-protein interfaces. Protein Engng. **10** (1997) 999–1012.
16. Eisenberg, D. and McLachlan, A.D.: Solvation energy in protein folding and binding. Nature **319** (1986) 199–203.
17. McDonald I. and Thornton J.: Satisfying hydrogen bonding potential in proteins. J. Mol. Biol. **238** (1994) 777–93.
18. Pace C., Shirley B., McNutt M. and Gajiwala K.: Forces contributing to the conformational stability of proteins. FASEB J. **10** (1996) 75–83.
19. Fersht A.: The hydrogen bond in molecular recognition. Trends Biochem. Sci. **12** (1987) 3214–3219.
20. Horovitz A., Serrano L., Avron B., Bycroft M. and Fersht A.: Strength and cooperativity of contributions of surface salt bridges to protein stability. J. Mol. Biol. **216** (1990) 1031–1044.
21. Akke M. and Forsen S.: Protein stability and electrostatic interactions between solvent exposed charged side chains. Proteins: Struct. Funct. Genet. **8** (1990) 23–29.
22. Page, M.I. and Jencks, W.P.: Entropic Contributions to Rate Accelerations in Enzymic and Intramolecular Reactions and the Chelate Effect. Proc. Natl. Acad. Sci. USA **68** (1971) 1678–1683.
23. McQuarrie, D.A. *Statistical Mechanics*. New York: Harper & Row, (1976).
24. Murray, C.W. and Verdonk, M.L.: The consequences of translational and rotational entropy lost by small molecules on binding to proteins. J. Comput.-Aided Mol. Design **16** (2002) 741–753.
25. Finkelstein, A.V. and Janin, J.: The price of lost freedom: entropy of bimolecular complex formation. Protein Eng. **3** (1989) 1–3.
26. Mammen, J., Shakhnovich, E.I., Deutch, J.M. and Whitesides G.M.: Estimating the Entropic Cost of Self-Assembly of Multiparticle Hydrogen-Bonded Aggregates Based on the Cyanuric AcidMelamine Lattice. J. Org. Chem. **63** (1998) 3821–3830.
27. Collaborative Computational Project, Number 4.: The CCP4 Suite: Programs for Protein Crystallography. Acta Cryst. D **50** (1994) 760-763.
28. Sayle, R. A., and Milner-White, E. J.: RasMol: Biomolecular graphics for all. Trends in Biochemical Sci. **20** (1995) 374-376.

Molecular Similarity Searching Using COSMO Screening Charges (COSMO/3PP)

Andreas Bender[1], Andreas Klamt[2], Karin Wichmann[2],
Michael Thormann[3], and Robert C. Glen[1]

[1] Unilever Centre for Molecular Science Informatics, Department of Chemistry,
University of Cambridge, Lensfield Road, Cambridge CB2 1EW, United Kingdom
{ab454, rcg28}@cam.ac.uk
http://www-ucc.ch.cam.ac.uk
[2] COSMOlogic GmbH&CoKG, Burscheider Strasse 515, 51381 Leverkusen, Germany
{klamt, wichmann}@cosmologic.de
[3] Morphochem AG, Gmunder Str. 37-37a, 81379 Muenchen, Germany
michael.thormann@morphochem.de

Abstract. We present a novel approach to define molecular similarity and its application in virtual screening. The algorithm is based on molecular surface properties in combination with a geometric encoding scheme. The molecular surface is described by screening charges calculated via COSMO. COSMO, the COnductor-like Screening MOdel, is a quantum-chemical molecular description originally developed and widely validated for solubilities and partition coefficients of molecules in the liquid state. The screening charges it calculates also define properties relevant to ligand-target binding such as hydrogen-bond donors and acceptors, positive and negative charges and lipophilic moieties from first principles. Encoding of properties is performed by three-point pharmacophores which were found to outperform other approaches. The similarity measure was validated on a dataset derived from the MDL Drug Data Report (MDDR) which comprises five classes of active compounds that are 5HT3 ligands, ACE inhibitors, HMG-CoA reductase inhibitors, PAF antagonists and TXA2 inhibitors. Compared to other approaches, the method presented here compares favorably with respect to the number of active compounds retrieved, finds different active scaffolds and is based on a solid theoretical foundation. Further work will be undertaken in order to find better shape and pharmacophoric feature encoding schemes as well as to make quantitative predictions of bioactivity.

1 Introduction

Molecular similarity searching is based on the "similar property principle" which states that structurally similar molecules tend to have similar properties more often than structurally dissimilar molecules [1,2,3]. While, generally speaking, this assumption holds true exceptions lurk behind every structural modification. In particular different behavior between physicochemical properties and bioactivity can be observed. While physicochemical properties such as logP show - to

M.R. Berthold et al. (Eds.): CompLife 2005, LNBI 3695, pp. 175–185, 2005.
© Springer-Verlag Berlin Heidelberg 2005

a certain extent - indifference as to the precise location of structural change, geometrically restrained binding sites impose additional, spatial constraints on the molecule. Identical substitutions may thus give rise to a wide range of changes in bioactivity, depending on the site of substitution, the particular ligand (and its binding mode) and the particular target. The effect of a modification on binding depends functionally on whether the group leads to steric repulsion (rapidly decreasing or even completely eliminating bioactivity), fills an additional hydrophobic region of the binding site (increasing activity to varying degrees due to induced dipole interactions), forms a charge interaction (greatly influencing activity, depending on whether the interaction is attractive or repulsive) or just points into surrounding bulk water, often having no profound effect on binding affinity. Not knowing the effect of structural modifications on ligand binding is one of the most severe limitations of the "similar property principle".

Molecular similarity searching involves at least two steps which are on the one hand the structural representation of the molecules to be compared in "descriptor space", and on the other hand a similarity or distance measure between representations to establish a numerical similarity value between structures. Optionally, feature selection [4,5,6,7] may be employed, whose aim is to increase the signal-to-noise ratio of molecular representations by focusing on features relevant to the searching task.

Conventionally, similarity measures are distinguished by their dimensionality. Thus, one-dimensional properties such as logP or molecular weight assign only a globally derived real- or integer valued number to the molecule [8]. Two-dimensional properties such as fragment-based approaches [6,9,10] or graph-based multiple point pharmacophores [11] derive the descriptor space representation from the connectivity table. Finally, three-dimensional descriptors use energy values assigned to points in space such as CoMFA [12], GRIND [13] and MOLPRINT 3D [14] or spatial three-point pharmacophores [15,16].

The controversy of employing 2D vs. 3D descriptors for similarity searching has been ongoing for years. Earlier work [17] found in particular MACCS keys to be information-rich descriptors for predicting a variety of molecular properties (mainly physicochemical properties such as logP and pKa). For diversity selection, 2D descriptors were found to outperform their 3D counterparts [18]. The relationship between 2D and 3D descriptors has also been analyzed systematically [19] with the finding that 2D descriptors in some cases neglect important features. Overall the general conclusion can be drawn that 2D descriptors perform well in cases where distinct topological features are responsible for biological activity (and, conversely, perform more poorly where those features are not available). 3D descriptors on the other hand often have the advantage of being able to retrieve more topologically diverse molecules which still show the same biological activity.

Of crucial importance when developing a similarity measure is its effectiveness which means the reliable identification of active compounds. In addition at least two other properties are desirable in molecular similarity searching, namely the ability to identify novel active scaffolds (often referred to as "scaffold hopping") and the identification of molecular features deemed to be responsible for biological activity ("back-projectability"). Scaffold hopping is desirable both to circumvent current patents, and to identify novel lead structures which might confer improved bioactivity or ADME/Tox properties. Back-projectability is a valuable source of information in the lead development process, where, among others, activity optimization is performed. The medicinal chemist, if told that a particular lipophilic (or other) moiety at a certain site is important, may exploit this knowledge to synthesize more active analogues in the next round of lead optimization.

According to the idea that ligand-target interactions are mediated via interactions of the two molecular surfaces, localized surface point environments [14] have previously been used in combination with GRID [13,20] derived energetic molecular surface properties for similarity searching. Several different molecular probes were employed in order to capture areas of the molecular surface which correspond to putative ligand-target interaction types, such as hydrogen bond donors and acceptors, positively and negatively charged surface areas and lipophilic moieties. Still, force-field (such as GRID) derived descriptors possess several serious shortcomings. A number of different probes need to be used to ensure that different interaction types are covered sufficiently. This increases the time needed for descriptor generation as well as introducing a degree of arbitrariness into the choice of probes. Also, force fields conceptually only employ approximations of molecular properties, such as point charges which do not account for the directionality of lone pairs. They also do not capture sufficiently other properties of the electronic structure such as polarizability and they depend on a parameterization performed using a particular (arbitrary) data set.

The sum of those shortcomings led us to believe that a quantum-mechanical method for the description of molecular surfaces may be more appropriate since it eliminates all of the points above, of course bought at higher computational expense. In this work screening charges of the molecular surface are calculated by the COSMO [21] methodology and capture potential intermolecular interactions via a calculation of screening charges on the molecular surface. COSMO provides a set of surface patches, typically in the order of several thousand patches for a small molecule, with associated surface screening charges. In the COSMO extension to Realistic Solvation (COSMO-RS) [22,23,24] the special importance of the surface screening charge den sity for electrostatic interactions, hydrogen bonding and interactions of lipophilic regions has been elucidated.

It should be noted that in the work described here only a single AM1-optimized low-energy conformation was employed to form the query. While conformationally flexible searching has recently become popular, the extent to which it improves searching performance is difficult to establish [25]. This is probably due to the fact that the addition of conformational information also introduces noise into the descriptor, overall not necessarily improving the signal-to-noise ratio. Following the most straightforward route we thus chose the AM1 optimized structure of every compound to derive its representation in descriptor space.

The following chapter describes briefly the background of COSMO and COSMO-RS, the encoding scheme used for descriptor generation and the molecular database employed. Results are presented and discussed in the subsequent chapter. Finally, we give our concluding remarks about the performance of the method and envisaged future work.

2   Material and Methods

2.1   COSMO-Derived Screening Charges and Their Relevance for Molecular Interactions

The COnductor-like Screening MOdel [24] is a very efficient and robust approximation of the dielectric continuum solvation models [26] which is available in many quantum chemical programs. Based on a cavity grid of m surface segments it calculates the polarization by the screening charges of dielectric continuum representing the solvent by the electrostatic field and solute from a scaled conductor boundary condition:

$$f(\varepsilon)\underline{\Phi}^{solute} + \underline{\underline{A}}^{-1}\underline{q} = 0$$

where $\underline{\Phi}^{solute}$ is the vector of the electrostatic solute potential arising on the m segments of the cavity, q is the vector of the screening charges on the m segments, A is the Coulomb matrix of the surface segments, and $f(\epsilon)$ is the a scaling factor depending on the dielectric constant of the medium:

$$f(\varepsilon) = \frac{\varepsilon - 1}{\varepsilon + 0.5}$$

The solute potential $\underline{\Phi}^{solute}$ is calculated by quantum chemical methods, where density functional methods have proved as most efficient and reliable. Since the screening causes back-polarization of the solute by the solvent, the dielectric screening has to be taken into account self-consistently in the quantum chemical calculation. If efficiently implemented as on DFT level in TURBO-MOLE [27,28], COSMO calculations can be performed with only small computational overhead compared with gas phase calculations, including consistent geometry optimization in the presence of the dielectric solvent.

Starting from a fundamental criticism of the oversimplified dielectric continuum solvation concept, Klamt developed a statistical thermodynamics extension of the COSMO model named COSMO-RS [24,29]. In this model the interactions of molecules in a liquid phase are expressed as local contact energies of the molecular surfaces. Here, the COSMO screening charge densities s, which are the surface charge densities resulting from the set of surface charges q in eq. 1, play the key role for the quantification of electrostatic, hydrogen bonding, and hydrophobic/lipophilic interactions. While originally developed and widely validated for environmental and chemical engineering mixture thermodynamics, the value of the σ-based COSMO-RS concept for the quantification of many

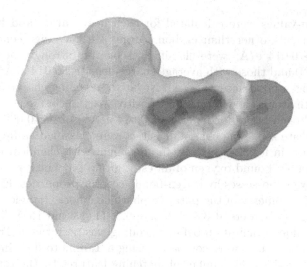

Fig. 1. COSMO-derived screening charge densities σ of an HMG-CoA reductase inhibitor. Hydrogen-bond acceptor features as well as negative charges are encoded in red while blue color denotes hydrogen-bond donor potential. Lipophilic moieties are colored in green.

ADME properties such as solubility, blood-brain barrier penetration, intestinal absorption, and even for pKa prediction has also been demonstrated [29,30]. Furthermore, first applications of the COSMO-RS concept to the evaluation of drug receptor binding are currently being developed. Thus COSMO screening charge densities are likely to provide a sound foundation for investigating ligand-target interactions from first principles.

The relevance of screening charges calculated by COSMO to molecular binding is illustrated in figure 1 for an HMG-CoA reductase inhibitor (statin). As known from crystal structures, binding of statins to HMG-CoA reductase is mediated by charge interactions of a carboxylic acid group of the ligand as well as hydrogen bond acceptor functions to the pyruvate-binding site of HMG-CoA. In addition a lipophilic function of the ligand is required which binds to a floppy lipophilic pocket of the target protein. All these features can be well distinguished from the COSMO screening charge densities σ, as illustrated in figure 1. The carboxylate function is shown on the right in red and purple, while hydrogen bond acceptor functions can readily be identified at the bottom of the same chain. Hydrogen bond donor functions point towards the viewer and are shown in blue while the lipophilic bulk of the structure is given in green color.

2.2   Encoding of Surface σ-Values as Three-Point Pharmacophores (3PP)

COSMO screening charge densities σ were encoded as atom-based three-point pharmacophores (3PP) [15,16]. By projection back on the associated atom centers average σ-values were calculated for each heavy atom and hydrogens attached to elements other than carbon. Atoms with average screening charge densities $\sigma > 0.014$ e/Å^2 were classified as bearing strongly negative partial charge (type N) and those with average charge densities 0.014 e/$\text{Å}^2 \geq \sigma > 0.009$ e/Å^2 as hydrogen-bond acceptors (A). Negative screening charge densities were associated with atoms showing strongly positive partial charge (P) at $\sigma < -0.014$ e/Å^2 and hydrogen-bond donors (D) at -0.014 e/$\text{Å}^2 \leq \sigma < -0.009$ e/Å^2. Atoms with intermediate screening charge densities were classified as lipophilic atoms (L). This results in features broadly in agreement with chemical intuition such as that the doubly bound oxygen of an ester group but not its neighboring sp^3 hybridized oxygen possesses hydrogen-bond acceptor properties. Eight bins were used to encode geometry of the putative pharmacophore triangles, starting at 2 and employing bin borders at 3.5, 5, 6.5, 8, 9.5, 11, 13 and 15 Å. Triangles were rotated to a unique orientation before encoding was performed. Triangle counts were kept and molecules were compared using a Tanimoto-like similarity coefficient [3] which divides the number of matching features by the total number of features present to give a similarity value in the range [0; 1] and also taking into account the size of the structure.

2.3 Molecular Database Used

For evaluation of the algorithm, 957 ligands extracted from the MDDR database [31] were used. The set [32] contains 49 5HT3 Receptor antagonists (from now on referred to as 5HT3), 40 Angiotensin Converting Enzyme inhibitors (ACE), 111 3-Hydroxy-3-Methyl-Glutaryl-Coenzyme A Reductase inhibitors (HMG), 134 Platelet Activating Factor antagonists (PAF) and 49 Thromboxane A2 antagonists (TXA2). An additional 547 compounds were selected randomly and did not belong to any of these activity classes. This dataset has previously been evaluated by Briem and Lessel [32] employing virtual affinity fingerprints, several 2D fingerprints and Feature Trees as well as on fragment-based descriptors (MOLPRINT 2D) [6,7] and force field-derived surface point environments (MOLPRINT 3D) [14,33]. Its wide use renders this dataset a suitable similarity searching benchmark, while restrictions are its rather small size, the comparatively small number of activity classes and the database they are derived from, the MDDR database, which includes a high number of analogue structures and no information about inactivity of compounds. Similarity searching performance was established as the number of structures from the same activity class as the query, as found among its ten nearest neighbors [32]. Similarity searching was repeated ten times and the hit rates obtained were averaged.

2.4 Computational Details

Structures were exported from the MDDR database in SD format. Protonation states were assigned using MOE [34], subsequently 3D structures were generated using CORINA [35]. Geometries were optimized with AM1/COSMO followed by a single-point BP-SVP-DFT/COSMO calculation using TURBOMOLE [28].

Screening charges of surface elements were translated into three-point pharma-cophores by a Perl script. Geometry optimization required approximately 10 minutes per compound on a 3 GHz CPU. While we chose COSMO calculations on such a high level for this initial study, it should be noted that a recently developed, very fast screening method COSMOfrag [30] can reduce the compu-tational cost to as little as one second per compound if required for large scale screening projects.

3 Results and Discussion

Hit rates and enrichments achieved on the five classes of active compounds are given in table 1, together with the standard deviation of the hit rate for the different queries. The hit rate is defined as the number of compounds from the

Table 1. Average hit rates, enrichment factors and standard deviations for the five classes of active compounds

Dataset	All Hetero-H			Hetero-H Except NH		
	Hitrate	Enrichment	σ (Hitrate)	Hit Rate	Enrichment	σ (Hitrate)
5HT3	4.9	14.0	3.5	5.6	11.2	3.3
ACE	5.1	7.6	4.1	5.6	13.7	2.8
HMG	5.2	5.8	3.8	9.1	7.9	0.9
PAF	5.1	5.2	3.1	7.0	5.0	1.9
TXA2	4.9	9.4	2.7	4.7	9.4	2.7
Mean	5.0	7.3	3.4	6.4	8.1	2.3

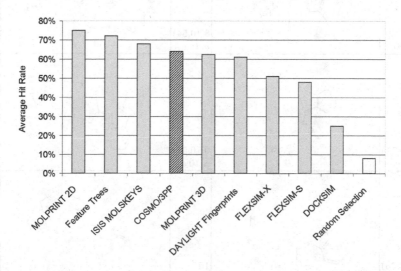

Fig. 2. Comparison of retrieval rates obtained by COSMO-based three-point pharma-cophores to established methods. While hit rates achieved are superior to other meth-ods, the added value of the method presented here lies in the type of active structures retrieved.

Table 2. Structures identified via COSMO screening charge densities σ in combination with a three-point pharmacophore encoding scheme. The query 5HT3 ligand is shown at the top.

Rank (T_c)	Structure	Rank (T_c)	Structure
1 (1.00) active		6 (0.56) inactive	
2 (0.55) active		7 (0.57) active	
3 (0.59) active		8 (0.54) active	
4 (0.57) active		9 (0.54) active	
5 (0.56) active		10 (0.51) active	

same activity class as the query among its ten most similar compounds (nearest neighbors). The enrichment given is the ratio of the hit rate obtained by employing COSMO/3PP, divided by the hit rate expected from random selection. In the first case all hydrogens bonded to heavy atoms except carbon were kept, in the second also hydrogen atoms bonded to nitrogen were neglected.

Considering all hydrogen atoms bound to heavy atoms except carbon as putative pharmacophore points the hit rate for all classes is around 5 (between 4.9 in case of the 5HT3 and TXA2 datasets and 5.2 in case of the HMG dataset), giving an average hit rate of 5.0 and enrichments of about 5- to 14-fold. If only hydrogen atoms bound to heavy atoms other than carbon and nitrogen are considered, performance improves to a mean hit rate of 6.4, corresponding again to enrichments of between 5- and 14-fold. Performance is compared to established methods in figure 2. The average hit rate of 6.4 active compounds in the first 10 structures of the ranked list is slightly better than Daylight fingerprints while not achieving performance of 2D-information based Feature Trees [36] and MOLPRINT 2D fingerprints [6,7]. This is in agreement with earlier results on the information content of 2D and 3D descriptors [17] although it should be noted that the difference remains small and probably not relevant in practice.

Still, the method presented here possesses a major advantage over many 2D based methods that is its ability to retrieve active compounds with widely different scaffolds. An example is given in table 2 for the 5HT3 dataset. If the query shown at the top of the table is used to rank the database of the remaining 956 structures, nine out of the ten most similar compounds retrieved indeed bind to at least one of the serotonin receptor subtypes. Shown are ranking positions with similarity scores and activity class. The compound retrieved at position 1 is identical to the query (but contained twice in the MDDR database) and the method gives a reproducible descriptor, assigning a similarity value of 1 to it. Compounds found at positions 1, 2 and 3 retain the scaffold, while those at positions 4 and 5 retain the amide and heterocyclic structure of the query. Structures retrieved at positions 7 and 8 already identify novel active scaffolds while those at positions 9 and 10 of the list display completely novel spiro- and bicyclic ring systems.

4   Conclusions

We have shown here that COSMO screening charge densities can successfully be employed for virtual screening and provide a conceptually sound as well as intuitively accessible scheme for calculating putative ligand-receptor interaction types. Retrieval rates are comparable to 2D fingerprints while considerably more diverse compounds are identified. Further work will employ a larger database for evaluating virtual screening performance. In addition novel encoding schemes are to be used since atom-based three-point pharmacophores can be assumed not to fully exploit the information content contained in the screening charge densities σ calculated on the molecular surface. One example are hydrogen-bond acceptor atoms, whose directionality is well captured by σ, yet neglected in the atom-based encoding scheme used here. In addition the possibility of quantitative activity predictions will be investigated.

References

1. Bender, A., Glen, R. C.: Molecular similarity: a key technique in molecular informatics. Org. Biomol. Chem. **2** (2004) 3204–3218
2. Johnson, M. A., Maggiora, G. M.: Concepts and Applications of Molecular Similarity (1990). New York: John Wiley & Sons.
3. Willett, P., Barnard, J. M., Downs, G. M.: Chemical similarity searching. J. Chem. Inf. Comput. Sci. **38** (1998) 983–996
4. Byvatov, E., Schneider, G.: SVM-based feature selection for characterization of focused compound collections. J. Chem. Inf. Comput. Sci. **44** (2004) 993–999
5. Wegner, J. K., Frohlich, H., Zell, A.: Feature selection for Descriptor based classification models. 1. Theory and GA-SEC algorithm. J. Chem. Inf. Comput. Sci. **44** (2004) 921–930
6. Bender, A., Mussa, H. Y., Glen, R. C., Reiling, S.: Molecular similarity searching using atom environments, information-based feature selection, and a naive bayesian classifier. J. Chem. Inf. Comput. Sci. **44** (2004) 170–178
7. Bender, A., Mussa, H. Y., Glen, R. C., Reiling, S.: Similarity searching of chemical databases using atom environment descriptors: evaluation of performance. J. Chem. Inf. Comput. Sci. **44** (2004) 1708–1718
8. Downs, G. M., Willett, P., Fisanick, W.: Similarity Searching and Clustering of Chemical-Structure Databases Using Molecular Property Data. J. Chem. Inf. Comput. Sci. **34** (1994) 1094–1102
9. Artymiuk, P. J., Bath, P. A., Grindley, H. M., Pepperrell, C. A., Poirrette, A. R., Rice, D. W., Thorner, D. A., Wild, D. J., Willett, P., Allen, F. H., et al.: Similarity searching in databases of three-dimensional molecules and macromolecules. J. Chem. Inf. Comput. Sci. **32** (1992) 617–630
10. Erlanson, D. A., McDowell, R. S., O'Brien, T.: Fragment-based drug discovery. J. Med. Chem. **47** (2004) 3463–3482
11. Schneider, G., Neidhart, W., Giller, T., Schmid, G.: "Scaffold-hopping" by topological pharmacophore search: A contribution to virtual screening. Angew. Chem.-Int. Edit. **38** (1999) 2894–2896
12. Cramer, R. D., Patterson, D. E., Bunce, J. D.: Comparative Molecular-Field Analysis (Comfa) .1. Effect of Shape on Binding of Steroids to Carrier Proteins. J. Am. Chem. Soc. **110** (1988) 5959–5967
13. Pastor, M., Cruciani, G., McLay, I., Pickett, S., Clementi, S.: GRid-INdependent descriptors (GRIND): a novel class of alignment-independent three-dimensional molecular descriptors. J. Med. Chem. **43** (2000) 3233–3243
14. Bender, A., Mussa, H. Y., Gill, G. S., Glen, R. C.: Molecular surface point environments for virtual screening and the elucidation of binding patterns (MOLPRINT). J. Med. Chem. **47** (2004) 6569–6583
15. Gund, P.: Three-dimensional pharmacophoric pattern searching. Prog. Mol. Subcell. Biol. **5** (1977) 117–143
16. Mason, J. S., Good, A. C., Martin, E. J.: 3-D pharmacophores in drug discovery. Curr. Pharm. Des. **7** (2001) 567–597
17. Brown, R. D., Martin, Y. C.: The information content of 2D and 3D structural descriptors relevant to ligand-receptor binding. J. Chem. Inf. Comput. Sci. **37** (1997) 1–9

18. Matter, H.: Selecting optimally diverse compounds from structure databases: a validation study of two-dimensional and three-dimensional molecular descriptors. J. Med. Chem. **40** (1997) 1219–1229
19. Thimm, M., Goede, A., Hougardy, S., Preissner, R.: Comparison of 2D similarity and 3D superposition. Application to searching a conformational drug database. J. Chem. Inf. Comput. Sci. **44** (2004) 1816–1822
20. Goodford, P. J.: A computational procedure for determining energetically favorable binding sites on biologically important macromolecules. J. Med. Chem. **28** (1985) 849–857
21. Klamt, A., Schuurmann, G.: Cosmo - a New Approach to Dielectric Screening in Solvents with Explicit Expressions for the Screening Energy and Its Gradient. J. Chem. Soc.-Perkin Trans. **2** (1993) 799–805
22. Eckert, F., Klamt, A. (2004). COSMOtherm (Version Version C1.2, Release 01.04). Leverkusen, Germany: COSMOlogic GmbH & Co. KG.
23. Klamt, A., Eckert, F., Hornig, M.: COSMO-RS: A novel view to physiological solvation and partition questions. J. Comput.-Aided Mol. Des. **15** (2001) 355–365
24. Klamt, A.: Conductor-Like Screening Model for Real Solvents - a New Approach to the Quantitative Calculation of Solvation Phenomena. J. Phys. Chem. **99** (1995) 2224–2235
25. Haraki, K. S., Sheridan, R. P., Venkataraghavan, R., Dunn, D. A., McCulloch, R.: Looking fo Pharmacophores in 3D-Databases: Does Conformational Searching Improve the Yield of Actives? Tetrahedron Comput. Methodol. **3** (1990) 565–573
26. Tomasi, J., Persico, M.: Molecular-Interactions in Solution - an Overview of Methods Based on Continuous Distributions of the Solvent. Chem. Rev. **94** (1994) 2027–2094
27. Schafer, A., Klamt, A., Sattel, D., Lohrenz, J. C. W., Eckert, F.: COSMO Implementation in TURBOMOLE: Extension of an efficient quantum chemical code towards liquid systems. PCCP Phys. Chem. Chem. Phys. **2** (2000) 2187–2193
28. Ahlrichs, R., Bar, M., Haser, M., Horn, H., Kolmel, C.: Electronic-Structure Calculations on Workstation Computers - the Program System Turbomole. Chem. Phys. Lett. **162** (1989) 165–169
29. Klamt, A. COSMO-RS, From Quantum Chemistry to Fluid Phase Thermodynamics and Drug Design (2005). Amsterdam: Elsevier.
30. Hornig, M., Klamt, A.: COSMOfrag: A Novel Tool for High Throughput ADME Property Prediction and Similarity Screening Based on Quantum Chemistry. J. Chem. Inf. Model. **submitted** (2005)
31. MDL Drug Data Report; MDL ISIS/HOST software, MDL Information Systems, Inc.
32. Briem, H., Lessel, U.: In vitro and in silico affinity fingerprints: Finding similarities beyond structural classes. Perspect. Drug Discov. Des. **20** (2000) 231–244
33. Bender, A., Mussa, H. Y., Gill, G. S., Glen, R. C.: Molecular surface point environments for virtual screening and the elucidation of binding patterns (MOLPRINT). IEEE Int. Conf. Syst. Man Cybern. **5** (2004) 4553–4558.
34. MOE (Molecular Operating Environment); Chemical Computing Group Inc.: Montreal, Quebec, Canada.
35. Sadowski, J., Gasteiger, J., Klebe, G.: Comparison of Automatic 3-Dimensional Model Builders Using 639 X-Ray Structures. J. Chem. Inf. Comput. Sci. **34** (1994) 1000–1008
36. Rarey, M., Dixon, J. S.: Feature trees: a new molecular similarity measure based on tree matching. J. Comput.-Aided Mol. Des. **12** (1998) 471–490

Increasing Diversity in In-silico Screening with Target Flexibility

B. Fischer, H. Merlitz, and W. Wenzel

Forschungszentrum Karlsruhe GmbH, Institut für Nanotechnologie,
Postfach 3640, D-76021 Karlsruhe, Germany
wenzel@int.fzk.de
http://www.fzk.de/biostruct

Abstract. We investigate the impact of receptor flexibility with the all-atom FlexScreen docking approach using the thymidine kinase (TK) receptor as a model system. We study the screening performance when selected side chains of the target are treated in a continuously flexible fashion in a screen of a database of 10000 compounds, which contains ten known substrates for the TK receptor. While the binding modes of the known substrates are not significantly affected as a function of receptor flexibility the mean binding energies of the database screen initially drop rapidly with increasing receptor flexibility but saturate when the number of target degrees of freedom is increased further. We demonstrate a dramatically increased diversity of the screen as 40% newly selected ligands appear in the top 500 ligands of the screen when receptor flexibility is taken into consideration.

1 Introduction

Virtual screening of a chemical database to targets of known three-dimensional structure is rapidly developing into a reliable method for finding new lead candidates in drug development [1,2,3]. Both better scoring functions [4] and novel docking strategies [5,6] contribute to this trend, although no completely satisfactory approach has been established yet [7]. This is not surprising since the approximations which are needed to achieve reasonable screening rates impose significant restrictions on the virtual representation of the physical system.

Three important ingredients of a reliable *in-silico* screening approach, based on the direct approximation of the affinity in an all-atom force field, can be identified: (1) The docking algorithm has to find the global minimum of the potential (or free) energy surface within the given conformational space in an accurate and reproducible manner. (2) All relevant conformations of the receptor-ligand complex in nature must be accessible in the virtual representation of that system. While ligand flexibility is now considered in most docking tools, the inclusion of receptor degrees of freedom has become the focus of recent investigations [8,9,10,11,12]. (3) The scoring function, which approximates the free energy change from the free ligand to the receptor-ligand complex, should approximate

M.R. Berthold et al. (Eds.): CompLife 2005, LNBI 3695, pp. 186–197, 2005.
© Springer-Verlag Berlin Heidelberg 2005

the experimental affinity or at least the relative ranking of a compound database accurately as possible.

Present day screening methods necessarily contain approximations for each of the above ingredients to permit the treatment of large ligand databases with an acceptable computational effort. One of the main obstacles to improve the overall reliability of in-silico screening methods is the difficulty associated with the consideration of target flexibility, which presently remains at an experimental stage. Several approaches have been suggested to include more than one receptor structure to the screening simulation: (1) Induced fitting: Once the ligand is docked into a (rigid) receptor, the side chains are allowed to relax and to find the optimum binding mode [10]. (2) A conformational ensemble of several different structures of the target is employed to either generate Boltzmann-weighted grids for the scoring function [11] or to define a united protein description [12]. In the latter approach, similar parts of the structures are merged whereas dissimilar areas are treated as separate alternatives and are combined during the docking process to form new overall structures. Although each of these techniques has been shown to deliver results superior to the rigid-target docking methods, their limitations are obvious: With the induced fitting approach only small structural variations can be taken into account. If the side chain would have to to be significantly moved from the starting structure to make the ligand fit, it might have been discarded during the docking stage. Using conformational ensembles, the proper choice of the different target structures is critical. This approach is best suited when several crystallographic structures of the receptor are available, otherwise it is difficult to define an ensemble which exhaustively covers the critical area of the conformational space, but remains small enough to avoid a combinatorial explosion of the number of possible conformations. Recent approaches have become more sophisticated in their construction of target conformations [13] and their combination [14] to address these problems.

In this investigation we moved toward the unconstrained consideration of receptor flexibility by allowing selected side chain bonds to rotate continously. Using our docking tool *FlexScreen* [15,16,17], we start with a rigid receptor and then gradually enable up to 15 selected side chain bonds to rotate, thereby investigating selectivity and accuracy of the screening simulations. We analyze the benefits of this approach and its impact on both screening reliability and computational cost.

By docking a database of 10000 molecules we demonstrate an increasing diversity of the selected ligands with the number of flexible target bonds. The fraction of binding ligands increased from 28% for the rigid receptor to 65% with 15 flexible bonds were released. We found that with increasing target flexibility stiffer ligands of larger size were able to fit into the receptor pocket and find competitive binding modes. The energy fluctuations of the docking simulations increased with the dimension of the conformational space, but the docking results appeared to saturate at about 10 receptor degrees of freedom. It is therefore advisable to balance the number of flexible bonds in the target against to computational cost of the screen. The FlexScreen methodology reported here removes one of the most significant physical limitations of present day screening methods and thus offers a platform for the improvement of lead selection with moderate additional computational cost.

2 Method

2.1 Docking Method

The screens in this investigation were performed with *FlexScreen*, an all-atom docking approach [15,16] based on the stochastic tunneling method [18]. In this approach receptor and ligand are represented in atomistic detail, the global minimum•of the energy surface is located using the stochastic tunneling method (STUN) [18,19].

We used a simple, first principle based atomistic scoring function:

$$S = \sum \left(\frac{R_{ij}}{r_{ij}^{12}} - \frac{A_{ij}}{r_{ij}^{6}} + \frac{q_i q_j}{\epsilon r_{ij}} \right) + \sum_{h-bonds} \cos \Theta_{ij} \left(\frac{\tilde{R}_{ij}}{r_{ij}^{12}} - \frac{\tilde{A}_{ij}}{r_{ij}^{10}} \right) \qquad (1)$$

which contains the empirical Pauli repulsion (first term), the Van de Waals attraction (second term), the electrostatic Coulomb potential (third term) and the angular dependent hydrogen bond potential (term four and five). The Lennard-Jones parameters R_{ij} and A_{ij} were taken from OPLSAA [20], the partial charges q_i were computed with InsightII and esff force field, and the hydrogen bond parameters $\tilde{R}_{ij}$, $\tilde{A}_{ij}$ were taken from AutoDock [21]. The dielectric constant was chosen as $\epsilon = 4$, which reduces the relative importance of Coulomb interaction against hydrogen bonds.

The scoring function in equation (1) is neglecting solvent related effects. Even in its simplest approximation the affinity of the ligand is the sum of the in-vacuo binding energy, the de-solvation energy of the ligand and the de-solvation energy of the receptor. Investigations are under way to extend the present scoring function in order to account for these effects.

Each simulation run consisted of a total number of 8×10^5 steps using the STUN optimizer. These simulation steps were partitioned following a cascadic strategy [17]: 100 short simulations of 5000 simulation steps led to 100 conformations, the 5 best of which were selected for another 30000 simulation steps. Finally, the 2 best conformations were further optimized using another 75000 steps each. This cascadic approach was shown to reduce the chance of false negatives, and as a side effect, it delivers two final conformations whose energies can be employed to estimate the accuracy of the docking simulation (Sec. 3.1).

2.2 Receptor

For this investigation the thymidine-kinase (TK) receptor in complex with its substrate ganciclovir (pdb entry 1ki2 [22]) was used as a model system. This system was used as a benchmark in several recent docking investigations [7,17] since crystal structures for 10 important substrates in complex with the receptor are available. Throughout this investigation the notation from Ref. [7] is used. The whole receptor is used for the screeing run: The protein was first protonated and partial charges were attached using InsightII and its *esff* forcefield. The proper treatment of conserved water molecules remains a disputed topic. Although the importance of specific water molecules involved in water-mediated contacts has been demonstrated for some TK ligands [23], it was found that the inclusion of these molecules during database screens did prevent docking of ligands which otherwise had displaced them [24]. We have therefore chosen to neglect the treatment of water molecules in this study until a proper treatment of solvation/de-solvation effects is available and the positional flexibility of the water can be taken in account.

We performed several simulations with an increasing degree of sidechain flexibility. A close inspection of the docking site of 1ki2, which is prototypical for the TK receptor, revealed that 14 side-groups were located within a 3 Å vicinity of the docked ligand gcv. Seven of them were identified to exhibit some structural variations when the crystallographic structures of various receptor conformations were compared. Based on these data, the five bonds most likely to account for conformational screening were made flexible in the first non-rigid screen (SIM5). To investigate convergence we performed further screens with 10 and finally with 15 free rotating bonds. More specifically, the following bonds were allowed to rotate in the respective simulations: simulation:

- SIM0; Fixed side-groups
- SIM5; 5 rotatable bonds: HIS A58 CB–CG, GLN A125 CG–CD, GLU A225 CG–CD, CB–CA, CG–CB
- SOM10; 10 rotatable bonds: SIM5 2 plus HIS A58 CA–CB, GLU A83 CG– CD, TYR A101 CZ–OH, TYR A172 CZ–OH, ARG A222 CA–CB
- SIM15; 15 rotatable bonds: SIM10 plus TYR A101 CB–CG, TYR A172 CB–CG, ARG A222 CB–CG, CG–CD, CD–NE

For each particular set of sidechains, all single bonds of the side chain are permitted to rotate continously. These rotations are achieved by random changes of the dihedral angles in each simulation step. Using standard off-the-shelf hardware (PC XEON 2.2 GHz), FlexScreen requires about 20s docking time per ligand. As the number of sidechain degrees of freedom rises, longer runs are required and the cost of an energy evaluation increases. For 15 flexible bonds a single simulation can take as much as 10 minutes. However, the present implementation of FlexScreen recomputes all interactions involving atoms on the flexible sidechains in each energy evaluations. Restricting the evaluation to just the atoms that actually move can reduce in computational savings by a factor of three on average.

2.3 Ligand Database

The ligands were taken from the open part of the National Cancer Institute (NCI) database, *nciopen3D* [25], which, in its latest version, contains 249061

Table 1. Ranks of the 10 tk-inhibitors when screened against 10000 compounds. Each ligand was docked 32 times, E_{bind} is the average binding energy, Δ its root-mean-square deviation (nd indicates non-docking ligands). E_{mean} denotes the average of binding energies over all docked inhibitors. The score of the screen (bottom row) is a measure for the selectivity of the screen.

	$N_{flex} = 0$			$N_{flex} = 5$			$N_{flex} = 10$			$N_{flex} = 15$		
	E_{bind}	Δ	Rank	E_{bind}	Δ	Rank	E_{bind}	Δ	Rank	E_{bind}	Δ	Rank
gcv	-129	3	7	-147	4	8	-172	6	7	-169	9	21
acv	-115	5	31	-119	2	118	-129	8	245	-142	6	190
dhbt	-116	6	29	-153	8	5	-160	7	18	-158	6	49
hpt	-113	1	42	-120	7	110	-134	6	186	-142	6	187
pcv	-98	10	178	-132	5	35	-133	9	189	-143	8	180
dt	-68	3	901	-96	3	549	-109	6	878	-123	7	606
idu	-58	3	1313	-104	3	341	-111	7	735	-122	6	624
ahiu	nd			-53	1	2380	-98	10	1588	-97	6	2113
mct	nd			nd			-73	10	3738	-66	6	4668
hmtt	nd			nd			-56	3	4929	-64	7	4793
E_{mean}	-99.6			-115.5			-117.5			-122.6		

compounds and represents the largest freely available ligand database. For this investigation 10000 ligands (which contain at least one ring [17]) with the following properties were randomly chosen from the database: (1) The number of atoms N was restricted to $20 \leq N \leq 80$. (2) The number of rotational degrees of freedom was restricted to $R \leq 12$. (3) Charged ligands were discarded.

3   Results

3.1   Screening Results

The experimentally determined conformation of inhibitor ganciclovir in docked position (pdb entry 1ki2 [22]) served as a measure of the docking accuracy. Each of the 128 simulations led to rms deviations below 1.5Å and thus to a successful docking, underlining the high level of reliability of the docking algorithm.

The degree of scattering in the binding energies of the docked conformations, however, increased with the target flexibility (see also Table 1). It therefore appears that these energy fluctuations are a more sensitive measure for the accuracy of the docking procedure. After all, in a database screen, it is the affinity which finally decides upon a ligand to become a suitable lead candidate or not.

The first striking result is that the number of docked compounds had significantly increased from 28% (rigid target) to around 64% (10 and more flexible bonds). With increasing flexibility, sterical restrictions became more and more relaxed, allowing compounds to dock which were previously unable to fit into the receptor pocket. Additionally, binding modes were optimized for a better fit:

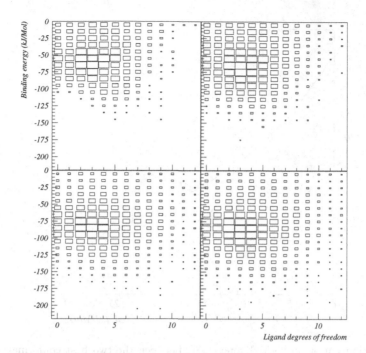

Fig. 1. Correlation of binding energies with compound flexibility for rigid target (upper left), 5, 10 and 15 target degrees of freedom (upper right, lower left, lower right). The area of the rectangle at each position is proportional to the number of ligands with the given number of degrees of freedom found in the screen with the corresponding binding energy.

The average binding energy of the docked compounds decreased as a function of receptor flexibility from -54 kJ/Mol (rigid receptor) to -77 kJ/Mol (10 flexible bonds), but appeared to saturate after that. The transition from 10 to 15 degrees of freedom caused a rather modest change of -4.5 kJ/Mol.

The two-dimensional plots of Fig. 1 show a breakup of the binding energies with respect to ligand flexibility. With a rigid receptor (upper left panel), a clear relation between both variables is visible: few inflexible ligands reached binding energies below -100 kJ/Mol. The best binding modes were achieved with ligands from 4 to 9 flexible bonds, beyond that a fast drop took place. At 10 and 15 target degrees of freedom (lower left and lower right) the binding energies not only improved in general, but their relation to ligand flexibility had also disappeared.

This behavior is related to the fact that a fixed target is highly selective: Only ligands with internal flexibility were able to fit into the receptor and find a proper binding mode, whereas for inflexible ligands the conformational space was too restricted for a tight binding. With increasing target flexibility, those restrictions were relaxed and this allowed ligands with few degrees of freedom to compete more successfully with rather flexible compounds. On the other side,

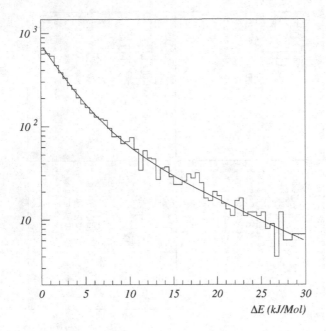

Fig. 2. Distribution of energy differences ΔE between the two best conformations of each docked ligand with 10 target degrees of freedom. Solid line: Fit of two negative exponentials Eq. (2).

the rigid target imposed a rather strict limitation to the size of the compounds. Only after inclusion of target flexibility, ligands with more than 40 atoms were docking well, and among those was the majority of ligands with 10 and more flexible bonds. This plot is an impressive demonstration on how target flexibility contributes to overcome sterical restrictions and increase the diversity of the database screen.

In order to determine the accuracy of the database screens, we analyzed their energy fluctuations. The cascadic docking approach delivers two different final conformations. The difference ΔE of their binding energies can be plotted for all docked ligands to yield the energy-difference distribution, the width of which is an indicator for the reliability of the screen. As has been argued elsewhere [17], this distribution can be interpreted as the energy-energy correlation function and fitted to two negative exponentials

$$F(\Delta E) = w_1 \, \lambda_1 \, e^{-\lambda_1 \, \Delta E} + w_2 \, \lambda_2 \, e^{-\lambda_2 \, \Delta E} \qquad (2)$$

yielding the inverse decay constants λ_i^{-1} and the weight factors w_i. Figure 2 displays such a fit for the simulation with 10 receptor degrees of freedom, and table 2 summarizes the resulting values for the fit parameters. According to this analysis, the set of docked ligands is formally partitioned into compounds of high docking accuracy (creating the fast decay λ_1^{-1}) and those of low docking accuracy (decaying with λ_2^{-1}). It is obvious how these energy fluctuations rapidly

Table 2. Median energy difference (kJ/Mol), inverse decay parameters λ_i^{-1} (kJ/Mol)(a measure of the screening error) and weight factors w_1/w_2 for a fit of Eq. (2) to the distribution of energy differences ΔE. N_{flex} is the number of flexible bonds of the target, N_{docked} the total number of docked compounds, E_{mean} denotes the mean binding energy of all docked compounds.

N_{flex}	0	5	10	15
Median	0.23	1.41	3.06	4.50
λ_1^{-1}	0.23	1.27	2.7	3.2
λ_2^{-1}	2.3	7.8	9.7	9.8
w_1/w_2	4.3	2.2	1.4	0.58
N_{docked}	2839	4047	6378	6540
E_{mean} (kJ/Mol)	-54.3	-60.0	-77.1	-81.6

increased with the number of target degrees of freedom. Not only the inverse decay constants were increasing, but also their relative weights shifted: Whereas in the fixed target the fast decaying compounds were dominating ($w_1/w_2 = 4.3$), the situation was reversed at 15 target degrees of freedom: Here the slow decay had a higher weight ($w_1/w_2 = 0.58$) and hence ligands of high docking accuracy were outnumbered by those of low docking accuracy. The decreasing accuracy could as well be expressed with the Median of the distribution, which was growing from 0.23 kJ/Mol (rigid target) to 4.5 kJ/Mol (15 flexible bonds).

The last two rows of Table 2 contain the total number of docked compounds and their mean binding energies. As mentioned above, these parameters were beginning to saturate beyond 10 target degrees of freedom, i.e. a further increase of target flexibility did not significantly improve the diversity of the screen any more. The accuracy, however, continued to decrease so that it seems to be wise to avoid any excessive flexibility of the target. Reasons for this decrease lie mostly in the dramatically increasing size of the search space. Discounting internal ligand degrees of freedom, the number of degrees of freedom doubles from five in SIM0 to ten in SIM10, with a corresponnding loss of efficiency of any applied search method- In order to form a proper hydrogen bond between a free sidechain and the ligand in a flexible receptor screen both fragments must coordinate properly. Even efficient optimization methods, such as the stoichastic tunneling method underlying FlexScreen, are reduced in their accuracy as the search space dimension increases.

3.2 Ranking of Known Substrates

Each of the 10 inhibitors was docked 32 times and the mean as well as the rms-deviations of the binding energies were evaluated. Table 1 contains the results for the different stages of target flexibility. The ranking was determined using the average binding energy of each inhibitor, not the best energy of the 32 runs. The general trend is defined with an increasingly tight binding of the inhibitors along with increasing target flexibility. This is shown in the lower row of the table (second to bottom), containing the mean energies of all docked inhibitors. It is

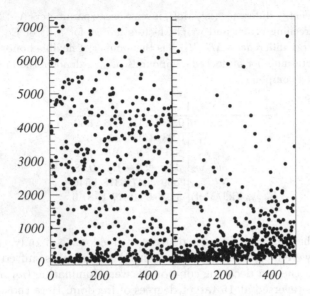

Fig. 3. Correlation of the ranks of the top 500 ranking database compounds of the $N_{flex} = 15$ simulation (x-axis) against the ranks of the same ligands in a $N_{flex} = 0$ simulation (left) and in a $N_{flex} = 10$ simulation (right). 212 (4) of these 500 best ligands did not dock at all during the $N_{flex} = 0$ ($N_{flex} = 10$) screen.

interesting to note how the average binding energy of the inhibitors is initially dropping quickly (from -97 kJ/Mol down to -116 kJ/Mol) and then rather slowly as a function of target flexibility.

Table (1) displays the binding energies of database compounds and inhibitors for the various target models. The overall ranking of the ligands had improved; expecially those which initially docked badly or not at all (ahiu, mct and hmtt) had significantly improved once the receptor became flexible. The latter two of them, however, never achieved a reasonably tight binding. It is striking how well the database ligands had improved their binding energies. This is the reason why the overall score of the inhibitors did not vary much, i.e. the selectivity of the screen remained essentially unchanged with increasing target flexibility.

The overall shift of the database toward higher affinity requires closer investigation. If the inclusion of target flexibility results in little more than a homogeneous shift of the binding energies of all ligands, there were little justification for the increased computational cost of such screens. Closer inspection of the data, however, reveals, that this is not the case. The data in the left panel of Fig. 3 demonstrates a lack of correlation of the ranks of the top 500 ligands of the $N_{flex} = 15$ in a screen with no target flexibility. We find that 212 of the 500 best ranking compounds in the flexible receptor screen did not dock at all when the rigid target was used. Only the inclusion of target flexibility permits more than 40% of the top ranking ligands to attain a competitive binding mode. All of these molecules would have been discarded in a rigid receptor screen. The data also implies that a rigid receptor screen necessarily results in comparatively low diversity. The particular receptor conformation used in such a screen, serves as a sieve which only allows a highly restricted subset of ligands to find a proper binding mode. As a result it may be misleading to interpret a particular good score for the known substrate as an indication of the selectivity of the screening methodology or scoring function. The right panel of Fig. 3 correlates the ranks of the best 500 ligands with their ranks during the $N_{flex} = 10$ screen. Here the correlations are much higher, indicating that the screening results saturate with approximately 10 receptor degrees of freedom. Only 4 (0.8%) of the top 500 at $N_{flex} = 15$ did not dock during the $N_{flex} = 10$ simulation.

4 Summary

Starting with a rigid target, we investigated the impact of increasing side-chain flexibility on the docking performance during a database screen of 10000 ligands. The target degrees of freedom were treated on the same continuous footing as the ligand degrees of freedom. Using the X-ray crystallographic structure of thymidine kinase complexed with the inhibitor gcv, as a blueprint for the exact binding mode, we demonstrated that the binding pose remained almost unchanged for various degrees of target flexibility.

By docking a database of 10000 molecules we demonstrate an increasing diversity of the screens with the number of flexible target bonds. The fraction of binding ligands increased from 28% for the rigid receptor to 65% with 15 flexible bonds were released. We found that with increasing target flexibility stiffer ligands of larger size were able to fit into the receptor pocket and find competitive binding modes.

The energy fluctuations of the docking simulations increased with the dimension of the conformational space. Beyond 10 target degrees of freedom the docking results appeared to saturate with the number of receptor degrees of freedom. It is therefore advisable to balance the number of flexible bonds in the target against to computational cost of the screen.

We demonstrated that increasing target flexibility results in a significant increase in the chemical variability of the ligands that find a competitive binding mode. While this increase of ligand variability in flexible receptor screens is highly desirable from the perspective of lead selection, it places a strong burden on the scoring function to discriminate between the large variety of different binding mechanisms. In the absence of perfect scoring functions, consensus scoring of the docked ligands after the screen may ameliorate this problem, but ultimately improved scoring functions may be required to further increase the reliability of large database screens.

The FlexScreen methodology reported here removes one of the most significant physical limitations of present day screening methods and thus offers a platform on which to develop such scoring functions. We noted that consideration of a relatively small number of receptor degrees of freedom resulted in a significant improvement of the screening results, a finding that we will validate for other receptors. Chemical intuition and the inspection of known binding modes may aid the selection of such target degrees of freedom and permit the improvement of lead selection with the FlexScreen approach with moderate additional computational cost.

Acknowledgments. We thank the Deutsche Forschungsgemeinschaft (grant WE 1863/13-1) and the Kurt Eberhard Bode Stiftung for financial support.

References

1. William L. Jorgensen. The many roles of computation in drug discovery. *Science*, 303:1813–1818, 2004.
2. W.P. Walters, M.T. Stahl, and M.A. Murcko. Virtual screening — an overview. *Drug Discovery Today*, 3:160, 1998.
3. J. Drews. Drug discovery: a historical perspective. *Science*, 287:1960, 2000.
4. G. Klebe and H. Gohlke. Approaches to the description and prediction of the binding affinity of small-molecule ligands to macromolecular receptors. *Angew. Chemie (Intl. Ed.)*, 41:2644, 2002.
5. G. Schneider and H.J. Boehm. Virtual screening and fast automated docking methods. *Drug Discovery Today*, 7:64, 2003.
6. Paul D. Lyne. Structure-based virtual screening: an overview. *Drug Discovery Today*, 7:1047, 2002.
7. C. Bissantz, G. Folkerts, and D. Rognan. Protein-based virtual screening of chemical databases. 1. evaluation of different docking/scoring combinations. *J. Med. Chem.*, 43:4759, 2000.
8. Maxim Totrov and Ruben Abagyan. Flexible protein-ligand docking by global energy optimization in internal coordinates. *Proteins*, 29:215–220, 1998.
9. C. W. Murray, C. A. Baxter, and A. D. Frenkel. The sensitivity of the results of molecular docking to induced fit effects: Application to thrombin, thermolysin and neuraminidase. *J. Comput.-Aided Mol. Design*, 13:547–562, 1999.
10. V. Schnecke and L.A. Kuhn. Virtual screening with solvation and ligand-induced complementarity. *Persp. Drug. Des. Discovery*, 20:171, 2000.
11. F. Osterberg. Automated docking to multiple target structures: incorporation of protein mobility and structural water heterogeneity in autodock. *Proteins*, 46:34, 2002.
12. H. Claußen, C. Buning, M. Rarey, and T. Lengbauer. Flexe: Efficient molecular docking consiudering protein structure variations. *J. Mol. Biol.*, 308:377–395, 2001.
13. Claudio N. Cavasotto and Ruben A. Abagyan. Protein flexibility in ligand docking and virtual screening to protein kinases. *J. Mol. Biol.*, 337:1161, 2004.
14. Binqung Q. Wei, Larry H. Weaver, Anna M. Ferrari, Brian W. Matthews, and Brian K. Shoichet. Testing a flexible-receptor docking algorithm in a model binding site. *J. Mol. Biol.*, 337:1161, 2004.
15. H. Merlitz and W. Wenzel. Comparison of stochastic optimization methods for receptor-ligand docking. *Chem. Phys. Lett.*, 362:271–277, 2002.
16. H. Merlitz, B. Burghardt, and W. Wenzel. Application of the stochastic tunneling method to high throughput database screening. *Chem. Phys. Lett.*, 370:68, 2003.
17. W. Wenzel H. Merlitz, T. Herges. Fluctuation analysis and accuracy of a large-scale in silico screen. *J. Comp. Chem.*, 25:1568–1575, 2004.

18. W. Wenzel and K. Hamacher. Stochastic tunneling approach for global optimization of complex potential energy landscapes. *Phys. Rev. Lett.*, 82:3003–3007, 1999.
19. H. Merlitz and W. Wenzel. Impact of receptor flexibility on in-silico screening performance. *Chem. Phys. Lett*, 390:500, 2004.
20. W.L. Jorgensen and N.A. McDonald. Development of an all-atom force field for heterocycles.properties of liquid pyridine and diazenes. *J. Mol. Struct.*, 424:145, 1997.
21. G.M. Morris, D.S. Goodsell, R. Halliday andR. Huey, W.E. Hart, R.K. Belew, and A.J. Olson. Automated docking using a lamarckian genetic algorithm and an empirical binding free energy function. *J. Comput. Chem.*, 19:1639, 1998.
22. J.N. Champness, M.S. Bennett, F. Wien, R. Visse, W.C. Summers, P. Herdewijn, E. de Clerq, T. Ostrowski, and R.L. Jarvestand M.R. Sanderson. Exploring the active site of herpes simplex virus type-1 thymidine kinase by x-ray crystallographyof complexes aciclovir and other ligands. *Proteins*, 32:350, 1998.
23. P. Pospisil, T. Kuoni, L. Scapozza, and G. Folkers. Methodology and open problems of molecular docking: Cases of dihydroorotate dehydrogenase, thymidine kinase and phosphodiesterase 4. *J. Recept. Signal Transduct. Res.*, 22:141, 2002.
24. Louise Birch, Christopher W. Murray, Michael J. Hartshorn, Ian J. Tickle, and Marcel L. Verdonk. Sensitivity of molecular docking to induced fit effects in influenza virus neuraminidase. *J. Comput. Aided Mol. Design*, 16:855, 2002.
25. G.W.A. Milne, M.C. Nicklaus, J.S. Driscoll, and and D. Zaharevitz S. Wang. National cancer institute drug information system 3d database. *J. Chem. Inf. Comput. Sci.*, 34:1219, 1994.

Multiple Semi-flexible 3D Superposition of Drug-Sized Molecules

Daniel Baum

Zuse Institute Berlin (ZIB), Germany
baum@zib.de

Abstract. A new algorithm for multiple semi-flexible superposition-ing of drug-sized molecules is described. It identifies structural similar-ities between two or more molecules. To account for the flexibility of a molecule, multiple conformers drawn from molecular ensembles gener-ated by conformational analysis are used. To address the varying degree of similarity among the molecules, similar substructures present in dif-ferent subsets of the molecules are identified.

All molecules are compared to a preselected reference molecule. Clique detection on the correspondence graph of two molecular structures is ap-plied to generate feasible start transformations, which are used to com-pute common substructures. The results of these pairwise comparisons are efficiently merged using binary matching trees.

Despite considering the full atomic information for identifying multi-ple structural similarities, the algorithm is well suited as an interactive tool for exploring drug-sized molecules, and has been integrated into the molecular visualization package AmiraMol. The algorithm's capabilities are demonstrated on two sets of molecules.

1 Introduction

In pharmaceutical drug design, one often faces the question, what properties a drug (ligand) must have to bind to a specific receptor. In the absence of the receptor structure, the only given information is a set of ligands for which we know or assume, that they all bind to the same receptor using similar binding modes. Beside the physico-chemical properties of the ligand, its form plays a major role in the binding process. Hence, it is not enough to consider the two-dimensional structure of the ligands, but one also needs to look at their three-dimensional forms. In general, the active forms of the ligands are not known. Therefore, the ligand's flexibility needs to be taken into account.

There exist two classes of algorithms that account for the flexibility of the ligands. The first class of algorithms keeps the ligands flexible during the com-parison stage [1,2,3,4]. The advantage of this approach is that the search space is not limited to a precomputed number of (generally) low-energy conformers. And indeed, active conformers often have a slightly higher energy. However, this ap-proach has two major disadvantages. First, the search space needs to be sampled for each comparison. Second, one might end up with statistically very unlikely

M.R. Berthold et al. (Eds.): CompLife 2005, LNBI 3695, pp. 198–207, 2005.

conformers. The second class of algorithms uses precomputed conformers to consider the flexibility of the ligands, hence they are called semi-flexible algorithms. Their advantage is that the conformers of each molecule need to be computed only once, independent of the comparison to be accomplished, and thus a more exhaustive conformational analysis can be carried out. Also, the development of comparison algorithms using multiple conformers is uncoupled from sampling the conformational space. The disadvantage is that the conformational space of a molecule might be very large and one can easily generate thousands of different conformers, posing the question, which conformers should be used for the comparison. This problem arises, since the conformers are generated before and not during the comparison.

Related work includes, e.g., rigid body alignment algorithms, trying to maximize some kind of volume overlap [5,6,7], and molecular surface alignment approaches [8,9]. Good coverage of publications predating 1990 can be found in [10,11]. Related work predating 2000 is excellently reviewed in [12].

The work most closely related to our approach is the multiple semi-flexible superposition algorithm by Martin et al. [13]. They specify a reference molecule to which all other molecules are pairwise aligned. Multiple conformers are considered separately. For the pairwise alignment they require pharmacophore points to be specified which are matched using a clique detection method. The results of the pairwise comparisons are finally merged to get multiple matchings. Since the number of pharmacophore points in each molecule is considerably smaller than the overall number of atoms, clique detection can easily be applied and merging the results is relatively easy. In parts, our algorithm is based on the work of Kirchner [14], who uses a greedy matching strategy to find the optimal matching of any two conformers of two molecules. Since he is only interested in the optimal matching between two molecules, he can apply a branch-and-bound method to prune large parts of the search tree.

With our approach we try to bridge the gap between the work of Martin et al. [13], which is fast, but uses only parts of the full structural information, and very detailed, expensive algorithms which explore the full structural information, but either are not able to handle flexibility or cannot do multiple superpositioning.

2   Algorithm

The algorithm requires to preselect a reference molecule R. Each reference conformer R_i is compared separately to all query molecules Q_k. This is done by computing pairwise matchings (Sect. 2.1) of all conformers Q_{kl} of the query molecule Q_k with R_i. The pairwise matchings between R_i and Q_k are stored in a matching tree T_{ik} (Sect. 2.2). Next, all matching trees T_{ik} are merged into a single final matching tree T_i containing all matching clusters corresponding to reference conformer R_i. In the last step, we compute the score values of all matching clusters. By interpreting these values as vectors, we can sort the matching clusters into Pareto sets (Sect. 2.3). The algorithm is sketched in Fig. 1.

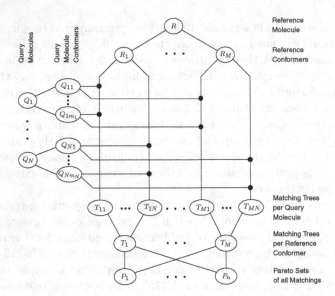

Fig. 1. Overview of the algorithm. The reference molecule is denoted by R, its conformers by R_j. The query molecules are denoted by Q_i, their conformers by Q_{ij}. The ● symbol denotes the computation of pairwise matchings. T_{ij} denotes the matching tree containing all pairwise matchings between molecule Q_i and reference conformer R_j. T_i is the final matching tree comprising all multiple matchings (matching clusters) with respect to R_i. Finally, P_i is a Pareto set containing multiple matchings of different reference conformers.

2.1   Computation of Pairwise Matchings

The computation of pairwise matchings between two conformers, a reference conformer R_i and a query conformer Q_{kl}, is done in two steps.

First, we compute a set of rather small substructures common to R_i and Q_{kl} by applying clique detection [15] to the correspondence graph of the two molecular structures [10]. Each clique in the correspondence graph represents a single common substructure. For each clique, we compute a rigid transformation by least-squares fitting [16] the respective substructures.

Before we describe the second step, we need to define the term *pairwise matching*. Let $m, n \in \mathbb{N} \setminus \{0\}$. A function $M : \{1, \ldots, m\} \longrightarrow \{0, \ldots, n\}$ is called a *pairwise matching*, or simply *matching*, if it meets the following property: $\forall i, j \in \{1, \ldots, m\} : M(i) = M(j) \Rightarrow i = j \vee M(i) = 0$. Furthermore, we define the set $M^* := \{i | M(i) \neq 0\}$.

In the second step, we use each transformation generated in the first step as start transformation for an iterative greedy point matching method [14]. In each iteration we compute the optimal matching with respect to the current rigid transformation T, i.e., the matching maximizing the weighted scoring function

$$score(A, B; M; T) := \frac{|M^*|}{\min(m, n)} \cdot e^{-rms(A, B; M; T)} , \qquad (1)$$

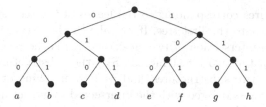

Fig. 2. Complete matching tree for a reference molecule with three atoms. For example, leaf a represents the empty substructure, whereas leaf d represents the substructure with atoms 2 and 3.

where A and B are the atom coordinates of R_i and Q_{kl}, respectively, M is a matching, and rms is defined as

$$rms(A, B; M; T) := \sqrt{\frac{\sum_{i \in M^*} w(i, M(i)) \cdot \|A_i - TB_{M(i)}\|^2}{|M^*|}} , \qquad (2)$$

where $w : \{1, \ldots, m\} \times \{1, \ldots, n\} \longrightarrow \mathbb{R}^+$ is a weight function, which allows us to favor atom pairs, such as donors or acceptors, or to penalize an atom pair consisting of, e.g., a donor and a hydrophobic atom. If $w(i,j) := 1.0, \forall i \in \{1, \ldots, m\}$ and $\forall j \in \{1, \ldots, n\}$, our scoring function is equal to the one used in [17]. The transformation for the next iteration is computed by least-squares fitting the substructures corresponding to the optimal matching.

2.2 Matching Tree

A *matching tree* is a binary tree used for storing matchings corresponding to a single reference conformer. A matching tree allows to find all matchings corresponding to a given substructure in the reference conformer in time $\mathcal{O}(m)$, where m is the number of atoms in the reference molecule. A matching tree has depth m. Level 0 of the matching tree represents the root node, level $i > 0$ represents the i'th atom in the reference molecule. Each leaf corresponds to a unique substructure of the reference molecule (see Fig. 2 for an example). The maximum number of leaves is 2^m. The path $p = [p_0, \ldots, p_m]$ from the root to a leaf uniquely defines the substructure associated with the leaf: If atom i of the reference molecule is contained in the leaf's substructure, p_i will be the right child of p_{i-1}, otherwise it will be the left child. Each matching is inserted into the matching tree according to the matched substructure of the reference conformer. Thus, two matchings M_x and M_y will be stored at the same leaf if $\forall i : M_x(i) \neq 0 \Leftrightarrow M_y(i) \neq 0$. For each leaf we maintain a sorted list of matchings. The substructure corresponding to a leaf can also be thought of as a bit string, where the i'th bit is set to 1 if the i'th atom of the reference molecule is in the substructure.

Two matching trees are merged in two steps. In the first step, we compute the intersections of all substructures corresponding to the leaves of the first tree

and all substructures corresponding to the leaves of the second tree and insert
these intersections into the first tree. If we take the bit string representation of
substructures, the intersection of two substructures is the result of the bitwise
AND operation of the two bit strings. If S is the substructure resulting from
the intersection of two substructures, and M is a matching of either matching
list of the two leaves associated with the intersected substructures, we define the
restricted matching M_S as

$$M_S(i) := \begin{cases} M(i) & \text{if } i \in S, \\ 0 & \text{otherwise.} \end{cases}$$

All restricted matchings resulting from intersecting two substructures are inserted in the first matching tree. In the second step, we insert all matchings of
the second tree into the first tree, if they are not already there.

Merging of the matching trees is done iteratively, i.e., we merge two trees
into one of them which is then merged with another one and so forth, until all
trees are merged into a single tree, the *final matching tree*.

The merging step drastically reduces the amount of work to be done when
bringing together matchings of different query molecules. Since in each merging
step the intersected substructures are inserted into the merged tree, substructures contained in multiple query molecules only need to be intersected once
during a single merging step.

2.3   Sorting of Matching Clusters

All restricted matchings gathered at one leaf of a matching tree build a *matching
cluster*. If there are several matchings of the same query molecule, we take the
one with the highest score. For each matching cluster, we compute a number
of measures, such as the number of query molecules in the cluster, the size of
the matching, and the averaged *score* value. If we want to sort the clusters
according to these measures using a single scoring function, we are faced with
the problem of assigning weights to the measures. We circumvent this problem
by considering the measure values as a vector and by using the concept of *Pareto
dominance* [18]. According to this concept, a vector $u = (u_1, \ldots, u_k)$ is said to
dominate vector $v = (v_1, \ldots, v_k)$ if and only if u is partially larger than v, i.e.,
$\forall i \in \{1, \ldots, k\}, u_i \geq v_i \land \exists i \in \{1, \ldots, k\} : u_i > v_i$. All vectors mutually not
dominating each other are assigned to the same so-called *Pareto set*. According
to the dominance of the vectors of different Pareto sets, we can define an order
on the Pareto sets.

3   Results

We tested the described algorithm on several groups of molecules. Unfortunately,
there are no benchmarks for superimposing molecules. Therefore, we chose two
groups that had previously been used for the assessment of superposition algorithms and thus seemed well suited [4,9]. The first group consisted of a set of
four angiotensin II antagonists. For this group the active conformations were not

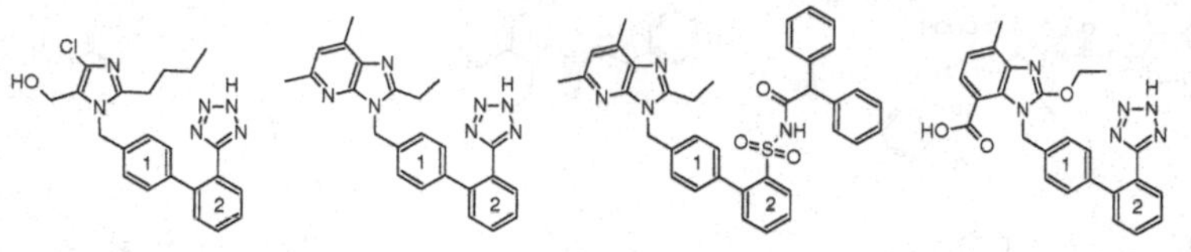

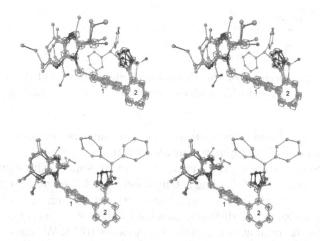

Fig. 3. Structural formulas of the angiotensin II antagonists: Losartan, L-158,809, L-159,894, and CV-11974 (from left to right)

Fig. 4. Stereo views of two matchings of four angiotensin II antagonists from the first (*top*) and the second (*bottom*) Pareto set, respectively. The red wire-frame cubes denote the common substructure. The numbers denote the same rings as in Fig. 3.

available. We therefore computed the metastable conformations [19] for each of these molecules and used representatives of these conformations as input to our algorithm. The second group consisted of seven thermolysin inhibitors. For this set of molecules the active conformations were available from the PDB (Protein Data Bank) [20], and thus we used these as input to the algorithm.

All computations were performed on a PC workstation with 3GHz processor. The implementation was done in C++, and the source code was integrated into the visualization software AMIRAMOL [21]. All images in this chapter were also made with AMIRAMOL.

3.1   Angiotensin II Antagonists

The four angiotensin II antagonists considered here are Losartan, L-158,809, L-159,894, and CV-11974. The two-dimensional structures as well as the nomenclature of these molecules were taken from [22]. The three-dimensional structures were generated using CORINA [23]. These structures were used as input to the program ZIBMol [19], which generated 20, 13, 13, and 25 metastable conforma-

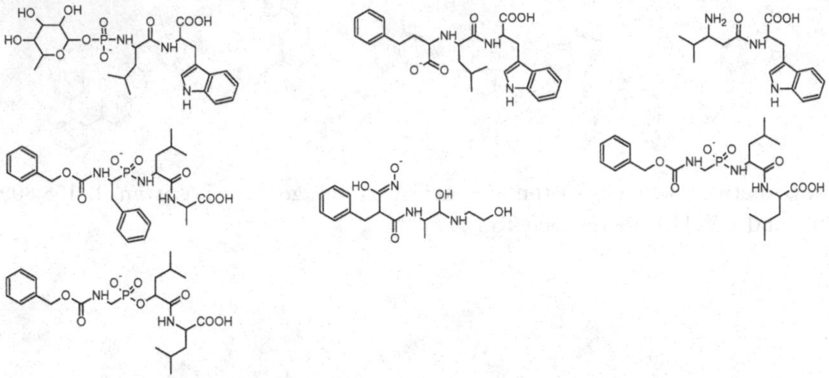

Fig. 5. Structural formulas of the thermolysin inhibitors of 1TLP, 1TMN, 3TMN, 4TMN, 5TLN, 5TMN, and 6TMN (from left to right and top to bottom)

tions, respectively. From these metastable conformations we chose the structures with minimum energy as input to our algorithm.

For the computation of the multiple structure superposition of the angiotensin II antagonists the following settings were used. The minimum clique size was set to 5, i.e., we only used substructures with a minimum size of 5 to generate start transformations. The distance threshold for the correspondence graph was set to 0.2, and the minimum matching size was set to 15. We chose Losartan as reference molecule.

325196 matching clusters were computed in 2 minutes. The large number of matching clusters is due to the large similarity between the molecules which leads to many slightly different matchings. Two matchings were selected, one from the first and one from the second Pareto set, which are shown in Fig. 4. The two matchings differ in the relative orientation of the imidazole ring of Losartan to the double-ring of the other three molecules. In the first matching the second nitrogen atom is not matched to the nitrogen atoms of the other molecules, in the second matching the nitrogen atoms are matched. In both matchings the imidazole ring as well as the aromatic rings are matched between all molecules. Furthermore, the tetrazole ring of Losartan, L-158,809, and CV-11974 is matched to the sulfonamide group of L-159,894. These groups are probably responsible for forming hydrogen bonds to the receptor.

3.2 Thermolysin Inhibitors

The active conformations of seven inhibitors of thermolysin (TLN, EC-number 3.4.24.27) were compared with each other. These inhibitors were extracted from the enzyme-inhibitor complexes found in the PDB: 1TLP, 1TMN, 3TMN, 4TMN, 5TLN, 5TMN, and 6TMN. The separated inhibitors were then parametrized using the Merck Molecular Force Field (MMFF) implemented in the program ZIB-Mol. The parametrization was needed for assigning atom types (donor, acceptor, hydrophobic, aromatic).

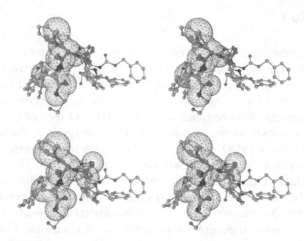

Fig. 6. Stereo views of two matchings of thermolysin inhibitors. The wire-frame surfaces denote common substructures. *Top:* 7 inhibitors with a matching of size 12. *Bottom:* 6 inhibitors (without 3TMN) with a matching of size 16.

For the computation of the multiple structure superposition of the seven thermolysin inhibitors we used the same settings as for the comparison of the angiotensin II antagonist except that we set the minimum matching size to 8. We chose the 4TMN inhibitor as reference molecule. However, this choice was somewhat arbitrary since we got similar results for the other inhibitors except for that of 3TMN. This is due to the size of the 3TMN inhibitor, which is considerably smaller and in particular misses the functional group responsible for binding to the zinc ion in the active site.

The computation of the superposition took less than a second. 308 matchings were computed which were sorted into 70 Pareto sets. Two superpositions from the Pareto optimal set, namely the one with all inhibitors and the one with 6 inhibitors are shown in Fig. 6. Due to the size of the 3TMN inhibitor and the absence of the functional group responsible for binding to the zinc ion, the first matching size is considerably smaller with 12 atoms in contrast to 16 atoms.

Of special interest in the second matching is the zinc ion binding group. For the 1TLP, 4TMN, 5TMN, and 6TMN inhibitors this region is identical, consisting of a phosphor atom which two oxygen atoms are bound to. The 1TMN and 5TLN inhibitors, however, differ in this region from the rest and between each other. At the position of the phosphor atom, in the 1TMN inhibitor we find a carbon which a COO- is bound to. This means that apart from the absence of the phosphor atom, there is also an additional carbon atom which slightly dislocates the oxygen atoms. This is similar to the 5TLN inhibitor, here, however, we have an additional nitrogen atom inserted between the carbon and the oxygen atoms, further dislocating one of the oxygen atoms. Despite this, the oxygen atoms were correctly matched.

4 Discussion

We presented a new algorithm for multiple semi-flexible superposition of drug-sized molecules. Our algorithm is similar to DISCO [13] in two points. (1) A reference molecule needs to be specified to which all other molecules are compared. (2) The results of these pairwise comparisons are then merged to get substructures present in all or many molecules. DISCO uses clique detection to identify common substructures. Clique detection, however, is limited to rather small sets of points. DISCO circumvents this problem by prespecifying a few pharmacophore points which are used instead of all atoms. Our approach, in contrast, uses the full atomic information. With clique detection we would only be able to identify rather small common substructures. We therefore use a two-step approach for the comparison of two molecular structures. In the first step, clique detection is used to generate good start transformations. The second step refines each start transformation and computes the matching. The matchings found using this two-step approach are, in general, considerably larger than those found by clique detection alone. However, the large matchings that we get impose a new problem. Merging of the pairwise results becomes much more expensive. We solve this problem by introducing matching trees.

We tested our algorithm on several groups of molecules. The results suggest, that the new approach is very well capable of identifying common substructures in a set of molecules. The algorithm depends on the computation of resonable molecular conformers. However, the algorithm is fault-tolerant to the absence of conformers by also computing substructures not present in all molecules.

The parameters of the algorithm allow the user to decide how thoroughly the search for common substructures should be performed. A trade-off has to be found between runtime and search detail.

In the future, the main focus will be on considering even more conformers per molecule while keeping short runtimes. The runtime of our algorithm, however, is mainly determined by the number of conformers per molecule. Exploiting similarities between a molecule's conformers might enable us to considerably reduce the algorithm's runtime, and, thus, it might allow for even more flexibility.

Acknowledgments. I would like to thank Johannes Schmidt-Ehrenberg and Frank Cordes for helpful discussions and Peter Deuflhard for his continuous support.

References

1. McMartin, C., Bohacek, R.: Flexible matching of test ligands to a 3d pharmacophore using a molecular superposition force field. J. Comput. Aid. Mol. Des. **9** (1995) 237–250
2. Jones, G., Willett, P., Glen, R.C.: A genetic algorithm for flexible molecular overlay and pharmacophore elucidation. J. Comput. Aid. Mol. Des. **9** (1995) 532–549
3. Lemmen, C., Lengauer, T.: FLEXS: a method for fast flexible ligand superposition. J. Comp. Chem. **41** (1998) 4502–4520

4. Handschuh, S., Wagener, M., Gasteiger, J.: Superposition of three-dimensional chemical structures allowing for conformational flexibility by a hybrid method. J. Chem. Inf. Comp. Sci. **38** (1998) 220–232
5. Good, A.C., Hodgkin, E.E., Richards, W.G.: Utilization of Gaussian functions for the rapid evaluation of molecular similarity. J. Chem. Inf. Comp. Sci. **32** (1992) 188–191
6. Grant, J.A., Gallardo, M.A., Pickup, B.T.: A fast method of molecular shape comparison: A simple application of a Gaussian description of molecular shape. J. Comp. Chem. **17** (1996) 1653–1666
7. Lemmen, C., Hiller, C., Lengauer, T.: RigFit: A new approach to superimposing ligand molecules. J. Comput. Aid. Mol. Des. **12** (1998) 491–502
8. Cosgrove, D., Bayada, D.M., Johnson, A.P.: A novel method of aligning molecules by local surface shape similarity. J. Comput. Aid. Mol. Des. **14** (2000) 573–591
9. Hofbauer, C.: Molecular Surface Comparison. A Versatile Drug Discovery Tool. PhD thesis, Technische Universität Wien (2004)
10. Brint, A.T., Willett, P.: Algorithms for the identification of three-dimensional maximal common substructures. J. Chem. Inf. Comp. Sci. **27** (1987) 152–158
11. Martin, Y.C., Bures, M.G., Willett, P.: Searching databases of three-dimensional structures. In Lipkowitz, K.B., ed.: Reviews in Computational Chemistry. Number 1. Elsevier Science Publishers B.V. (1990) 213–263
12. Lemmen, C., Lengauer, T.: Computational methods for the structural alignment of molecules. J. Comput. Aid. Mol. Des. **14** (2000) 215–232
13. Martin, Y.C., Bures, M.G., Danaher, E., DeLazzer, J., Lico, I.: A fast new approach to pharmacophore mapping and its application to dopaminergic and benzodiazepine agonists. J. Comput. Aid. Mol. Des. **7** (1993) 83–102
14. Kirchner, S.: Ein Approximationsalgorithmus zur Berechnung der Ähnlichkeit dreidimensionaler Punktmengen. Diploma Thesis, Department of Computer Science, Humboldt University Berlin (2003)
15. Bron, C., Kerbosch, J.: Algorithm 457: Finding all cliques of an undirected graph. Communications of the ACM **16** (1973) 575–577
16. Kabsch, W.: A discussion of the solution for the best rotation to relate two sets of vectors. Acta Crystallographica A **34** (1978) 827–828
17. Thimm, M., Goede, A., Hougardy, S., Preissner, R.: Comparison of 2d similarity and 3d superposition. Application to searching a conformational drug database. J. Chem. Inf. Comp. Sci. **44** (2004) 1816–1822
18. Veldhuizen, D.A.V.: Multiobjective Evolutionary Algorithms: Classification, Analyses, and New Innovations. PhD thesis (1999)
19. Fischer, A., Schütte, C., Deuflhard, P., Cordes, F.: Hierarchical uncoupling-coupling of metastable conformations. Volume 24 of LNCSE Series., Berlin, Springer (2002) 235–259
20. Berman, H., Westbrook, J., Feng, Z., Gilliland, G., Bhat, T., Weissig, H., Shindyalov, I., Bourne, P.: The Protein Data Bank. Nucleic Acids Res. **28** (2000) 235–242
21. AmiraMol – User's Guide and Reference Manual. Zuse Institute Berlin (ZIB) and Indeed - Visual Concepts GmbH, Berlin, http://www.amiravis.com (2002)
22. Wexler, R.R., Greenlee, W.J., Irvin, J.D., Goldberg, M.R., Prendergast, K., Smith, R.D., Timmermans, P.B.M.W.M.: Nonpeptide Angiotensin II Receptor Antagonists. J. Comp. Chem. **39** (1996) 625–656
23. Sadowski, J., Gasteiger, J.: From Atoms and Bonds to Three-Dimensional Atomic Coordinates: Automatic Model Builders. In: Chemical Reviews. (1993) 2567–2581

Efficiency Considerations in Solving Smoluchowski Equations for Rough Potentials

Polina Banushkina[1], Olaf Schenk[2], and Markus Meuwly[1]

[1] Department of Chemistry, University of Basel, Klingelbergstrasse 80
[2] Department of Computer Science, University of Basel, Klingelbergstrasse 50,
4056 Basel, Switzerland

Abstract. We present an efficient and numerically robust algorithm for solving the Smoluchowski equation (SE) to follow diffusive processes on smooth and rough potential energy surfaces. The hierarchical nature of the algorithm (hierarchical discrete approximation or HDA) allows to fully explore the fine- and coarse-grained structure of the free energy surface and can be extended to multidimensional problems. It is shown that for free energy surfaces where the minima are separated by considerable barriers the reaction kinetics can be captured using only a small number of eigenvalues of the corresponding rate matrix which leads to a considerable speedup of the computation. This technique, in combination with HDA, is applied to study the rebinding of carbon monoxide (CO) to native myoglobin (Mb) and a mutated protein (L29F), a process of fundamental importance in biophysics.

1 Introduction

The notion of smooth and structured (rough) energy landscapes is an important concept in the discussion of dynamical processes in complex systems. Examples include the folding of proteins, the reaction of two end-groups in a polymer chain or the motion in bistable, metastable or periodic potentials. The stochastic dynamics on these landscapes is governed by the heights of the barriers between the local minima which can be large. In protein dynamics one is usually interested in the way how an initial population of conformational substates relaxes towards a steady state. A useful way to determine the reaction kinetics of such a system is to follow the temporal and spatial relaxation of an initial distribution $p(x,0)$ to the final, steady-state (equilibrium) distribution $p_{eq}(x)$. To this end the Smoluchowski equation is solved for the particular potential energy surface. There exist several methods to solve SEs. They include finite-difference schemes in space (x) and time (t) [1], finite-differences in x with time propagation based on the formal solution of the time-dependent part [2,3], basis set expansions [4], and path integral methods [4]. Although the first two methods are conceptually simple and appealing, their applicability to problems with realistic potential energy curves is limited. With increasing number of intermediate states described by the potential $V(x)$ the distance between two points Δx must decrease and the number of grid points required increases. With increasing number of grid

M.R. Berthold et al. (Eds.): CompLife 2005, LNBI 3695, pp. 208–216, 2005.

points the computational effort significantly increases due to storage requirements and execution time. Other methods including approximations based on analytical mean first passage are restricted to one-dimensional (1D) systems and generalizations to multidimensional dynamics are difficult and in fact seem not to have been successful so far.

Based on a recently developed hierarchical approach (HDA) to solve the SE in one and two spatial dimensions we discuss a procedure to increase the efficiency of the algorithm. The main concern is to limit the number of eigenvalues that have to be calculated for an underlying free energy surface and the conditions that allow such an approximation.

In the first section we present an efficient and numerically robust algorithm to follow diffusive processes on rough potential energy surfaces. Because of the hierarchical structure of the method, HDA can be extended to multidimensional problems. In the second section we consider a more efficient way to solve the SE by calculating only a small number of eigenvalues instead of the entire spectrum. This modification considerably speeds up calculations while retaining accuracy. In the last section we apply the method to investigate migration of carbon monoxide (CO) between different internal binding sites in native myoglobin and the L29F mutant along a well-defined reaction coordinate. From the behavior of $p(x, t)$ the rebinding time for CO diffusing between metastable states is calculated and compared with experiment.

2   Hierarchical Discrete Approximation Method

Brownian motion or diffusion of the system in a potential $V(x)$ is described by the Smoluchowski equation,

$$\frac{\partial p(x,t)}{\partial t} = \frac{\partial}{\partial x} D(x) e^{-\beta V(x)} \frac{\partial}{\partial x} \left[e^{\beta V(x)} p(x,t) \right] \tag{1}$$

Here, $D(x)$ is a space-dependent diffusion constant, $\beta = 1/kT$ is the Boltzmann factor and $p(x,t)$ is the space and time-dependent probability distribution. Equation (1) can be recast as a Master equation [2]

$$\frac{\partial p(x_n, t)}{\partial t} = l(n|n+1)p(x_{n+1}, t) + l(n|n-1)p(x_{n-1}, t) \tag{2}$$
$$- (l(n+1|n) + l(n-1|n))p(x_n, t)$$

Here $l(n|n \pm 1)$ are the rates to move to the left and to the right from the starting point x_n.

The time evolution of $p(x, t)$ in the equation (1) can be followed numerically whereby the space coordinate is discretized such that the continuous variable x takes discrete values x_n, $n = 1, \ldots, N+1$. The solution to (1) may be obtained by solving equations (2). In discrete approximation method (DA) the rate coefficient $l(m|n)$ for making a transition $x_n \to x_m$ is given by

$$l(m|n) = \frac{D(n) + D(m)}{2d^2} \exp(-\frac{\beta(V(m) - V(n))}{2}), \tag{3}$$

with $d = x_{n+1} - x_n$ (see Ref. [2]).

For rough potentials $V(x)$ the number of grid points n required to resolve the roughness increases rapidly. This leads to a large rate matrix that has to be diagonalized. In the present work we explore the possibility to solve the SE on a hierarchy of grids by defining a coarse grid (N) and a subgrid (M) by dividing each interval (x_n, x_{n+1}) into $j = 1, \ldots, M + 1$ points (see Fig.1).

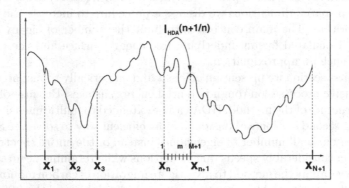

Fig. 1. Discretization of the grid $[1..N + 1]$. Each interval $[n..n + 1]$ is subdivided into $[1..M + 1]$ points for which a local rate constant is calculated using DA.

The mean passage time $\tau(n + 1|n)$ from x_n to x_{n+1} is given by [5]

$$\tau(n + 1|n) = \int_0^\infty \sum_{j=1}^M p(x_j, t)dt = \int_0^\infty (1 - p(x_{M+1}, t))dt. \qquad (4)$$

and we calculate the corresponding rate coefficients as $l_{HDA}(n + 1|n) = 1/\tau(n + 1|n)$. For finding $\boldsymbol{p_M}(x, t)$ ($\boldsymbol{p_M}(x, t)$ is the vector of probabilities in points $x_1, \ldots, x_{M+1}$ on each subinterval) we use the DA method and calculate the rate coefficients for making transitions in inner points by formula (3). Finally, the solution of equation (2) can be determined from

$$\boldsymbol{p_M}(x, t) = U \exp(\boldsymbol{\lambda} t)U^{-1}\boldsymbol{p_M}(x, 0) \qquad (5)$$

The elements of the matrix U and the vector $\boldsymbol{\lambda}$ are the eigenvectors and eigenvalues of the rate matrix, respectively. The boundary conditions on each subinterval (see Fig. 1) are

$$l(1|0) = l(0|1) = 0, \qquad l(M + 2|M + 1) = l(M + 1|M + 2) = 0,$$
$$l(M|M + 1) = 0$$

and the initial condition for the probability distribution is a δ-function at the first point of each subinterval. Taking $\boldsymbol{p_M}(x, 0) = \{\delta_j p_1\}$, the probability $p(x_{M+1}, t)$ at grid point x_{M+1} (required for $\tau(n + 1|n)$) is

$$p(x_{M+1}, t) = \sum_k U_{M+1,k} \exp(\lambda_k t)U_{k,1}^{-1},$$

where $U_{M+1,k}$ and $U_{k,1}^{-1}$ are the $(M+1,k)$-th and $(k,1)$-th elements of the matrices U and U^{-1}, respectively. Substituting this expression into equation (4) the integral can be evaluated analytically to yield

$$\tau(n+1|n) = \sum_k U_{M+1,k} \frac{1}{\lambda_k} U_{k,1}^{-1}$$

The coefficients $l_{\mathrm{HDA}}(n|n+1)$ are determined from the condition of detailed balance

$$l_{\mathrm{HDA}}(n|n+1) = \frac{l_{\mathrm{HDA}}(n+1|n)p_e(n)}{p_e(n+1)} \tag{6}$$

where $p_e(x)$ is the Boltzmann equilibrium probability distribution for the potential $V(x)$. It should be noted that this definition of $l_{\mathrm{HDA}}(n|n+1)$ is not necessary but it increases the numerical stability of the procedure. The coefficients $l_{\mathrm{HDA}}(n|n+1)$ can also be calculated from the mean first passage time without imposing detailed balance.

3   Solving SE Using Only Smallest by Module Eigenvalues of Rate Matrix

For the following we consider a master equation in matrix form

$$\frac{\partial \boldsymbol{p}(t)}{\partial t} = L\boldsymbol{p}(t), \tag{7}$$

where L is the rate matrix. For finding the probabilities by formula (5) one has to calculate all eigenvalues λ_i, the corresponding eigenvectors u_i (matrix U) of the matrix L and its inverse U^{-1}. The computational effort for calculating the inverse U^{-1} and the eigenvalues of L is proportional to N^3. The size of L depends on the discrete spatial grid. With increasing number of grid points the computational effort significantly increases also due to storage requirements. The HDA method allows to use fewer grid points compared to other DA methods[6] but even with HDA for multidimensional problems the size of L grows exponentially. For example, a two-dimensional (2D) coarse 100×100 grid leads to a matrix L with dimensions 10000×10000. It is possible to drastically reduce computing times by addressing the above-mentioned points in the following way.

First, for symmetrical matrices the equality $U^{-1} = U^T$ is valid. The matrix L is not symmetrical but can be symmetrized. Replacing the elements $p(x_i, t)$ by $p(x_i, t)/\sqrt{p_e(i)}$ in equation (7) gives a new rate matrix L_s with the following symmetrized elements:

$$l_s(i,j) = l(i,j)\sqrt{p_e(j)/p_e(i)},$$

where $l(i,j)$ are the coefficients of the matrix L, $p_e(i)$ is the equilibrium probability in point x_i. The equality $l_s(i,j) = l_s(j,i)$ is satisfied using detailed balance

(6) for the coefficients $l(i,j)$. The matrices L and L_s have the same eigenvalues and the eigenvectors u_i of L can be expressed by the eigenvectors u_{i_s} of L_s through

$$u_i = u_{i_s} \sqrt{p_e(i)}.$$

Second, we consider strategies to more efficiently solving the SE by calculating just a small number of eigenvalues instead of diagonalizing the entire rate matrix. In diffusion problems all eigenvalues obey $\lambda_i \leq 0$ where $\lambda = 0$ corresponds to the equilibrium probability distribution. If the product $\|\lambda_i t\|$ is large enough then $\exp(\lambda_i t) \to 0$ in formula (5) and in the long time limit only the smallest by module eigenvalues contribute to the solution. If the spectrum of eigenvalues λ_i separates into small and large eigenvalues by module (see example) it is sufficient to consider only the smallest by module eigenvalues and their corresponding eigenvectors to find the solution of the SE at large times. For such a case only the 10 to 15 smallest by module eigenvalues of a sparse, real, symmetric matrix are sought [7], and thus iterative methods can be used (for example ARPACK)[8]). The rate matrix L is 3-diagonal for a 1D problems and 5-diagonal for 2D problems. To reduce memory requirements the sparse matrices can be stored such that instead of $N \times N$ elements only about $5 \times N$ or $7 \times N$ elements have to be kept for 1D or 2D problems, respectively [8]. In the following this strategy is applied to a problem relevant in biophysical chemistry: the rebinding time scales of CO in native and mutant myoglobin.

4   Application to Rebinding Time Scales in Myoglobin

We apply our method to calculate the rebinding time for CO diffusing from the internal cavity (Xe4 pocket) to the primary binding site. The systems contained a total of 2532 heme protein atoms (native Mb) and 2533 heme protein atoms (L29F), respectively, the CO ligand and 181 water molecules. The ligand motion and the associated free energy barriers can be described along suitably chosen (progression) coordinates. For these systems free energy profiles (see Fig. 2) were calculated using molecular dynamics (MD) simulations and umbrella sampling along the reaction coordinate, which is the distance between iron (Fe) and carbon (C) atoms.[9] The FEPs were calculated along escape paths from 1ns molecular dynamics simulations previously calculated. In particular, the electrostatic interaction between the CO and the environment is accurately described by using the recently developed fluctuating point charge model for CO.[10] This is important because CO only interacts via non-bonded (electrostatic and van der Waals) interactions with the surrounding. Since the total charge of CO is zero and its dipole moment is small and changes sign around the equilibrium structure, the contribution of the quadrupole moment to the total interaction is essential. The combination of umbrella sampling and stochastic simulations provides new and fundamental insight into the ligand rebinding Mb·CO $\to$ MbCO for native and mutant myoglobin. Direct simulations for converged barrier crossing statistics is computationally demanding since the time scales involved in the rebinding process are long and sufficient sampling by standard MD simulations can be difficult.

Solving SE on a constructed FE profiles allows not only to calculate rebinding times but also investigate the dynamics of CO migration and its population in different pockets for the early stages following photodissociation of the ligand from the heme. The binding site (B) (see Fig.2) is of major importance since after photodissociation it is rapidly populated. From there, the ligand can either rebind directly (A) or it follows a largely unknown path within the protein to diffuse towards the solvent from where it rebinds at much longer time scales. One possible, secondary binding site in the neighborhood of site B is the Xe4 pocket.

We solved the SE equation using the free energy profiles ($V(x)$ in eq. (1)) with an initial distribution of CO in the Xe4 pocket with $T = 300$ K and a constant diffusion coefficient $D = 2.2$ Å ^{2}/ps. D was calculated in the distal pocket from mean square displacements at $T = 300$ K. Corresponding to the notation of the HDA method the grid was discretized by $N + 1 = 205$ and $M + 1 = 6$ points. The discretization on the coarse and fine grid are 0.05 and 0.01, respectively. The inset in Fig.2 shows that the FEPs are quite rough. For example, to fully explore the fine structure of the potential by the DA method, 1021 grid points are required that corresponds to the step discretization 0.01.

In Table 1 the ten smallest by module eigenvalues of the rate matrix L_{HDA} for native MbCO are given.

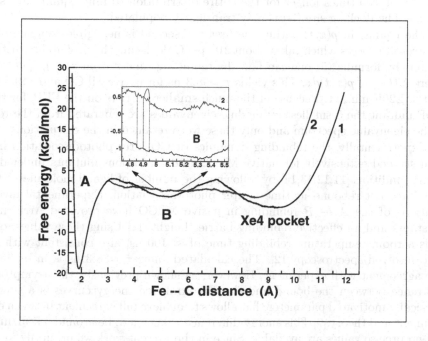

Fig. 2. Escape path along which the free energy profile was calculated. The path follows the motion of the center of mass of CO from the heme binding site (A) via the a metastable position (docking site (B)) to the Xe4 pocket: trace 1 in native myoglobin, trace 2 in the L29F mutant.

Table 1. Smallest by module eigenvalues of the rate matrix for native MbCO

i	196	197	198	199	200	201	202	203	204	205
λ_i	-74.07	-63.41	-49.18	-42.13	-31.55	-24.94	-22.87	-0.014	-0.000044	0

The three smallest by module eigenvalues differ from the remaining eigenvalues by at least 3 orders of magnitude. At time $t = 1$ ps the value $\exp(\lambda_{202}t) \approx 1.16 \cdot 10^{-10}$ and after 1 ps only λ_{203}, λ_{204} and λ_{205} contribute to the solution. Typical rebinding time scales are known to be of the order of 100 ns for native Mb. Thus, the calculation of $p(x,t)$ for $t > 10$ ps starting from $p(x_0, t_0)$ and using the three smallest by module eigenvalues is sufficient. The probability distributions calculated with the entire spectrum of λ_i and with 3 eigenvalues are the same at $t \geq 1$ ps. As an example, Figure 3 shows $p(x,t)$ calculated at $t = 100$ ps where the solid line and line with circles represent the distributions obtained with the entire spectrum of λ_i and with only λ_{203}, λ_{204}, λ_{205}, respectively. The probability distributions are virtually indistinguishable but the effort to calculate them differs considerably. It can also be seen that $p(x,t)$ is sensitive to the roughness of the free energy profile. Finally, even after 100 ps some of the population has migrated to the bound state (A, see Fig. 2) although it will take about 100 times longer for the entire distribution to fully equilibrate (see below). The docking site (B) is only transiently populated.

The change in $p(x,t)$ within the first picosecond is negligible compared to the overall process which takes about 10^5 ps. Calculating the probability distributions by formula (5) one can find the rebinding time τ from $\tau = \int_{t=1}^{\infty} dt \Sigma(t)$ where $\Sigma(t) = \int p(x,t)dx$. This yields $\tau = 38.3$ ns for native MbCO and 119.5 ps for the L29F mutant. Because of the much smaller barriers on the FEP, for the L29F mutant the 5 smallest by module eigenvalues are separated from the rest of the eigenvalue spectrum and only these were retained in the calculations.

Experimentally, the rebinding dynamics of CO after photodissociation has been studied extensively for native MbCO and for various mutants under different conditions[11,12,13,14] by following the number of ligand molecules $N(t)$ that have not rebound at time t after photodissociation. Experimental investigations of the $A \leftarrow B$ rebinding in native MbCO have provided Arrhenius constants and an effective rebinding barrier height.[11] Using these values, one finds a room temperature rebinding time of ≈ 100 ns, also consistent with ns time resolved spectroscopy[12]. The calculated value $\tau = 38.3$ ns is in qualitative agreement with experiment. However, as shown previously, the asymptotic difference between the bound and the unbound free energy curves is a single, physically motivated parameter that allows to achieve full agreement between experiment and theory.[9] This energy difference is known to be around 5 kcal/mol, but no precise values are available. Since in the present work we primarily focus on efficiency considerations in solving SE, further investigations of this aspect have been omitted. For the L29F mutant the rebinding time is not experimentally determined. From time-resolved X-ray experiments[15] it is known that ligand migration between B and Xe4 is considerably more rapid than in native

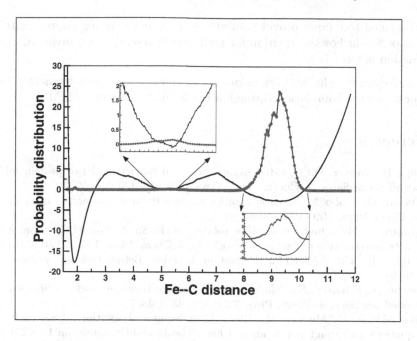

Fig. 3. Probability distributions at time $t = 100$ ps for CO diffusing in Mb: solid line presents the distribution calculated using 205 eigenvalues, line with circles – using 3 smallest by module eigenvalues. The magnifications show the transient population in the docking site (B) and the rough profile of the free energy surface around the Xe4 pocket ($p(x,t)$ rescaled for better visibility).

Mb. In the L29F mutant the CO escapes on a ns time scale from B to the Xe4 pocket after dissociation. This is in good agreement with the constructed free energy profile where free energy barriers are much smaller than for native Mb (see Fig. 2) and CO rebinding occurs on a ns time scale. Thus, the flat FEP for CO migration for the L29F mutant reflects the faster dynamics. Again, the precise rebinding time will depend upon the asymptotic separation between the two free energy profiles.

Summary. The HDA method for solving SEs on rough potentials was discussed and analyzed. The method presented here is particularly useful for processes taking place on long-time scales. In such processes it may be possible that the eigenvalues separate into two groups of different magnitude and only the smallest by module eigenvalues and their eigenvectors are required for accurate time propagation of $p(x,t)$. Here, for the first time, HDA is applied for a realistic free energy profile to calculate rebinding time for CO diffusing in Mb and L29F mutant. Comparisons of HDA with alternative methods on model potentials can be found in[6]. The results were obtained using only 3 and 5 eigenvalues (for native MbCO and its mutant, respectively) which considerably speeds up the calculations but without affecting the accuracy of the propagation. In summary, the described technique provides an attractive alternative for approximate solutions of Smoluchowski equation for multidimensional systems involving rough interaction potentials.

Acknowledgment. The authors acknowledge financial support from the Swiss National Science Foundation through a Förderungsprofessur to MM.

References

1. Jun, B., Weaver, D.: One-dimensional potential barrier model of protein folding with intermediates. J. Chem. Phys., **116** (2002) 418–426
2. Bicout, D., Szabo A.: Electron transfer reaction dynamics in non-Debye solvents. J. Chem. Phys. **109** (1998) 2325
3. Ansari, A.: Mean first passage time solution of the Smoluchowski equation: Application to relaxation dynamics in myoglobin. J. Chem. Phys. **112** (2000) 2516–2522
4. Risken, H.: The Fokker-Planck equation. Springer, Berlin, Heidelberg, New York (1989)
5. Szabo, A., Schulten, K., Schulten, Z.: 1st passage time approach to diffusion controlled reactions. J. Chem. Phys. **72** (1980) 4350–4357
6. Banushkina, P., Meuwly, M.: Hierarchical Numerical solution of Smoluchowski equations with rough potentials. J. Chem. Theory and Computation **1** (2005) 208–214
7. Saad, Y.: Iterative Methods for Sparse Linear Systems. SIAM Publications, Philadelphia (2003)
8. Lehoucq, R., Sorensen, D., Yang, Ch.: ARPACK users guide. SIAM, Philadelphia (1998)
9. Banushkina, P., Meuwly, M.: Free energy barriers in MbCO rebinding. (submitted)
10. Nutt D., and Meuwly M.: Theoretical Investigation of Infrared Spectra and Pocket Dynamics of Photodissociated Carbonmonoxy Myoglobin. Biophys. J. **85**, (2003) 3612-3623.
11. Steinbach P., Ansari A., Berendzen J., Braunstein D., Chu K., Cowen B., Ehrenstein D., Frauenfelder H., Johnson J., Lamb D., Luck S., Mourant J., Nienhaus G., Ormos P., Philipp R., Xie A. and Young R.: Ligand-binding to heme-proteins – connection between dynamics and function. Biochem. **30**, (1991) 3988-4001.
12. Ansari A., Jones C., Henry E., Hofrichter J. and Eaton W.: Conformational relaxation and ligand-binding in myoglobin. Biochem. **33**,(1994) 5128-5145.
13. Ostermann A., Waschipky R., Parak F. and Nienhaus G.: Ligand binding and conformational motions in myoglobin. Nature **404**,(2000) 205-208.
14. Srajer V., Ren Z., Teng T., Schmidt M., Ursby T., Bourgeois D., Pradervand C., Schildkamp W., Wulff M. and Moffat K.: Protein conformational relaxation and ligand migration in myoglobin: A nanosecond to millisecond molecular movie from time-resolved Laue X-ray diffraction. Biochem. **40**, (2001) 13802-13815.
15. Schotte, F., Lim M., Jackson T., Smirnov A., Soman J., Olson J., Phillips Jr., Wulff M. and Anfinrud P.: Watching a protein as it functions with 150-ps time-resolved X-ray crystallography. Science **300** (2003) 1944-1947

Fast and Accurate Structural RNA Alignment by Progressive Lagrangian Optimization[*]

Markus Bauer[1,2], Gunnar W. Klau[3], and Knut Reinert[1]

[1] Algorithmic Bioinformatics Group, Institute of Computer Science,
Free University of Berlin, Germany
[2] International Max Planck Research School on Computational Biology
and Scientific Computing
[3] Mathematics in Life Sciences Group, Institute of Mathematics,
Free University of Berlin, Germany

Abstract. During the last few years new functionalities of RNA have been discovered, renewing the need for computational tools for their analysis. To this respect, multiple sequence alignment is an essential step in finding structurally conserved regions in related RNA sequences. In contrast to proteins, many classes of functionally related RNA molecules show a rather weak sequence conservation but instead a fairly well conserved secondary structure. Hence, any method that relates RNA sequences in form of multiple alignments should take structural features into account, which has been verified in recent studies.

Progress has been made in developing new structural alignment algorithms, however, current methods are computationally costly or do not have the desired accuracy to make them an everyday tool. In this paper we present a fast, practical, and accurate method for computing multiple, structural RNA alignments. The method is based on combining a new pairwise structural alignment method with the popular program T-Coffee. Our pairwise method is based on an integer linear programming (ILP) formulation resulting from a graph-theoretic reformulation of the structural alignment problem. We find provably optimal or near-optimal solutions of the ILP with a Lagrangian approach. Tests on a recently published benchmark set show that our Lagrangian approach outperforms current programs in quality and in the length of the sequences it can align.

1 Introduction

Recently, it has become clear that RNA molecules perform additional functions that were previously thought of being carried out by proteins. Many more of these *functional RNAs* have yet to be discovered. Computing multiple alignments to detect structural features is usually the first step in analyzing sequences

---

[*] Supported by the DFG Research Center MATHEON "Mathematics for key technologies" in Berlin, the German Federal Ministry of Education and Research, (grant no. 0312705A 'Berlin Center for Genome Based Bioinformatics'), and the IMPRS for Computational Biology and Scientific Computing.

M.R. Berthold et al. (Eds.): CompLife 2005, LNBI 3695, pp. 217–228, 2005.

of biomolecules. Unfortunately, and unlike proteins, many functional classes of RNA show little sequence conservation, but rather a conserved secondary structure which is formed by folding in space and forming hydrogen bonds between its bases. Among such RNAs are tRNA, rRNA, snoRNAs, and SRP RNA [1].

Hence, algorithms to compute multiple alignments ought to take not only the sequence, but also the secondary structure into account. Washietl and Hofacker [2] support this consideration by showing that sequence based alignments are significantly worse than sequence-structure based alignments if their pairwise sequence identity sinks below ≈ 60%. This observation is confirmed by Gardner and coworkers [3] in a paper that also benchmarks numerous multiple alignment programs.

Thus, the problem of producing RNA alignments that find a common structure has become the bottleneck in the computational study of functional RNAs. To date, the available tools for computing structural alignments are often incapable of handling reasonable input sizes or produce alignments of low quality. With this work we present a *multiple* RNA sequence-structure alignment tool that computes fast and accurate alignments. Our method uses a new pairwise structural alignment algorithm based on Lagrangian relaxation in combination with the progressive alignment tool T-Coffee.

Previous Work. The computational problem of considering sequence and structure of an RNA molecule simultaneously was first addressed by Sankoff [4] who proposed a dynamic programming algorithm that aligns a set of RNA sequences while at the same time predicting their common fold. The running time of this algorithm is $O(n^{3m})$ where m is the number of sequences and n their length. Algorithms similar in spirit were proposed later for the problem of comparing one RNA sequence to one or more of known structure. Corpet and Michot [5] align simultaneously a sequence with a number of other, already aligned, sequences using both primary and secondary structure. Their dynamic programming algorithm requires $O(n^5)$ running time and $O(n^4)$ space and thus can handle only short sequences. Current implementations modify Sankoff's algorithm by imposing limits on the size or shape of substructures, *e.g.*, `Dynalign` [6,7], `Foldalign` [8,9], `pmcomp` [1], and `Stemloc` [10,11].

Bafna *et al.* [12] gave an algorithm that simultaneously aligns the primary and secondary structure of two sequences that runs in time $O(n^4)$ which still does not make it applicable to instances of realistic size. Common motifs among several sequences are searched by Waterman [13]. Eddy and Durbin [14] describe probabilistic models for measuring the secondary structure and primary sequence consensus of RNA sequence families. They present algorithms for analyzing and comparing RNA sequences as well as database search techniques. Since the basic operation in their approach is an expensive dynamic programming algorithm, their algorithms cannot analyze sequences longer than 150-200 nucleotides. Instead of folding and aligning sequences simultaneously, Hofacker *et al.* [1] present a dynamic programming approach to align the corresponding *base pair probability matrices*, computed by McCaskill's partition function algorithm [15]. Their approach PMcomp—also a variant of Sankoff's algorithm—takes time $O(n^6)$ and space $O(n^4)$, but can be reduced by solving a banded version of the problem to $O(n^4)$ time and $O(n^3)$ space complexity. Their tool, `pmmulti`, which we also use to compare our new approach with, obtains multiple structural alignments by aligning consensus base pair probability matrices in a progressive fashion.

The base pair probabilities can be directly used to weight edges in the structural alignment graph introduced in Lenhof *et al.* [16] where the authors presented a branch-and-cut algorithm for structurally aligning two RNA sequences. The underlying graph-theoretical formulation is flexible and allows for pseudoknots. Previous work on contact map overlap in the area of proteomics by Caprara and Lancia [17] and for the two-sequence case of the structural alignment problem by Bauer and Klau [18] indicates, however, that Lagrangian relaxation is better suited to obtain provably optimal or near-optimal solutions to the corresponding integer linear programming (ILP) formulation than a direct branch-and-cut approach in terms of running time. Bauer, Klau, and Reinert extend these ideas to multiple sequences [19]. Currently, however, the approach is applicable only to few sequences and small instance sizes.

Contribution. Our goal is to devise a fast method to compute high-quality, multiple structural alignments for a large number of possibly long RNA sequences.

Our key idea is to use the program T-Coffee [20], a successful multiple sequence alignment program that conducts a progressive alignment similar to ClustalW [21] but additionally incorporates *local* alignment information in form of so called *libraries*. This idea is not new by itself. Siebert and Backofen [22] already employ it in their program MARNA, which we also use in our experimental comparisons. The difference lies in the way the pairwise alignments are computed: MARNA takes fixed RNA structures as input and minimizes their pairwise edit distances as proposed by Jiang *et al.* [23]. In the general case of unknown structures, MARNA uses a whole ensemble of suboptimal structures in order not to overlook structurally conserved motifs.

We use the implementation of Bauer and Klau [18] (Lara) and improve it in several ways, such that the obtained pairwise, structural alignments are very accurate. das hier weggenommen und oben ein wenig ausfuerhlicher beschrieben while Siebert and We will show that T-Lara is better than or competitive to other, more costly, structural alignment programs and can handle much longer sequences while maintaining a running time of only a couple of minutes. Furthermore, T-Lara yields better results than MARNA in terms of structural conservation, in particular for test instances that are longer than ≈ 300 bases.

2   Lagrangian Structural Alignment of Two Sequences

We have described the theoretical framework of the Lagrangian approach to structural sequence alignment elsewhere (see [18,19]). We therefore provide only a short summary of the basic approach and focus on recent practical improvements such as the incorporation of affine gap costs and a more sophisticated selection of candidate edges.

2.1   Terminology and Basic Approach

Let S be a sequence $s_1, \ldots, s_n$ of length n over the alphabet $\Sigma = \{A, C, G, U\}$. A paired base (i, j) is called an *interaction* if (i, j) forms a Watson-Crick-pair. The set P of interactions is called the *annotation* of sequence S. Two interactions are said to be in *conflict*, if they share one base; they form a *pseudoknot* if they cross each other. A pair (S, P) is called an *annotated sequence*. Note that a structure where no pair of interactions is in conflict with each other forms a valid secondary structure of an RNA sequence, possibly with pseudoknots.

We are given two annotated sequences (S_1, P_1) and (S_2, P_2) and model the input as a graph $G = (V, L \cup I)$. The set V denotes the vertices of the graph, in this case the bases of the sequences. The set L contains *alignment edges* between vertices of the two input sequences (for sake of better distinction called *lines*) whereas the set I codes the two annotations by means of *interaction edges* between vertices of the same sequence. A subset $\mathcal{L} \subset L$ corresponds to an *alignment* of the two sequences if $\mathcal{L}$ does not contain crossing lines, since those correspond to ordering conflicts of the letters in the sequences. Two interaction edges $(i_1, i_2) \in P_i$ and $(j_1, j_2) \in P_j$ are said to be *realized* by an alignment $\mathcal{L}$ if and only if $\mathcal{L}$ contains the alignment edges $l = (i_1, j_1)$ and $m = (i_2, j_2)$. The pair (l, m) is called an *interaction match*. Note that we define (l, m) as an ordered tuple, that is, (l, m) is distinct from (m, l). Figure 1 illustrates the above definitions by means of an example.

Note that the graph-theoretical model gives absolutely no restriction on the annotation of the sequences, *i.e.*, one can align two known structures, infer a known structure to an unknown or compare two unknown structures.

We assign positive weights w_l and w_{ij} to each line l and each interaction match (i, j), respectively, that represent the benefit of realizing the line or the match. The weights are given, for example, by mutation score matrices or—in the case of interaction matches—by the base pair probabilities as computed by McCaskill's algorithm [15].

The structural alignment problem now corresponds to finding a maximally weighted subset of lines and interaction edges in the input graph such that

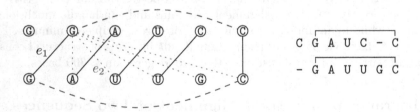

Fig. 1. Graph-theoretic concept of alignment. The right side shows a structural alignment of two annotated sequences, the left side the corresponding graph G. Solid lines represent alignment edges in $\mathcal{L}$, dotted lines represent additional candidate edges from L (only a subset shown). Replacing, *e.g.*, $e_1 \in \mathcal{L}$, by e_2 creates a crossing. Lines $\mathcal{L}$ realize two interaction matches (remember that interaction matches are ordered tuples).

no lines cross each other, each interaction match is realized, and no vertex is incident to more than one interaction edge. We define binary variables x_l for each alignment edge l and y_{lm} for each interaction match (l, m) and rewrite the problem as the following integer linear program:

$$\max \quad \sum_{l \in L} w_l x_l + \sum_{l \in L} \sum_{m \in L} w_{lm} y_{lm} \tag{1}$$

$$\text{s.t.} \quad \sum_{l \in I} x_l \leq 1 \qquad \qquad \forall \text{ sets of crossing lines } I \tag{2}$$

$$y_{lm} = y_{ml} \qquad \qquad \forall l, m \in L \tag{3}$$

$$\sum_{m \in L} y_{lm} \leq x_l \qquad \qquad \forall l \in L \tag{4}$$

$$0 \leq x \leq 1, \quad 0 \leq y \leq 1 \qquad \qquad \text{integer} \tag{5}$$

We have shown in [18] that dropping constraints (3) leads to a much easier problem (the *relaxed problem*). Figure 2 shows possible solutions for the original and the relaxed problem. Since each alignment edge l is free to choose its own best interaction match (l, m) regardless of what interaction match line m chooses (recall that we dropped the equality constraint $y_{lm} = y_{ml}$), the relaxed problem can be reduced to a classical primary sequence alignment problem that is in turn solvable in polynomial time.

We follow the iterative Lagrangian optimization method and move the complicating constraints into the objective function with a penalty term for their violation. An iteration consists of solving an instance of the relaxed problem and adapting the penalty terms. As a by-product we obtain a feasible solution in each iteration by interpreting the solution of the relaxed problem as an input graph for a maximum weighted matching problem. With an increasing number of iter-

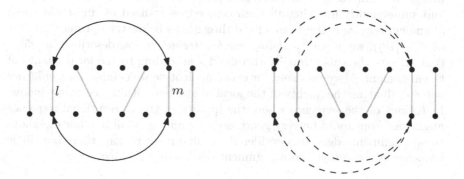

Fig. 2. The left part shows an interaction match that is realized from both sides, where the constraint $y_{lm} = y_{ml}$ ist satisfied. Dropping the equality constraints allows solutions like the one shown in the right part: These solutions have a higher score, but do not necessarily yield a valid solution for the original problem.

ations, these generated solutions become better and better and finally converge to a provable optimal or near-optimal solution of the above ILP. The advantages of this approach over alternative methods based on dynamic programming are threefold: First, the Lagrangian method is fast and applicable to sequence lengths where dynamic programming must fail computationally. Secondly, the graph-theoretical formulation allows for the formation of pseudoknot structures that cannot be handled by alternative approaches. Finally, as we show in the next section, we are able to incorporate an affine gap cost model, which results in a more realistic gap distribution.

2.2   Practical Improvements

We have implemented two major modifications of the basic approach described in the preceding section in order to increase its applicability to RNA data from practical applications.

– The basic approach does not consider gap costs and alignments computed with an early version of our implementation suffered from this drawback. We have therefore replaced the recurrence relation in the standard dynamic programming algorithm for classical primary sequence alignment by a version that takes into account affine gap scores (see, *e.g.*, [24]). We have also modified the backtracking in the dynamic programming matrix in order to account for a different treatment of gaps occurring at the beginning or the end of the sequence.
– We achieved a speed-up compared to the basic approach by providing another way we select the candidate edges. Note that only the complete bipartite graph models all possible alignments of two sequences. In practice, this is computationally too expensive, and we resort to a heuristic selection of promising candidate edges:
  Instead of computing a conventional sequence alignment with affine gap costs and subsequently inserting all alignment edges realized by any suboptimal alignment scoring better than a fixed threshold s below the optimal score (as used in [18]), we provide a *sliding window technique*—as described in [25]—that adjusts the suboptimality threshold s according to the local quality of the alignment. More precisely, for every nucleotide we compute a *confidence value* evaluating the quality of the local alignment within a certain window. In regions of the sequence where the quality of the conventional sequence alignment appears to be very good, none or only a small number of suboptimal alignment edges are considered. In alignment regions that show little sequence conservation, more alignment edges are generated.

3   Extension to Multiple Sequences

We have shown how to extend the formulation (1)-(5) and the Lagrangian relaxation technique to the multiple sequence case in [19]. Here, we follow a different approach, since the inherent computational complexity of the multiple structural sequence alignment problem impedes the use of exact methods for instances with many sequences. We wish to remark that we are following two different lines of research: on the one hand, we investigate the structure of truly optimal multiple alignments and aim at solving instances of three or four sequences to provable optimality. On the other hand, we wish to provide a fast and practical—although possibly suboptimal—tool based on the good results of the pairwise algorithm. For this reason, we decided to integrate our pairwise algorithm into a multiple alignment framework.

3.1 Progressive Alignment with T-Coffee

T-Coffee uses a progressive sequence alignment approach similar to the one of ClustalW [21]. Progressive methods build multiple alignments from pairwise alignments. The pairwise distances are usually used to compute a guide tree which in turn determines the order in which the sequences are aligned to the evolving multiple alignment.

Progressive approaches often suffer from their sensitivity to the order in which the sequences are chosen during the alignment process. T-Coffee reduces this effect by making use of local alignment information from *all* pairwise sequence alignments during its progressive alignment phase. This local information, however, is computed with Lalign [26] and therefore considers only sequence-based information.

A nice feature about the T-Coffee implementation is that the user can supply such local alignment information. Therefore, we compute all pairwise structural alignments using Lara [18], assigning a high score to conserved interaction matches. The structural information is subsequently passed on to T-Coffee that computes a multiple alignment, taking into account the additional structural information.

4   Computational Results

4.1   Materials and Methods

We took a subset of data from the recently published BRaliBase dataset[4] [3] and used two different scores: the *sum-of-pairs score* SPS and the *structure conservation index* SCI.

If reliable reference alignments were available, we compared the computed alignments to the reference alignments and computed the SPS score: the SPS score is a value between 0 and 1 indicating the number of sequences correctly aligned (in case of a SPS score of 1, the two compared alignments are identical).

On the other hand, the SCI value compares the minimum free energies of the single sequences in an alignment with a "consensus energy" imposed by the alignment, which is computed by incorporating covariation terms into a

---

[4] Freely available from http://www.binf.ku.dk/users/pgardner/bralibase/

Table 1. Average SCI scores computed over a test set of 242 instances with different programs

Program	Av. SCI	Av. SPS
clustal	0.6076	0.7345
MUSCLE	0.6069	0.7666
T-Coffee	0.5972	0.7543
T-Lara	0.71	0.77

Table 2. Average T-Lara SCI scores for the different groups of test instances

Group (# of instances)	Av. SCI	Av. SPS
5S rRNA (39)	0.84	0.903
U5 spliceosomal (101)	0.60	0.765
Group II introns (72)	0.73	0.696
tRNA (30)	0.84	0.800

free energy minimization computation. More technically, let $\hat{E}$ be the consensus energy value of the alignment and let E_n be the mean of all MFE (*minimum free energy*) values of n sequences, respectively. Then the SCI is defined as

$$\mathrm{SCI} = \frac{\hat{E}}{E_n}$$

An SCI close to zero indicates that there is no conserved structure within the alignment, whereas SCI > 1 exhibits a perfectly conserved structure, additionally supported by compensatory mutations. Therefore, the SCI assesses in particular the structural quality of an alignment.

As a first test, we took all instances with low homology (that is with sequence identity < 55%) of the first dataset that was used by Gardner *et al.* in [3]: we computed a structural alignment of all 242 instances, with one instance being a set of either five Group II introns, 5S rRNA, tRNA, or U5 spliceosomal RNA sequences. The entire computation took 345.93 minutes on an AMD Opteron server running at 2Ghz. Table 1 shows the average SCI scores of the three best-scoring sequence-based programs on the low homology data. It should be noted that the alignment program (clustalW) computing the best SCI score of the first dataset reached an average SCI score of only 0.6076. Table 2 gives a more detailed view of T-Lara's performance on the different subgroups.

The big gap between T-Lara and the other programs is easily explained by the fact that due to the extensive computational demands of structure alignment programs, Gardner and colleagues only used sequence based approaches for the first dataset. T-Lara removes In case of sequences with low sequence identity (say below 50%), structure alignment programs compute significantly better alignments in terms of conserving structural motifs.

For comparing structure alignment programs, Gardner *et al.* chose a subset of tRNA instances consisting of only two tRNA sequences (some programs tested in their survey are only capable of computing pairwise structural alignments).

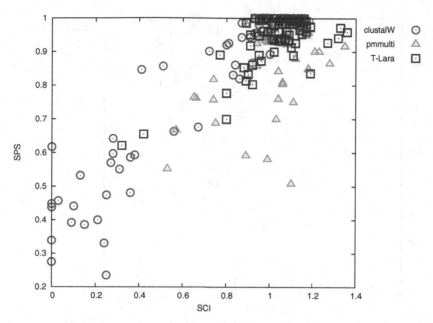

Fig. 3. 2D plots of 97 tRNA instances taken from BRalibase. The x- and y-axis are labelled with the corresponding SPS and SCI scores.

Since our approach can handle multiple sequences, we augmented this dataset and calculated all tRNA instances (consisting of five sequences) from the first dataset and compared them to pmmulti and clustalW. Over a set of 97 instances of five tRNA sequences (pmmulti failed on one instance) the average SCI score of clustalW—one of the best sequence-based alignment programs from the first dataset—is 0.82, whereas pmmulti and T-Lara reach average SCI scores of 1.044 and 1.047 at a running time of 101.91 and 75.16 minutes, respectively. The average SPS scores of clustalW, pmmulti, and T-Lara are 0.854, 0.926, and 0.954, respectively. Figures 3 and 4 show the SPS and SCI scores for all 97 tRNA instances.

It is interesting to observe that a high SPS score does not necessarily imply a high SCI score (and vice versa): pmmulti, for instance, computes alignments that have a high SCI score, whereas on average the SPS is worse compared to T-Lara. Consequently, considering the SCI score alone is a reasonable first indication for the structural quality of the alignment: a more thorough analysis, however, should take the SPS score into account as well if reference alignments are available.

To illustrate our ability to handle long sequences, we randomly chose 14 instances of three SRP RNA sequences with low sequence identity from BRaliBase and compared the alignments computed by T-Lara to those of clustalW and MARNA.

Table 3 shows the computed SCI scores of clustalW, MARNA, and T-Lara, respectively. We were able to calculate only such a small number of instances,

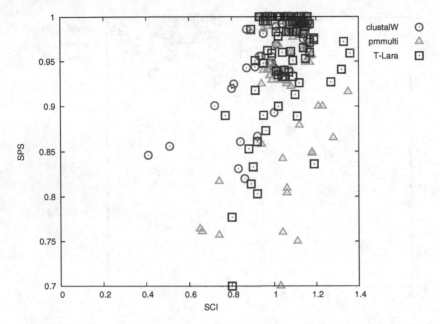

Fig. 4. Detailed view on alignment instances with a SPS score of 0.7 or higher

since MARNA can be accessed only by a web interface which makes the evaluation tedious. Since the reference alignment from BRaliBase "(...) were found to
be untrustworthy (...)" (quote from the BRaliBase website), we refrained from
considering the SPS score for the instances. For the instances computed, however, the table shows that T-Lara clearly outperforms clustalW and MARNA in
terms of conserving structural elements. Furthermore, it takes T-Lara just ≈ 40
minutes in total to compute the alignments of the 14 instances.

5   Discussion

In this paper we presented the new multiple structural alignment program
T-Lara. Our experiments show that T-Lara computes structural alignments
comparable or better than those computed by variants of Sankoff's algorithm.
Yet, our approach can also be applied to longer sequences (*e.g.*, 16S rRNA sequences of length ≈ 1600 nucleotides) since we do not suffer from the restrictive
demands in terms of CPU time and memory imposed by Sankoff's dynamic programming algorithm. Additionally, our algorithm does not restrict the secondary
structure of a given sequence in any way (*i.e.*, the approach allows arbitrary pseudoknots). Therefore, we plan to integrate more accurate base pair probabilities
based on pseudoknot energy parameters (like, for example, [27]).

   In the future we will extend our Lagrangian approach with our own progressive code (similar in spirit to pmmulti), and incorporate better scoring matrices
(*e.g.*, RIBOSUM matrices) should additionally enhance the quality of the align-

Table 3. SCI scores of `clustalW`, `MARNA`, and `T-Lara` on SRP sequences

SeqID	Instance	clustalW	MARNA	T-Lara
0.49	aln38	0.55	0.57	**0.66**
0.50	aln58	0.86	0.68	**1.00**
0.50	aln27	0.54	0.40	**0.58**
0.51	aln16	0.54	0.29	**0.62**
0.51	aln34	0.84	0.62	**0.93**
0.52	aln11	0.48	0.13	**0.54**
0.53	aln6	0.62	0.36	**0.66**
0.53	aln7	0.63	0.55	**0.70**
0.54	aln20	0.66	0.71	**0.78**
0.54	aln28	0.62	0.41	**0.69**
0.54	aln5	0.63	0.56	**0.73**
0.58	aln43	0.66	0.42	**0.79**
0.59	aln48	**0.78**	0.33	0.76
0.60	aln21	0.49	0.36	**0.54**

ments. Furthermore, a web service providing access to our algorithm is currently developed. A public-domain version of the program will follow in the next weeks.

Acknowledgments. The authors thank Veronika Gamper for implementing `T-Lara`'s T-Coffee library support and the gap score modifications, and Rolf Backofen for pointing us to the possible discrepancies between SPS and SCI scores.

References

1. Hofacker, I.L., Bernhart, S.H.F., Stadler, P.F.: Alignment of RNA base pairing probability matrices. Bioinformatics **20** (2004) 2222–2227
2. Washietl, S., Hofacker, I.L.: Consensus folding of aligned sequences as a new measure for the detection of functional RNAs by comparative genomics. J. Mol. Biol. **342** (2004) 19–30
3. Gardner, P., Wilm, A., Washietl, S.: A benchmark of multiple sequence alignment programs upon structural RNAs. Nucl. Acids Res. **33** (2005) 2433–2439
4. Sankoff, D.: Simultaneous solution of the RNA folding, alignment, and proto-sequence problems. SIAM J. Appl. Math. **45** (1985) 810–825
5. Corpet, F., Michot, B.: RNAlign program: alignment of RNA sequences using both primary and secondary structures. CABIOS **10** (1994) 389–399
6. Mathews, D.H., Turner, D.H.: Dynalign: An algorithm for finding secondary structures common to two RNA sequences. J. Mol. Biol. **317** (2002) 191–203
7. Mathews, D.: Predicting a set of minimal free energy RNA secondary structures common to two sequences. Bioinformatics **21** (2005) 2246–2253
8. Gorodkin, J., Heyer, L.J., Stormo, G.D.: Finding the most significant common sequence and structure motifs in a set of RNA sequences. Nucl. Acids Res. **25** (1997) 3724–3732
9. Hull Havgaard, J., Lyngsø, R., Stormo, G., Gorodkin, J.: Pairwise local structural alignment of RNA sequences with sequence similarity less than 40%. Bioinformatics **21** (2005) 1815–1824

10. Holmes, I.: A probabilistic model for the evolution of RNA structure. BMC Bioinformatics **5** (2004) 166
11. Holmes, I.: Accelerated probabilistic inference of RNA structure evolution. BMC Bioinformatics **5** (2004) 73
12. Bafna, V., Muthukrishnan, S., Ravi, R.: Computing similarity between RNA strings. In: Proc. of CPM'95. Number 937 in LNCS, Springer (1995) 1–16
13. Waterman, M.S.: Consensus methods for folding single-stranded nucleic adds. Mathematical Methods for DNA Sequences (1989) 185–224
14. Eddy, S.P., Durbin, R.: RNA sequence analysis using covariance models. Nucl. Acids Research **22** (1994) 2079–2088
15. McCaskill, J.S.: The Equilibrium Partition Function and Base Pair Binding Probabilities for RNA Secondary Structure. Biopolymers **29** (1990) 1105–1119
16. Lenhof, H.P., Reinert, K., Vingron, M.: A polyhedral approach to RNA sequence structure alignment. Journal of Comp. Biology **5** (1998) 517–530
17. Caprara, A., Lancia, G.: Structural alignment of large-size proteins via Lagrangian relaxation. In: Proc. of RECOMB'02, ACM Press (2002) 100–108
18. Bauer, M., Klau, G.W.: Structural alignment of two RNA sequences with Lagrangian relaxation. In: Proc. of ISAAC'04. Number 3341 in LNCS, Springer (2004) 113–123
19. Bauer, M., Klau, G.W., Reinert, K.: Multiple structural RNA alignment with Lagrangian relaxation. In: Proc. of WABI 2005. LNBI (2005) To appear.
20. Notredame, C., Higgins, D.G., Heringa, J.: T-Coffee: A novel method for fast and accurate multiple sequence alignment. Journal of Molecular Biology (2000)
21. Thompson, J.D., Higgins, D.G., Gibson, T.J.: Clustal w: improving the sensitivity of progressive multiple sequence alignment through sequence weighting, positions-specific gap penalties and weight matrix choice. Nucl. Acids Res. **22** (1994) 4673–4680
22. Siebert, S., Backofen, R.: MARNA: Multiple alignment and consensus structure prediction of RNAs based on sequence structure comparisons. Bioinformatics (2005) In press.
23. Jiang, T., Lin, G.H., Ma, B., Zhang, K.: A general edit distance between RNA structures. J. of Computational Biology **9** (2002) 371–388
24. Gotoh, O.: An improved algorithm for matching biological sequences. Journal of Molecular Biology (1982) 705–708
25. Kececioglu, J., Lenhof, H.P., Mehlhorn, K., Mutzel, P., Reinert, K., Vingron, M.: A polyhedral approach to sequence alignment problems. Discrete Applied Mathematics **104** (2000) 143–186
26. Huang, X., Miller, W.: A time efficient, linear space local similarity algorithm. Adv. Appl. Math. **12** (1991) 337–357
27. Dirks, R., Pierce, N.: An algorithm for computing nucleic acid base-pairing probabilities including pseudoknots. Journal of Computational Chemistry **25** (2004) 1295–1304

Visual Analysis of Molecular Conformations by Means of a Dynamic Density Mixture Model

Johannes Schmidt-Ehrenberg and Hans-Christian Hege

Zuse Institute Berlin (ZIB)
{schmidt-ehrenberg, hege}@zib.de

Abstract. We propose an approach for transforming the sampling of a molecular conformation distribution into an analytical model based on Hidden Markov Models. The model describes the sampled shape density as a mixture of multivariate unimodal densities. Thus, it delivers an interpretation of the sampled density as a set of typical shapes that appear with different probabilities and are characterized by their geometry, their variability and transition probabilities between the shapes. The gained model is used to identify atom groups of constant shape that are connected by metastable torsion angles. Based on this description an alignment for the original sampling is computed. As it takes into account the different shapes contained in the sampled set, this alignment allows to compute reasonable average shapes and meaningful shape density plots. Furthermore, it enables us to visualize typical conformations.

1 Introduction

Molecules are flexible structures. They move, vibrate, and interact with other molecules and their environment. Understanding these movements and interactions is essential for the complete comprehension of structure-function relationships, including many aspects of drug design and intermolecular interactions. Information about possible shapes of a molecule is carried by a density in the molecule's state space. Given some reasonably sized molecule, Hybrid Monte Carlo (HMC) simulation techniques allow to compute a set of molecular configurations that approximates this probability density of molecular shapes in thermodynamic equilibrium (see, e.g., [1]). Thanks to the molecular dynamics step in the HMC algorithm, the resulting sequence of states (trajectory) also contains dynamic information. Thus, the molecule's space of shapes can be analyzed [2] for *metastable subsets,* i.e. for regions in shape space that will be left by the molecule with very low probability.

These metastable subsets of shape space can be understood as the different rough shapes the molecule typically can take. However, the subsets can still contain more than one mode of the shape density. These modes can again be understood as typical shapes between which the molecule can change easier than between metastable sets. Overall, this constitutes a hierarchy of metastable sets.

For the visualization of metastable conformations, a partitioning of the shape space is of interest that also separates the modes inside a metastable subset. A

M.R. Berthold et al. (Eds.): CompLife 2005, LNBI 3695, pp. 229–240, 2005.

further challenge is to find a suitable alignment of the single configurations to each other. In particular, this is important for reasonable averaging and visualization of shapes.

In this paper we will present a method for analyzing molecular dynamics trajectories with respect to typical molecular shapes. After shortly surveying related work (Sect. 2), we will give some background on the treatment of circular coordinates (Sect. 3), in particular dihedral angles, on Hidden Markov Models (Sect. 4), and on the concept of Perron cluster cluster analysis (Sect. 5). Based on this we present

- a technique for partitioning the shape variations of a molecular dynamics trajectory into long-time changes and thermal fluctuations using Hidden Markov Models (Sect. 6)
- a method to determine a hierarchical decomposition of the molecule into rigid sets of interconnected atoms that are connected by metastable degrees of freedom (Sect. 7)
- a new alignment strategy that clearly separates the long-time shapes in cartesian coordinates while minimizing the variance induced by thermal fluctuations (Sect. 8).

Application of the approach is demonstated in Sect. 9. Conclusions and future work are presented in Sect. 10.

2   Related Work

Visual molecular analysis is well established and widely used in research and industry. The multifarious demands in chemical, biochemical and pharmaceutical applications have been addressed by commercial and academic software packages, offering a variety of visual representations of molecules as well as editing functions. However, specific tools for visual shape analysis based on molecular trajectories seem not to exist.

Identification of molecular conformations is a current research topic, see e.g. [3,4,5]. Recently, in [6] a method has been proposed - similar to our analysis step - for identification of the most important conformations of a biomolecular system from Metropolis Monte Carlo time series. The authors, however, do not aim at alignment and visualization of long-time shapes.

Alignment is a classical task in molecular science. When two molecules are to be compared in 3D space, alignment is necessary in order to eliminate differences caused by global rigid transformations. Kabsch [7,8] gives a straightforward method for computing an alignment between two point sets. Pennec [9] develops an approach to align multiple point sets iteratively. Both methods are characterized in more detail in Sect. 8. Huitema and van Liere [10] describe techniques for interactive visualization of protein dynamics, utilizing the concept of essential dynamics [11]. They interpret the dynamics of a protein as a trajectory in a high dimensional space and employ covariance analysis to filter out large concerted motions. Results on visual analysis of metastable molecular conformations on base of time series have been presented in [12].

3   Statistics of Molecular Shapes

For the analysis of molecular shapes some coordinate system is needed that represents the essential aspects of a molecular shape. Since we are not interested in a molecule's absolute position or orientation in 3D space, Cartesian coordinates of the atom positions are not appropriate. Neglecting the need to distinguish between mirror symmetric molecular shapes, it would be sufficient to consider intra-molecular distances between atoms. However, it turns out that the triple of bond lengths, bond angles and dihedral angles is more suitable. Bond lengths and angles can be regarded as nearly constant with respect to the shape variations of interest here. The interesting changes thus can be expressed via *dihedral angles*, which are defined by a sequence of four atoms where the respective angle is the angle between the two planes spanned by the first three and the last three atoms. Regarding the three bonds that sequentially connect the four atoms, the dihedral describes the rotation of the third bond relative to the first around the axis defined by the middle bond.

As bond lengths and angles are not of interest, we can describe molecular shapes in a coordinate space build by dihedral angles, which have a bounded and periodic value range $[0, 2\pi)$. For statistical analyses this periodicity has to be taken into account. To get statistical informations about our data we cannot apply standard techniques. Naive averaging of angular values may lead to invalid results, because periodicity is ignored. To overcome this problem, we can interpret every angular value α as a point $z(\alpha) = e^{i\alpha}$ on the unit circle in the complex plane. This representation intrinsically reflects the periodicity of angular values and is independent of the choice of an interval of periodicity. Averaging this set of points in the complex plane, we can define a reasonable mean angle $\bar{\alpha}$ by

$$\bar{R}\,e^{i\bar{\alpha}} = \frac{1}{N}\sum_{j=1}^{N} e^{i\alpha_j}. \tag{1}$$

As we want to set up a Hidden Markov Model in the space of dihedral angles, we need a probability density function for circular variables. The circular analogon to the normal distribution is the *von-Mises-* or *circular normal* distribution:

$$f_{1D}(\alpha;\phi,\kappa) = \frac{1}{2\pi I_0(\kappa)} \exp\left\{\kappa\cos(\alpha - \phi)\right\} \quad \text{with } \phi \in [0, 2\pi) \quad \text{and } \kappa \geq 0 \tag{2}$$

where I_n is the modified Bessel function of order n. f_{1D} is a unimodal distribution with maximum at $\alpha = \phi$ and is symmetric on the interval $[\phi - \pi, \phi + \pi]$. The mean angle $\bar{\alpha}$ from (1) turns out to be a maximum likelihood estimator for the *mean direction* ϕ. The respective maximum likelihood estimator for the *concentration parameter* κ is based on the amplitude $\bar{R}$ of the complex mean in Eq. (1): $I_1(\kappa)/I_0(\kappa) = \bar{R}$. High values of κ correspond to narrow distributions, while the minimal value ($\kappa = 0$) makes the von-Mises distribution uniform. For description of multidimensional distributions of angles we use a tensor product of von Mises distributions.

4   Hidden Markov Models

A *Markov chain* is a sequence of random variables $S_1, S_2, S_3, \ldots$ with state space I that fulfills the so called *Markov property*:

$$P(S_{n+1} = i_{n+1}|S_1 = i_1, \ldots, S_n = i_n) = P(S_{n+1} = i_{n+1}|S_n = i_n) \qquad (3)$$

with $i_1, \ldots, i_{n+1} \in I$. If this conditional probability is independent of n, it can be described by a stochastic matrix $T = \{t_{ij}\}$ with $t_{ij} := P(S_{n+1} = j|S_n = i)$ where $i, j \in I$ and $\sum_{j \in I} t_{ij} = 1$. Such a Markov chain is called *homogeneous*.

Hidden Markov Models (HMM), see e.g. [13], are a two-stage probabilistic concept for explaining the course of a time series. The primary assumption of a HMM is that a given time series is based on the realization of a homogeneous Markov Chain with finite state space I. This realization is not directly observable, but only by its influence on the second stage of the model. The HMM associates every state of its Markov chain with a probability density function defined on the value space of the time series to be explained. Depending on the state of the Markov chain realization at a given instant in time, a sample of the respective probability density is drawn as the observable value of the time series. A HMM is completely specified by the following parameters:

1. Markov chain start distribution $\pi_i = P(S_1 = i)$, $(i \in I)$,
2. Markov chain transition matrix $T = \{t_{ij}\}$, $(i, j \in I)$, and
3. probability density functions associated to the states of the Markov chain.

Dealing with HMMs two questions are typically of interest: First, given a sequence of observations, what are the optimal parameters of the HMM to explain this sequence? And second, given a sequence of observations and the parameters of the HMM, what is the underlying sequence of Markov states? Both question are answered by maximum likelihood estimation, i.e. by choosing the unknown parameters such that the likelihood of the observation sequence gets maximal.

For the first question this leads to the Baum-Welch-algorithm, which is a special case of the iterative Expectation-Maximization algorithm [14]. The iteration assumes a given set of model parameters. In a first step probabilities for the hidden parts of the model, in our case the states of the Markov chain realization, are computed. In a second step new model parameters are computed based on these probabilities. In every iteration cycle the likelihood of the observation given the model parameters increases. The iteration is terminated when the amount of increase drops under a threshold.

Maximum likelihood estimation for the second question is done via the direct Viterbi algorithm. The estimated sequence of states is called *Viterbi path*.

The construction of a HMM for a given series of observations requires the choice of the form of the probability density functions. We use tensor products of von Mises distributions, Eq. (2).

Further, the number of states of the hidden Markov chain has to be determined. In general, there is no way to measure whether a used number of states is appropriate. The achievable likelihood, which is the optimization criterion of the Baum-Welch-algorithm, monotonously increases with the number of states. Thus, it does not have a local maximum that would define an optimal number. In Sect. 6 we will use the concept of metastability to find a suitable number of states.

5 Perron Cluster Analysis

The phrase *metastable conformation* indicates a dynamic aspect of molecular behavior: it denotes approximate molecular geometries that survive the fast oscillations of molecular dynamics. In mathematical terms a metastable conformation is an *almost invariant set of the ensemble*, i.e. a subset of the molecular state space, that a molecular trajectory will only leave after a long time.

To find these metastable subsets of the state space, molecular dynamics is described using a transfer operator approach [15]. The state space is decomposed into subsets and a transfer operator is constructed, that specifies transition probabilities between these sets. Due to the reversibility of the dynamics, spectral analysis of the transfer operator leads to a real valued spectrum with maximal eigenvalue $\lambda_{max} = 1$, while the corresponding eigenvector is constant. If the state space contains l metastable subsets, the $l - 1$ next largest eigenvalues are very close to 1. This so called *Perron Cluster* of eigenvalues can be identified by a spectral gap that separates it from the remaining smaller eigenvalues.

If l has been determined, the metastable subsets can be constructed using the l corresponding eigenvectors, which define a mapping of the states to an approximate simplex in l-dimensional Euclidean space (cf. [5] for details). We can associate the l simplex vertices with the l metastable subsets we are looking for. Applying a linear transformation in the l-dimensional space mapping the simplex vertices onto the l vectors of an orthonormal basis, we get components for all the mapped sample points with respect to the orthonormal basis that approximately lie between 0 and 1 [16]. These can be interpreted as measures of membership to the respective metastable set. To turn this fuzzy and thereby ambiguous assignment into a definite one, we define a state space element to belong to the metastable set with the maximal membership value.

6 Adapting Hidden Markov Models

As the effort of fitting a HMM to a time series depends quadratically on the number of hidden states, we are interested in models with small numbers of states. Therefore we try to find as small as possible groups of dihedral angles that can be treated as independent from the rest of the molecule. In the first step we combine all dihedral angles into one HMM that rotate around the same bond, as these typically have a strong coupling.

To estimate the number of hidden states we start with a definite overestimate. After the Baum-Welch-iteration, we have a probabilistic decomposition of the molecule's shape space that is defined by the probability densities of the

HMM. The transition matrix of the HMM defines a transfer operator on this decomposition and we can apply Perron cluster analysis (cf. Sect. 5). This groups the Markov states into metastable sets. In [6], these sets were reduced to single states with mixture densities. In contrast to that, we replace the mixtures by single von Mises distributions. The resulting HMM is again optimized using the Baum-Welch-algorithm.

To find further correlations, we determine the Viterbi paths of all HMMs and compute for every two paths x and y the following entropy based measure of association [17]:

$$U(x,y) = 2 \frac{H(x) + H(y) - H(x,y)}{H(x) + H(y)} \tag{4}$$

where $H(x)$ and $H(y)$ are the state distribution entropies of the single paths and $H(x,y)$ denotes the entropy of the combined state distribution. The value of $U(x,y)$ will range between 0 and 1, with 0 representing complete independence; $U(x,y) = 1$ on the other hand indicates complete dependence. Thus, pairs of HMMs with high values of U are merged into one common HMM. An intial value for this merged HMM can be generated by building all possible combinations of states from both original HMMs. After the optimization, this HMM is again reduced by Perron Cluster Analysis.

7   Rigid Substructures

In the following, we will propose a policy to divide the shape variations of a molecular dynamics trajectory into long-time changes that lead to substantially different shapes, and thermal fluctuations around those shapes. We will use information from the HMMs that describe the various groups of dihedral angles in the molecule.

We distinguish between trivial HMMs with only one state and HMMs with multiple states. In case of a single state HMM, no hidden Markov chain exists and any shape variability is expressed by the variance of the corresponding probability density of this single state. For our purposes, we consider dihedral angles that are described by a single state HMM to be of constant shape, i.e. we interpret their complete shape variation as thermal noise.

If a dihedral angle is described by a HMM with multiple states, it changes between different shapes that are characterized by the corresponding probability densities. Therefore, the four atoms of the dihedral cannot be altogether in one rigid structure. Nevertheless, if we consider parts of the trajectory where the Viterbi path remains in the same state, the dihedral can be treated as constant in these subtrajectories.

In order to perform an alignment that takes these insights about rigid substructures into account, we build up a tree that specifies for every step of the trajectory to which other steps it has to be aligned and with respect to which atoms. Every node specifies a set of atoms, every edge corresponds to a HMM. The atoms of a node are considered to build a substructure of constant shape. The atoms of a child node can be added to this structure, if the trajectory is resolved into subtrajectories whose Viterbi paths with respect to the connecting node's HMM stay constant.

After the preliminary determination of all maximal rigid sets of interconnected atoms, we perform the following steps:

1. Choose a rigid structure containing central atoms of the molecule to be the root node.
2. For all leaf nodes of the tree:
   - Follow the root path of the current leaf node and collect all atoms contained in the nodes along the path. We will call this the *current rigid structure*.
   - Check all unused HMMs for a dihedral angle that overlaps in three atoms with the current rigid structure. If a HMM meets this criterion, collect all atoms of the dihedrals described by this HMM and remove those atoms, that are already contained in the tree. From these atoms build a child node of the current leaf node and associate the connecting edge with the HMM.
   - Check all unused rigid substructures for an overlap of at least 3 atoms with the union of the current rigid structure and one of the newly created nodes. If you find such a rigid substructure add it to the respective newly created node.
3. While unused rigid structures exist, repeat step 2.

8   Alignment

Since visualization of conformations takes place in Cartesian coordinates, it is necessary to assign global positions and orientations to the geometries and thereby to define a relative alignment between them. As has been detailed for example in [12], this can be done by superimposing the atomic positions via rigid-body transformations. On the one hand, a reference shape can be chosen to which all other shapes are pairwise aligned. On the other hand, an iterative algorithm can be used that in every step aligns a shape to the current mean of all other shapes. Although requiring some higher computational effort, this approach is superior to the first one, as it does not depend on the arbitrary choice of a reference.

In the following, we will introduce an extension of the second approach that uses the hierarchy of rigid structures constructed in Sect. 7. The tree of rigid structures specifies a hierarchy of atom sets together with sets of time steps in which the respective structure is considered to be constant. Therefore, we keep one mean for the atoms connected to the root node. If the HMM that connects a child node with the root has three possible states, we compute three means for the atoms corresponding to the child node. Only those time steps contribute to one of these means that have the same state in the Viterbi path of the HMM. If a grandchild node of this child node is connected by another HMM that again has three states, we have to deal with nine different mean structures for the atoms of the grandchild, because now two Viterbi paths with three states each have to be considered and, as the HMMs are independent, all nine combinations can arise.

In the following, M will denote the number of atoms in the molecule and N the number of time steps in the trajectory. Further, we write T for the tree constructed in Sect. 7 and $C_1, \ldots, C_L$ for the *rigid groups of atoms* that correspond to the nodes of T. It holds: $C_k \cap C_l = \emptyset$, $(k \neq l)$ and $\bigcup_{k=1}^{L} C_k = \{1, \ldots, M\}$.

If C_k is a rigid group of atoms, then C_k^* is the union of all groups of atoms corresponding to the nodes of T that build the connecting path from the root of T to the node corresponding to C_k. G_k is the number of subtrajectories for which C_k^* is considered to be of constant shape. We denote with S_{kg} ($g \in \{1, \ldots, G_k\}$, $k \in \{1, \ldots, L\}$) the set of time steps that are in the respective subtrajectories. It holds $S_{kg} \cap S_{kh} = \emptyset$, $(g \neq h;\ g, h \in \{1, \ldots, G_k\})$ and $\bigcup_{g=1}^{G_k} S_{kg} = \{1, \ldots, N\}$. With $S_k(t)$ we denote the set of time steps that contains a step t with respect to a C_k^*.

Let the original cartesian coordinates of all time steps be $\boldsymbol{x}_a^{(t)}$, where $t \in \{1, \ldots, N\}$ indicates the time step and $a \in \{1, \ldots, M\}$ the atom. Associated with every time step t we assume a weight factor w_t and define $W_{kg} = \sum_{t \in S_{kg}} w_t$ and $W_k(t) = \sum_{t \in S_k(t)} w_t$. We also define the weighted barycentric coordinates

$$
\hat{\boldsymbol{x}}_a^{(t)} = \boldsymbol{x}_a^{(t)} - \frac{\sum_{k=1}^{L} (1 - \frac{w_t}{W_k(t)}) \sum_{b \in C_k} \boldsymbol{x}_b^{(t)}}{\sum_{k=1}^{L} (1 - \frac{w_t}{W_k(t)}) \cdot |C_k|}, \tag{5}
$$

where $|C_k|$ is the number of atoms in C_k.

The aligned coordinates are $\tilde{\boldsymbol{x}}_a^{(t)} = R^{(t)} \hat{\boldsymbol{x}}_a^{(t)} + \boldsymbol{q}^{(t)}$, with $R^{(t)}$ a rotation matrix and $\boldsymbol{q}^{(t)}$ a translation vector. The determination of $R^{(t)}$ and $\boldsymbol{q}^{(t)}$ for $t \in \{1, \ldots, N\}$ will be described in the following. We seek an alignment that minimizes

$$
V = \sum_{k=1}^{L} \sum_{a \in C_k} \sum_{g \in G_k} W_{kg} \sum_{\substack{t,s \in S_{kg} \\ t \neq s}} \frac{w_t w_s}{W_{kg}^2} (\tilde{\boldsymbol{x}}_a^{(t)} - \tilde{\boldsymbol{x}}_a^{(s)})^2. \tag{6}
$$

This is the sum of variances of all atoms of the molecule, but, for rigid groups of atoms where the trajectory decomposes into different sets of time steps ($G_k > 1$), the variances are computed per set (S_{kg}) and then summed up using the set weights W_{kg}. Solving $\nabla_{\boldsymbol{q}^{(r)}} V = 0$ for $r \in \{1, \ldots, N\}$ we get

$$
\boldsymbol{q}^{(r)} = \frac{\sum_{k=1}^{L} \sum_{a \in C_k} \sum_{\substack{t \in S_k(r) \\ t \neq r}} \frac{w_t}{W_k(r)} \tilde{\boldsymbol{x}}_a^{(t)}}{\sum_{k=1}^{L} \sum_{a \in C_k} \left(1 - \frac{w_r}{W_k(r)}\right)} \qquad (r \in \{1, \ldots, N\}) \tag{7}
$$

In order to determine $R^{(r)}$, we isolate those parts of V that depend on $R^{(r)}$:

$$-2w_r \sum_{k=1}^{L} \sum_{a \in C_k} \left[\sum_{\substack{t \in S_k(r) \\ t \neq r}} \frac{w_t}{W_k(r)} (\tilde{\boldsymbol{x}}_a^{(t)} - \boldsymbol{q}^{(r)}) \right] R^{(r)} \hat{\boldsymbol{x}}_a^{(r)}. \tag{8}$$

To minimize this term we have to perform a pairwise alignment of the barycentric coordinates of time step r to the barycentric coordinates of the mean of all other time steps that belong to the same alignment groups $S_k(r)$. On this basis, we perform the following algorithm:

1. Set $R^{(1)} = \mathbb{1}$ and $\boldsymbol{q}^{(1)} = 0$.
2. Initialize all $R^{(r)}$ and $\boldsymbol{q}^{(r)}$ for $r \in \{2, \ldots, N\}$ by aligning time step r to time step 1 with respect to C_1.
3. Loop over all time steps $r \in \{1, \ldots, N\}$ and recompute $\boldsymbol{q}^{(r)}$ using (7) and $R^{(r)}$ by minimizing Eq. (8), but assuming $L = 1$ in Eq. (7) and Eq. (8).
4. Repeat step 3, thereby continuously increasing the influence of atoms group C_k with $k > 1$, but keeping the maximal change of an atom position under a threshold by increasing the influence slow enough.
5. Stop the iteration when all atoms have full influence and the maximal change of an atom position drops under another, lower threshold.

The computation time of a single iteration depends linearly on the number of time steps and the number of atoms.

9   Results

To demonstrate the described algorithm we use a Hybrid Monte Carlo sampling of a pentane molecule with 15,000 samples and another sampling of trialanine with about 500,000 samples. The density of pentane describes nine shape clusters of which seven are metastable conformations. The relevant shape variations of pentane are described by two dihedral angles, each of which consists of four of the five carbon atoms. Both have three typical values 0.4π, π, and 1.4π which are found by fitting an HMM as described in Sect. 4. The analysis by HMMs took about 15 minutes. Hence, the first three atoms of the first dihedral angle can be considered to be a rigid structure. Fig. 1 shows a comparison between (1) an alignment that only superposes these three atoms and (2) our new approach that takes the three atoms as root structure and aligns the other two carbons with respect to their shape groups.

The first alignment looks only on three atoms and shifts most of the trajectory's shape variance to the disregarded rest of the molecule. Thus, it gives a clearly defined geometry for the aligned carbons, but blurs the rest of the molecule. Our new method results in more variance for the positions of the three base atoms, but delivers a much clearer image of the rest of the molecule. The structure of all the nine shape clusters is clearly visible. The computation of

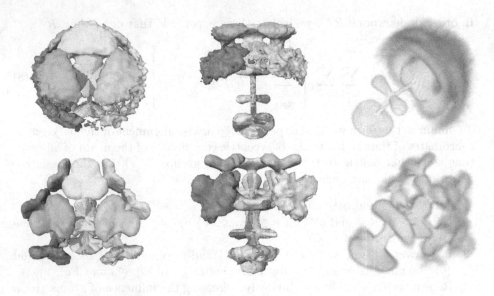

Fig. 1. Comparison of alignment strategies for a trajectory of the pentane molecule with 15,000 time steps that can be divided into 9 shape clusters. All images of a row depict the same 3D geometry from different viewing perspectives, while the two images in a column show approximately the same view on the geometries resulting from the two alignment strategies. In the left and middle column, the clusters are visualized by isosurfaces of their corresponding configuration densities. In the right column, the complete density is visualized by direct volume rendering. *Top row:* alignment by minimizing the positional variance of three carbon atoms. *Bottom row:* the new approach taking the same three carbons as the root structure which has constant shape over the whole trajectory. For the fourth carbon, three possible positions relative to the root structure are assumed and nine for the fifth atom.

the new alignment took about 15 seconds on a Pentium4 1.8 GHz Notebook. To visualize the aligned trajectories we accumulate the density over all geometries [12], where a geometry is the wireframe representation of an aligned time step. In accumulating the density, we count for every node of a uniform grid, how many geometries overlap with it. The densitiy is then visualized using isosurfaces or direct volume rendering (Fig. 1). The respective visualization for the bigger molecule trialanine is depicted in Fig. 2.

10   Conclusion and Future Work

We have presented an algorithm that divides the shape variations of a molecular dynamics trajectory into long-time changes which lead to substantially different shapes and thermal fluctuations around those shapes. Groups of dihedral

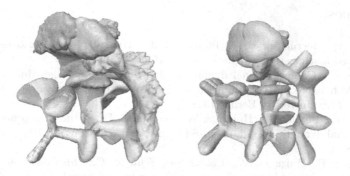

Fig. 2. Comparison of alignment strategies for a trajectory of trialanine with about 500,000 time steps. The 4 most important clusters are visualized by isosurfaces of their corresponding configuration densities. *Left image:* alignment by minimizing the positional variance of 5 selected atoms. *Right column:* the new approach taking the same 5 atoms as the root structure which has constant shape over the whole trajectory.

angles are analyzed by fitting Hidden Markov Models. In contrast to Perron cluster analysis based on uniform discretizations of dihedral angles, HMMs allow metastable clusters with fuzzy borders. We identified the fluctuation of the HMMs' distribution functions with thermal fluctuations of the molecule, while the state changes of multistate HMMs were interpreted as long-time changes. Thus, the combination of the different states of the multistate HMMs defines classes of different shapes. Using this decomposition we defined an alignment strategy that tries to mimimize the variance induced by thermal fluctuations. Thus, we got a clear depiction of the different shapes that the molecule takes on in its long-time changes. For the entire aligned trajectory, as well as the subtrajectories belonging to the long-time shapes, we accumulated configuration densities. These were visualized using isosurfaces and direct volume rendering. All the described techniques are integrated in the visualization system Amira [18]. Application to larger biochemically relevant molecules will be subject of further investigation. The algorithmic complexity of the analysis by HMMs depends linearly on the trajectory length and the number of atoms. The number of necessary Markov states increases with the number of atoms and quadratically increases the computational effort. While we succeeded in lower dimensions by using random start condition we expect this to be problematic in higher dimensions. Regarding the alignment, we are not expecting relevant problems, as it has a linear dependence on the problem size.

Acknowledgments

We would like to thank Frank Cordes for kindly providing the data for the example molecules as well as for detailed discussions and Peter Deuflhard for continuous support.

References

1. Mehlig, B., Heermann, D.W., Forrest, B.M.: Hybrid Monte Carlo method for condensed-matter systems. Phys. Rev. B **45** (1992) 679–685
2. Schütte, C.: Conformational dynamics: Modelling, theory, algorithm, and application fo biomolecules. Habilitation Thesis, Dept. of Mathematics and Computer Science, Free University Berlin (1998)
3. Schütte, C., Fischer, A., Huisinga, W., Deuflhard, P.: A direct approach to conformational dynamics based on hybrid Monte Carlo. J. Comput. Phys. **151** (1999) 146–168
4. Deuflhard, P., Huisinga, W., Fischer, A., Schütte, C.: Identification of almost invariant aggregates in reversible nearly uncoupled markov chains. Lin. Alg. Appl. **315** (2000) 39–59
5. Deuflhard, P., Weber, M.: Robust Perron cluster analysis of conformation dynamics. Lin. Alg. Appl. (Special Issue on Matrices and Mathematical Biology) **398 C** (2005) 161–184
6. Fischer, A., Waldhausen, S., Schütte, C.: Identification of biomolecular conformations from incomplete torsion angle observations by hidden markov models. Preprint, FU Berlin, Dept. of Mathematics and Computer Science (2004)
7. Kabsch, W.: A solution for the best rotation to relate two sets of vectors. Acta Crystallographica A **32** (1976) 922–923
8. Kabsch, W.: A discussion of the solution for the best rotation to relate two sets of vectors. Acta Crystallographica A **34** (1978) 827–828
9. Pennec, X.: Multiple registration and mean rigid shape - Application to the 3D case. In Mardia, K., Gill, C., I.L., D., eds.: Image Fusion and Shape Variability Techniques (16th Leeds Annual Statistical Workshop), University of Leeds, UK (1996) 178–185
10. Huitema, H., van Liere, R.: Interactive visualization of protein dynamics. In: Proceedings of IEEE Vis2000. (2000) 465–468
11. Amadei, A., Linssen, A.B.M., Berendsen, H.J.C.: Essential dynamics of proteins. Proteins: Structure, Function and Genomics **17** (1993) 412–425
12. Schmidt-Ehrenberg, J., Baum, D., Hege, H.C.: Visualizing dynamic molecular conformations. In: Proceedings of IEEE Visualization 2002. (2002) 235–242
13. Rabiner, L.R.: A tutorial on hidden markov models and selected applications in speech recognition. Proc. of IEEE **77** (1989) 257–286
14. Dempster, A.P., Laird, N.M., Rubin, D.B.: Maximum likelihood from incomplete data via the em algorithm. J. Roy. Stat. Soc. **39** (1977) 1–38
15. Deuflhard, P., Schütte, C.: Molecular conformation dynamics and computational drug design. In Hill, J.M., Moore, R., eds.: Applied Mathematics Entering the 21st Century. Proc. ICIAM 2003, Sydney, Australia (2004) 91–119
16. Weber, M.: Clustering by using a simplex structure. Report 04-03, Zuse Institute Berlin (2004)
17. Press, W.H., Teukolsky, S.A., Vetterling, W.T., Flannery, B.P.: 14. In: Numerical Recipes in C - Second Edition. Cambridge University Press (1992) 632–636
18. Stalling, D., Westerhoff, M., Hege, H.C.: Amira: A highly interactive system for visual data analysis. In Hansen, C.D., Johnson, C.R., eds.: The Visualization Handbook. Elsevier (2005) 749–767

Distributed BLAST in a Grid Computing Context

Micha M. Bayer and Richard Sinnott

National e-Science Centre, e-Science Hub, Kelvin Building,
University of Glasgow, Glasgow G12 8QQ
{michab, ros}@dcs.gla.ac.uk

Abstract. The Basic Local Alignment Search Tool (BLAST) is one of
the best known sequence comparison programs available in bioinformatics. It is used to compare query sequences to a set of target sequences,
with the intention of finding similar sequences in the target set. Here, we
present a distributed BLAST service which operates over a set of heterogeneous Grid resources and is made available through a Globus toolkit
v.3 Grid service. This work has been carried out in the context of the
BRIDGES project, a UK e-Science project aimed at providing a Grid
based environment for biomedical research. Input consisting of multiple
query sequences is partitioned into sub-jobs on the basis of the number of
idle compute nodes available and then processed on these in batches. To
achieve this, we have implemented our own Java-based scheduler which
distributes sub-jobs across an array of resources utilizing a variety of
local job scheduling systems.

1 The BRIDGES Project

The BRIDGES project (Biomedical Research Informatics Delivered by Grid En-
abled Services [1]) is a core project of the UK's e-Science Programme [28] and is
aimed at developing Grid-enabled bioinformatics tools to support biomedical research. Its primary source of use cases is the Cardiovascular Functional Genomics
Project (CFG) [30], a large collaborative study into the genetics of hypertension
(high blood pressure). Hypertension is partly genetically determined, and investigation of the genes and biological mechanisms responsible for high blood pressure is of great importance. BRIDGES aims to aid and accelerate such research
by applying Grid-based technology. This includes security focused data access
and integration tools [2] but also Grid support for compute intensive bioinformatics applications such as BLAST. The service described in this paper is based
primarily on the use case of microarray chip annotation, in which microarray
reporter sequences have to be compared against annotated sequence data (e.g.
from the human genome). These are highly compute-intensive tasks, involving
several hundred thousand input sequences and very large target databases, and
may take of the order of several weeks to compute on a single processor machine. An additional use case are BLAST runs of small batches of several tens
of sequences against standard databases such as the NCBI *nt* database [5].

M.R. Berthold et al. (Eds.): CompLife 2005, LNBI 3695, pp. 241–252, 2005.
© Springer-Verlag Berlin Heidelberg 2005

2   Parallelising BLAST

BLAST - the Basic Local Alignment Search Tool [3] is a widely used search algorithm that is used to compute alignments of nucleic acid or protein sequences with the goal of finding the n closest matches in a target data set. BLAST takes a heuristic (rule-of-thumb) approach to a computationally highly intensive problem and is one of the fastest sequence comparison algorithms available, yet it still requires significant computational resources. It does therefore benefit greatly from being run in a Grid computing context providing it is parallelised, i.e. single jobs must be partitioned into independent sub-jobs that can be run on remote resources concurrently. In the case of BLAST (and possibly similar sequence comparison programs) parallelisation can be achieved at three different levels [6,7]:

1. A single query can be compared against a single target sequence using several threads in parallel, since there are $O(nm)$ possible alignments for a query sequence of length n and a target sequence of length m. This approach is implemented by default by the BLAST executable itself.
2. Input files with multiple query sequences can be parsed to provide individual query sequences or blocks of sequences, and these can then all be compared against multiple identical instances of the target data file in parallel.
3. With input files containing only a single query sequence, the target data can be segmented into n copies for n available compute nodes, and then multiple identical instances of the query sequence can each be compared to a different piece of the target data in parallel.

There are several existing implementations which take approach 2 [7,8] or 3 [9,10,11] but none of these suited the particular requirements in this project. One of the better known implementations of approach 3 is mpiBLAST [10,16], an MPI-based implementation of BLAST in which the target database is segmented into a number of fragments that the input is then compared against concurrently. It was initially considered to include mpiBLAST on our back end resources so that single query input could benefit from parallelisation but after preliminary tests it was excluded from the design for the following reasons:

- The speedup described by the authors of mpiBLAST [16] appears to be closely linked to the particular set of conditions their tests were run under and could not be reproduced under the conditions on our clusters. Instead, execution times actually increased when more than 10 processors were used in the trial runs.
- mpiBLAST requires the target database to be segmented and the fragments to be formatted ahead of the actual BLAST run. This is a compute-intensive task which takes approximately 30 minutes for the NCBIs nt database [5], a standard nucleotide database widely used for BLAST searches. This is a significant overhead which is unsuitable for single, short jobs that may take only minutes to compute. The alternative of having a predetermined number of preformatted database fragments ready for computation on the cluster is unworkable because of the requirements of mpiBLAST itself.

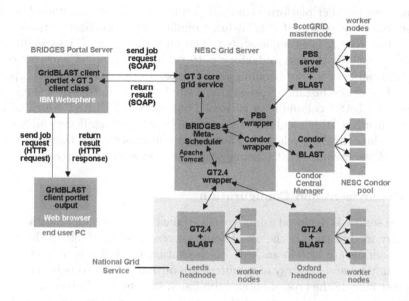

Fig. 1. Schematic of system architecture

- The software requires $n + 2$ processors to be available if n database fragments are to be used, and in practice this means that jobs will have to be queued until n processors are available, leading to very significant delays which usually outweigh any potential performance gains.

The final design therefore only included parallelization of BLAST at the level of the input data.

3   BRIDGES GT3 BLAST Service

There is are a growing number of Grid based implementations of BLAST [12,13,14,15], based on various Grid middleware but we here present the first all-Java, Open Grid Service Architecture (OGSA)-based [22] implementation, based on version 3 of the Globus toolkit. Globus v.3 (GT3) is based on the web services programming model but services have been extended to operate in a Grid computing context, with service data, client notification mechanisms and statefulness as added functionality. Our GT3 based implementation of BLAST is used in conjunction with our own meta-scheduler that allows jobs to be farmed out to remote clusters and the results to be combined.

3.1   Basic Design

A GT3 core based Grid service is used as a thin middleware layer on top of our application (Figure 1). The service itself has deliberately been kept basic to allow easy porting to other platforms since Grid computing is still very much in flux. It therefore only uses limited GT3 functionality, with no client notification or service data having been implemented. The only significant GT3-specific functionality needed is the service factory pattern. This is a useful feature because it provides an easy way of handling concurrency in the context of this application, with a service instance representing a single job submitted by a single user. The actual BLAST computation is carried out by NCBI BLAST which is freely available from NCBI [5], and BLAST executables are preinstalled on all compute resources since these are relatively unchanging components and therefore regular stage-in of the executable at runtime would be a waste of resources.

3.2   Deployment and Configuration

The Grid service is deployed using Apache Tomcat running on a Linux server. To allow easy modification of the set of compute resources available to the service, resource details are held in an XML configuration file which is read by the service at runtime and an array of resources are initialised accordingly. The details of interest include the type of batch submission system, the domain name of the resource's head node, the number of compute nodes, memory etc. There is currently no provision of dynamic resource discovery. The system has been designed to be easily extendible however and simply requires the XML configuration file to be extended with new resource details.

3.3   User Interface to Service

A web portal has been designed for the BRIDGES project to allow users easy web based access to resources and in order to avoid having to install Grid software on end user machines. This has been implemented using IBM Websphere, currently one of the more sophisticated portal packages. Users log in with standard username and password pairs and are then presented with a job submission portlet.

3.4   Scheduler and Input Data Segmentation

For the purpose of this project we decided to provide our own scheduler since at the time of designing the application Grid meta-schedulers were rare and none were available that satisfied our requirements with respect to the particular combination of OS and batch submission system on our resources (n.b. "resource" here denotes a computational back end such as a compute cluster).

Our scheduler can take an unlimited number of resources as an argument and uses the following algorithm to distribute subjobs across these:

```
parse input and count no. of query sequences
poll resources and establish total no. of idle nodes
set no. of sub-jobs to be run to equal no. of idle nodes
calculate no. of sequences n to be run per sub-job (= no. of
  idle nodes/no. of sequences)
```

```
while there are sequences left
        save n sequences to a sub-job input file
if the number of idle nodes on the whole Grid is 0
        make up small, predetermined number of sub-jobs and
            evenly distribute these into queues across resources
else
    for each resource
                send i subjobs to the resource(i = no. of idle
                nodes at the resource)
when all subjob results are complete, combine them into a single
  result file using the original input sequence order
return combined output file to the user
```

Thus, the system will always make use of the maximum number of idle nodes across resources if multiple query sequences exist. Load balancing is achieved by assigning only as many sub-jobs to a resource as there are idle nodes, and by making all input files roughly the same size.

3.5 Compute Resources and Wrapper Classes

eGrid computing environments usually feature a heterogeneous mixture of back end resources [23] and their associated operating systems, and the resources available to this project are typical in this respect. The requirement arising from this is that the Grid meta-scheduler must be capable of submitting jobs to the

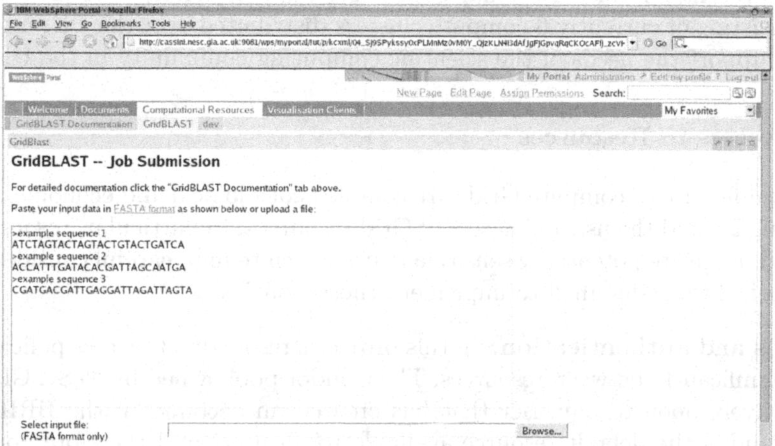

Fig. 2. BRIDGES Web portal showing the job submission portlet for the distributed BLAST service

resources via a number of wrapper components that feature shared functionality. The minimum functionality that a basic job submission system requires consists of 3 elements: job submission, job monitoring and job cancellation.

Our design satisfied these requirements through the use of an abstract *Resource* class which was then extended with wrappers that provided the above functionality for each type of back end batch system. We implemented Java wrappers for the Condor [19] and PBS [18] batch systems as well as for Globus 2 server side installations [24]. The *Resource* class contains abstract methods for job submission, job monitoring and cancellation, with the wrappers providing concrete implementations of these for the specific back end systems. In the case of the PBS and Condor batch systems we used the respective client packages provided with PBS and Condor and wrapped the relevant commands there as native processes in Java, using the java.lang.Runtime and java.lang.Process classes. In the case of the wrapper for Globus 2, we used the Java Commodity Grid Kit version 1.1 [25], which provides a convenient and easy-to-use high-level API for job submission to GT2 resources.

The actual resources available to the project included:

- Our local Condor pool at the National e-Science Centre Glasgow, a small cluster of 21 single processor desktop machines (Intel Celeron processor, 512 mb RAM)
- ScotGRID [17], a 250 processor compute cluster located at Glasgow University comprising IBM xSeries nodes (X330, X335, X340, X370) and Dell Poweredge 2650 machines with 15TB disk storage
- The recently formed National Grid Service (NGS) of the UK [26], a Grid consisting of currently 6 compute clusters distributed over the UK, intended to support the needs of the scientific computing community in the UK

3.6   Access to Resources

A typical feature of compute Grids are complex sociological and economic issues revolving around the use and access of Grid resources. In particular, resource access and usage are potentially sensitive issues which require careful consideration and a set of tools designed to implement those policies.

Policies and authentication. In this project's resource set, access policies differed significantly between resources. The Condor pool owned by NeSC Glasgow is effectively open to any user that has created an account on the BRIDGES portal and is the default resource available to all users and therefore does not require any authentication mechanism. There are, however, restrictions on the kind of jobs that can be executed there, which are enforced through providing application-specific rather than generic user interfaces (see section 3.3 on user front ends above). ScotGRID requires user registration through the support staff, and users then log on to the service using their ssh credentials. The UK National Grid Service allows any person in academia access after having registered with a local representative of the certification authority. The identity of the user is

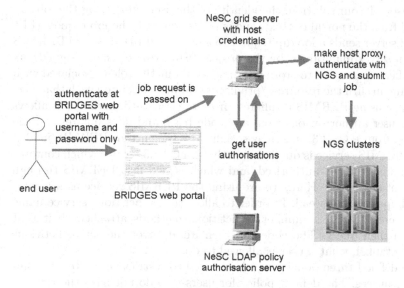

NeSC grid server
with host
credentials

make host proxy,
authenticate with
NGS and submit
job

authenticate at
BRIDGES web
portal with
username and
password only

job request is
passed on

get user
authorisations

NGS clusters

end user

BRIDGES web portal

NeSC LDAP policy
authorisation server

Fig. 3. BRIDGES security infrastructure for job submission onto the NGS nodes

backed up by a digital certificate and private key issued by the certification authority, and access to the resources is permitted through a user proxy that has been created from the users credentials.

One of the project requirements was that user authentication should not cause any additional learning or usability overheads for the users. Biology end users range widely in computer literacy and therefore systems providing a single mechanism for users of all abilities should aim at the lowest level of literacy. The process of obtaining and caring for digital certificates for the National Grid Service is currently still rather involved and requires familiarity with command line tools and a generally advanced level of computer literacy. It was therefore decided to remove digital certificates from the end user environment altogether and replace them with simple username and password authentication at a central project web portal (see section 3.6 below). GSI authentication at NGS resources is instead being carried out by means of a host proxy generated from the Grid servers host credentials (Fig. 3). The host's identity is then mapped locally to a project account in the local gridmap file. Thus, all jobs run under the project's identity on the NGS resources, and the logging and monitoring of user activity are the responsibility of the BRIDGES support staff.

Authorisation. Once a user is logged on, they have access to the complete set of tools available on the project portal. The finer grain control of what back end resources associated with a tool are accessible for a given user is implemented through the Grid authorisation software PERMIS [27]. PERMIS provides an enhanced Grid authorization infrastructure which allows to define policies about what resources users are allowed to access and subsequently use. This extends most Grid based security infrastructures which focus predominantly on authentication of users. In our solution, the identity of the user submitting the job can be extracted from the portal context, and is passed on with the job request (Fig. 3). The Grid server sends a lookup request to a dedicated LDAP based PERMIS authorisation server maintained by the project team, where secure (signed) attribute certificates are used to store information about the roles associated with given users and in turn the resources/privileges that are associated with that role.

This way of using PERMIS is different from the standard method but allows us to obtain user authorisations from any code base, and does not rely on the query coming from a GT3.3 service as specified by the current PERMIS model. There, the PERMIS service is queried directly by a GT3.3 service which contains the actual methods to be authorised, and which is tied into PERMIS through its deployment descriptor. Here, we are using the PERMIS service as a loosely coupled lookup service instead. In order to allow this, a fictitious service name needs to be created and a number of fictitious methods attached to it that PERMIS can then query. The code can then iterate over the set of fictitious methods to establish what a user is allowed to do.

We have defined three policies which are used to restrict access to the computational resources. The default policy for users who do not have the right to access ScotGrid or NGS resources, is that they are only able to access the NeSC Glasgow Condor pool. A policy is defined for users with a ScotGrid account which provides those users access to the ScotGrid and additionally the Condor pool. A third policy has also been defined which extends the authorized resources to include the National Grid Service.

It is important to note that the checks on user access to these resources are completely transparent to users. They simply select the target database and input sequences used in the BLAST search, and the selection of resources is done automatically.

3.7   Target Databases

Grid based file transfer is potentially time-consuming and the favourable option is to keep frequently used data cached locally, close to the computation. In the case of BLAST this is readily achievable since most users blast against a small number of publicly available target databases (available as flat text files for ftp download, e.g. from NCBI [5] or EBI [4]). Typically, the databases are in the order of gigabytes, with some of them growing exponentially [21]. The data can be stored on the local NFS of the compute resource and updated regularly through automated scripts.

4   Performance Evaluation

One of the main objectives of creating this service was, along with the ability to execute batch BLAST jobs remotely, to improve BLAST execution times for large jobs. BLAST is computationally costly and large batch jobs such as those supported in this project can take of the order of weeks to compute. To evaluate the speedup that can be obtained we conducted a simple experiment which compared the execution times of different sized jobs between a single CPU and our compute Grid.

4.1 Experimental Setup

We randomly selected a set of 100 nucleotide sequences from the mouse EST database from NCBI [5]. From this, three input files containing 1, 10 and 100 sequences were generated. These files were run against the *nt* nucleotide database, also obtainable from NCBI. This is a large database (currently approx. 12 gb uncompressed) that most researchers compare their sequences against, and contains entries from all traditional divisions of GenBank, EMBL and DDBJ. The BLAST options used were as follows:

-e 0.01 (set e-value to 0.01)
-w 7 (set word size to 7)
-b 0 (suppress alignments)
-T (HTML output)
-p blastn (use blastn program)

Each run was repeated three times on each a single CPU machine and on the BRIDGES compute grid. The single CPU machine has a single Intel Celeron 2.20GHz processor and 512 mb of RAM, and runs Fedora core 2. The resources utilised on the compute grid included:

- the NeSC Condor pool (see section 3.5 above) – specs identical to those of the single CPU machine
- ScotGRID (see section 3.5 above) – mixed specs, see http://www.scotgrid.ac.uk/equipment/ for details
- the NGS (see section 3.5 above) compute nodes at Oxford and Leeds – worker node specs dual 3.06GHz Intel Xeon CPUs, 2 gb memory

For the grid jobs, it was ensured prior to the run that at least the same number of processors were available on the grid as the number of input sequences, which results in the maximum number of subjobs (1, 10 and 100 respectively) as input sequences and therefore maximum performance. If fewer processors are available than input sequences, the scheduler will package several input sequences into each input file and execution time will then increase.

4.2 Results

The results of the benchmarking runs are shown in Fig. 4. For the single sequence job, there was no difference in execution times between repeat runs or between Grid and single CPU – all of these took 6 minutes to execute each. Repeat runs on the single CPU machine consistently produced the same execution time across all three sizes of job, with jobs taking 6, 55 and 668 minutes respectively

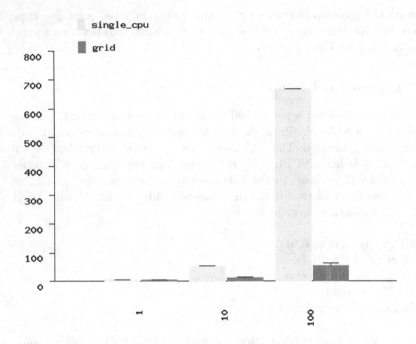

Fig. 4. Results from benchmarking experiments. Figure shows BLAST execution times in minutes, plotted against the number of sequences in the input file. The left column of each pair represents the result obtained with the single CPU machine, the right hand column that of the Grid job. Error bars = +1 SD

for the 1, 10 and 100 input sequence files. For the Grid jobs, there was some variation in execution times, with values ranging from 10 to 14 minutes and 47 to 66 minutes for the 10 and 100 sequence jobs respectively.

4.3   Discussion

The results clearly show a marked performance increase when jobs were sched-uled on the Grid, with the speedup factor (= single CPU execution time divided by Grid execution time) ranging from 3.9 to 5.5 and 10.1 to 14.2 for the 10 and 100 sequence jobs respectively. However, in theory the values for the speedup should be much closer to the number of input sequences used since each subjob only contained a single sequence and all subjobs were intended to be processed concurrently. In practice, however, the job submission policies implemented at the back end compute clusters degrade performance very significantly. On most resources policies are in place which limit the number of jobs that a single user can run at the same time, and the resources used here are no exception. This leads to a loss of concurrency, despite the apparent availability of free processors, with jobs starting at time offsets and thereby increasing overall execution time.

In addition to this there are time penalties incurred by the process of instan-tiating a service object (approx. 15 s with the Globus toolkit v.3), uploading and parsing the input, configuring and scheduling the job, staging input files in to the resources and combining the result at the end of the overall job. For the most minimal of jobs (running a synthetic sequence of 10 bp against a small bacterial genome such as that of *E. coli*) this amounts to a total of approx. 40s of overheads, and this amount increases proportionally with both the size of the input and the size of the target database. However, as job size increases the proportion of the overall execution time taken up by these overheads diminishes and becomes increasingly insignificant.

The major loss of performance observed here exemplifies one of the major challenges in Grid computing, which is of a logistic and sociological nature rather than technical. In devising scheduling policies, a trade-off must be struck between availability of resources to all users, and performance for the individual user. If individual users are allowed to submit unlimited numbers of jobs to a resource this may lead to the resource becoming unavailable to all other users, potentially for an extended period of time. Conversely, restricting resource uses for the individual user can be a frustrating experience if no other jobs are intended to be submitted any time soon by others but submission policies enforce a rigid load limit nevertheless.

5   Future Work

Grid computing is still in its early stages and the middleware toolkits available to developers are still undergoing major architectural changes [20]. Since the beginning of this project, the Globus Alliance has released several minor versions of GT3 and now a new major version, GT4 [29]. The current service should now be ported to GT4 to avoid problems regarding support etc. GT4 is a move towards greater homogeneity in the field of web/Grid services and has the support of several major industry partners. We plan to explore these solutions later within the recently funded Scottish Bioinformatics Research Network [31] and in the on-going efforts to establish a Scottish Grid Service which we expect to become another full node of the UK National Grid Service in 2006.

References

1. BRIDGES project. http://www.brc.dcs.gla.ac.uk/projects/bridges/
2. Sinnott, R., Atkinson, M., Bayer, M., Berry, D., Dominiczak, A., Ferrier, M., Gilbert, D., Hanlon, N., Houghton, D., Hunt, E., White, D.: Grid Services Supporting the Usage of Secure Federated, Distributed Biomedical Data. Proceedings of the UK e-Science All Hands Meeting, Nottingham, UK (2004)
3. Altschul, S.F., Gish, W., Miller, W., Myers, E.W., Lipman, D.J.: Basic Local Alignment Search Tool. J. Mol. Biol. **215** (1990) 403-410
4. EBI BLAST. http://www.ebi.ac.uk/blastall/index.html
5. NCBI BLAST website: http://www.ncbi.nlm.nih.gov/BLAST/
6. Pedretti, K.T., Casavant, T.L., Braun, R.C., Scheetz, T.E., Birkett, C.L., Roberts, C.A.: Three Complementary Approaches to Parallelization of Local BLAST Service on Workstation Clusters (invited paper). Proceedings of the 5th International Conference on Parallel Computing Technologies (1999) 271-282

7. Braun, R.C., Pedretti, K.T., Casavant, T.L., Scheetz, T.E., Birkett, C.L., Roberts, C.A.: Parallelization of local BLAST service on workstation clusters. Future Generation Computer Systems **17** (2001) 745-754
8. Clifford, R., Mackey, A.J.: Disperse: a simple and efficient approach to parallel database searching. Bioinformatics **16** (2000) 564-565
9. Mathog, D.R.: Parallel BLAST on split databases. Bioinformatics **19** (2003) 1865-1866
10. Darling, A.E., Carey, L., Feng, W.: The design, implementation and evaluation of mpiBLAST. Proceedings of ClusterWorld Conference, Expo and the 4th International Conference on Linux Clusters: The HPC Revolution (2003)
11. Hokamp, K., Shields, D.C., Wolfe, K.H., Caffrey, D.R.: 2003. Wrapping up BLAST and other applications for use on Unix clusters. Bioinformatics **19** (2003) 441-442
12. GridBlast at Keck BioCenter, University of Wisconsin. http://bioinf.ncsa.uiuc.edu/
13. GridBlast at A-Star Bioinformatics Institute Singapore: http://www.bii.astar.edu.sg/infoscience/dcg/gridGridblast/index.asp
14. North Carolina BioGrid. http://www.ncbiogrid.org/tech/apps.html
15. RIKEN GridBlast. http://big.gsc.riken.jp/big/Members/fumikazu/Activity_Item.2004-02-02.0425
16. mpiBLAST. http://mpiblast.lanl.gov/
17. ScotGRID. http://www.scotgrid.ac.uk/
18. OpenPBS. http://www.openpbs.org/
19. Condor. http://www.cs.wisc.edu/condor/
20. Globus WSRF. http://www.globus.org/wsrf/default.asp
21. GenBank statistics. http://www.ncbi.nih.gov/Genbank/genbankstats.html
22. OGSA Open Grid Services Architecture. http://www.gridforum.org/documents/GWD-I-E/GFD-I.030.pdf
23. Foster, I. , Kesselman, C., Tuecke, S.: The Anatomy of the Grid: Enabling Scalable Virtual Organizations. International J. Supercomputer Applications, **15(3)** (2001)
24. Globus toolkit version 2.4.3. http://www-fp.globus.org/gt2.4/
25. Java Cog kit version 1.1. http://www-unix.globus.org/cog/java/1.1/
26. UK National Grid Service. http://www.ngs.ac.uk/
27. PERMIS Grid authorisation software. http://www.permis.org
28. The UK e-Science Programme. http://www.rcuk.ac.uk/escience/
29. Globus toolkit. http://www.globus.org/toolkit/
30. Cardiovascular Functional Genomics project, http://www.brc.dcs.gla.ac.uk/projects/cfg
31. Scottish Bioinformatics Research Network project. http://www.nesc.ac.uk/hub/projects/sbrn

Parallel Tuning of Support Vector Machine Learning Parameters for Large and Unbalanced Data Sets

Tatjana Eitrich[1] and Bruno Lang[2]

[1] John von Neumann Institute for Computing,
Central Institute for Applied Mathematics,
Research Centre Juelich, Germany
t.eitrich@fz-juelich.de
[2] Applied Computer Science and Scientific Computing Group,
Department of Mathematics,
University of Wuppertal, Germany
Bruno.Lang@math.uni-wuppertal.de

Abstract. We consider the problem of selecting and tuning learning parameters of support vector machines, especially for the classification of large and unbalanced data sets. We show why and how simple models with few parameters should be refined and propose an automated approach for tuning the increased number of parameters in the extended model. Based on a sensitive quality measure we analyze correlations between the number of parameters, the learning cost and the performance of the trained SVM in classifying independent test data. In addition we study the influence of the quality measure on the classification performance and compare the behavior of serial and asynchronous parallel parameter tuning on an IBM p690 cluster.

1 Introduction

Support vector machines (SVMs) are one of the well accepted machine learning methods [1]. Numerous experiments have confirmed that the linear learning approach in combination with problem adapted implicit feature mappings leads to highly reliable nonlinear classification functions. Much work has been done to make SVM algorithms run very fast [2].

In recent years, however, a significant number of nontrivial problems has surfaced in the context of SVMs. The size of the classification problems increases rapidly, while at the same time better results are desired. The quality issue is particularly important if the data sets are unbalanced, which means that either the number of positive and negative data differ significantly, or the cost of a false positive classification differs significantly from the cost of a false negative, or both. Often, the differing costs of misclassifications have been neglected, and the success of a particular approach has been measured only by totaling the number of incorrectly classified test points.

M.R. Berthold et al. (Eds.): CompLife 2005, LNBI 3695, pp. 253–264, 2005.

Support vector machine classification involves a learning phase, in which the training data are used to adjust the classification parameters. This procedure, which can be formulated as a quadratic programming problem, is controlled by a—typically very small—set of learning parameters. Usually these have to be set by the user. Quoting [3], *"There is a lot of papers published about the SVM algorithms and kernel methods, but very few of them address the parameters tuning to get the high quality results usually presented [...] these results are difficult to reproduce because of the influence of the parameter settings."* In addition, it is often not clear which quality measure had been used.

In this paper we address the classification of *large unbalanced* data sets with SVMs, taking the differing costs of misclassifications into account. Large data sets feature two properties that are important for the training. On the one hand, they allow considering models with a higher number of learning parameters, so that nonlinearity can be captured more precisely than with models involving only a few parameters. (For smaller data sets, the number of parameters is limited by overfitting effects.) Finding appropriate values for many parameters, however, can no longer be done by hand or simple grid search, but must be automated. To this end we embed the learning in a numerical optimizer, which repeatedly trains an SVM with different settings of the learning parameters and strives to find parameters that optimize a suitable quality measure; see Section 3 for details. The overall procedure for adjusting the *learning parameters and classification parameters* is summarized in Figure 1. Note that evaluating the quality measure involves validating the SVM on data different from the training data.

The negative effect of large data sets is the high computational complexity. To reduce the overall learning time, each SVM training is done with a highly efficient quadratic program solver, and a parallelized optimizer is used for tuning the learning parameters; see Sections 2 and 4. In Section 5 numerical experiments with a large, hard classification problem will show that our automated approach is able to yield good results in a reliable manner. This is important for users from other fields because our method requires no human interaction and no familiarity with the underlying SVM theory to tune the SVM to a particular classification problem.

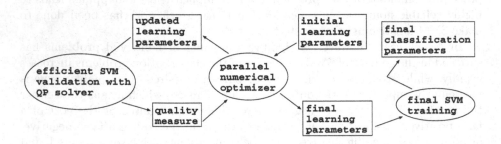

Fig. 1. Main components of the SVM system

2   Support Vector Learning

The task of support vector learning is to determine functions that can be used to classify data points. In this paper we consider only binary problems and leave out multi-class learning and regression. In the binary case support vector learning is the process of using so-called reference data of given input–output pairs $\{(\boldsymbol{x}_i, y_i) \in \mathbb{R}^n \times \{-1, 1\},\ i = 1, \ldots, l\}$ to find an optimal separating hyperplane $\boldsymbol{w}^T \boldsymbol{x} + b = 0$. Using assumptions of statistical learning theory the desired classifier is then defined as $h(\boldsymbol{x}) = \text{sgn}(f(\boldsymbol{x}))$ with the linear decision function $f(\boldsymbol{x}) = \boldsymbol{w}^T \boldsymbol{x} + b$; see [4,5] for details.

If the data are not linearly separable then a *kernel* $K : \mathbb{R}^n \times \mathbb{R}^n \to \mathbb{R}$ is used to learn a nonlinear decision function

$$f^*_{\text{nonlin}}(\boldsymbol{x}) = \sum_{i:0<\alpha_i} y_i \alpha_i^* K(\boldsymbol{x}_i, \boldsymbol{x}) + b^*.$$

Here the classification parameters α_i^* and b^* are given by the unique global solution of a suitable (dual) quadratic optimization problem [5]

$$\min_{\boldsymbol{\alpha} \in \mathbb{R}^l}\ g(\boldsymbol{\alpha}) := \frac{1}{2} \boldsymbol{\alpha}^T H \boldsymbol{\alpha} - \sum_{i=1}^{l} \alpha_i \tag{1}$$

with $H \in \mathbb{R}^{l \times l}$, $H_{ij} = y_i K(\boldsymbol{x}_i, \boldsymbol{x}_j) y_j$ ($1 \le i, j \le l$), constrained to

$$\boldsymbol{\alpha}^T \boldsymbol{y} = 0, \quad \boldsymbol{0} \le \boldsymbol{\alpha} \le C. \tag{2}$$

The kernel function K must be provided by the user.

Note that the the Hessian H is usually dense, and therefore the complexity of evaluating the objective function g in (1) scales quadratically with the number l of training pairs, leading to very time-consuming computations. A well-known method for the solution of such problems is the decomposition algorithm [6] that repeatedly selects a subset of the free variables and optimizes (1) over these variables. Its main advantage is the flexibility concerning the size of the subproblems. Decomposition provides a framework for handling large SVM training tasks but it does not define how to solve the reduced quadratic programming problems. To obtain good overall times it is necessary to have efficient QP solvers for the subproblems. We use our own implementation of the projection method described in [7]. This method is suitable for large data sets. It defines problems with diagonal matrices and solves them iteratively with a fast inner solver [8]. Thus a single optimization step of the decomposition method becomes very fast.

3   Learning Parameters and Quality Management

The constraints (2) involve learning parameters C_i, $i = 1, \ldots, l$, which have to be chosen before SVM training. Often a single value $C_i \equiv C$ is used for simplicity. In [9] the authors gave evidence that for unbalanced data sets at least two values should be used: $C_i = C^+$ if the ith training point is positive ($y_i = +1$), and $C_i = C^-$ otherwise ($y_i = -1$). In addition to correcting different sizes of the two classes, the (C^+, C^-) model can also capture different costs of false positive and false negative classifications. Since the data set treated in Section 5 is even more unbalanced than the example in [9], with a very small number of positive points, we again used the (C^+, C^-) model.

In addition to this weighting approach we consider generalizing the kernel function. One of the most commonly used functions is the Gaussian kernel,

$$K^G(\boldsymbol{x}, \boldsymbol{z}) = \exp\left(-\frac{\sum_{k=1}^{n}(x_k - z_k)^2}{2\sigma^2}\right) \quad (\boldsymbol{x}, \boldsymbol{z} \in \mathbb{R}^n), \tag{3}$$

where the standard deviation $\sigma > 0$ is chosen identically for all features of the data. The reason is again that hand tuning of the learning parameters requires their number to be very small.

If the learning parameters can be adjusted automatically then their number can be increased, and in the extreme case we may assign a different standard deviation to each feature [9]:

$$K^G(\boldsymbol{x}, \boldsymbol{z}) = \exp\left(-\sum_{k=1}^{n} \frac{(x_k - z_k)^2}{2\sigma_k^2}\right). \tag{4}$$

As a reasonable compromise between (3) and (4), one might divide the features into different groups (such as "binary" and "continuous") and assign one σ value to each group.

Comparing SVMs trained with these extended models to SVMs trained with the usual uniform approach confirms that the added complexity indeed leads to better classification results. Interestingly, allowing different σ values for the features can also yield additional information. Based on the optimized value σ_k one can estimate the relevance of the corresponding feature k, and thus one gets an implicit *feature selection* mechanism for free. To our knowledge, however, the option of tuning different σ values in the context of support vector learning has not been considered elsewhere.

We also work on other generalized kernels and on other parameters that are relevant for the training phase. For example, the decomposition method and the QP solver use several internal parameters, which may be tuned to enhance the performance [6,7]. Both issues cannot be discussed here due to space limitations.

The tuning of the learning parameters can be implemented by optimizing a certain quality measure, which is obtained in validation steps. Optimizing a nontrivial parameter model is almost impossible if a discrete quality measure is used, e.g., the number of validation errors. Following the ideas in [9] we use the continuous effectiveness measure

$$E_\beta = 1 - \frac{(\beta^2 + 1)\mathrm{pr} \cdot \mathrm{se}}{\beta^2 \cdot \mathrm{pr} + \mathrm{se}} \in [0, 1], \tag{5}$$

which we have to *minimize*. The sensitivity se (which percentage of the positive data have been recognized ?) and the precision pr (which percentage of the points that have been classified "positive" are indeed positive ?) are computed with a special smooth error measure. The quantity β can be used to enforce or diminish the influence of sensitivity. In Section 5 we will present results achieved with (5) for different values of β and discuss the problem of defining the quality measure.

4   Automatic Parallel Parameter Optimization

Tuning a nontrivial number of parameters can be very time consuming, and therefore it is reasonable to use parallel computing resources. There are three ways to insert parallelism during the SVM model selection stages: parallelizing the training of a single SVM, training several SVMs in parallel, and using a parallel algorithm for parameter optimization. Concerning the first option, promising parallelization techniques for decomposition methods exist [10], whereas parallel SMO [2] methods are currently investigated, but seem not to be reliable yet. The second option has also been addressed with mostly straight-forward approaches, e.g., parallel mixture of SVMs [11], parallel training of binary SVMs for multiclass problems [12], and parallel cross validation models [13]; see [3] for a short overview.

Concerning parallel parameter optimization, ongoing work is on parallel grid search techniques [14]. Grid search uses a predefined set of values for each parameter and determines which combination of these values yields the best results. Thus parallel grid search is an easy and perfectly scalable method that needs no communication at all. Unfortunately this approach scales *exponentially* in the number of parameters and therefore is applicable only for very simple models.

Since we are interested in tuning complex models with larger numbers of parameters, we rely on an efficient numerical optimizer instead. We decided to use the *APPSPACK* [15] software for this task because it does not require derivatives of the objective function and because an MPI-based parallel version is available. Parallelism is achieved by assigning evaluations of the objective function E_β to different processors, the so-called workers. Note that each evaluation of E_β requires a complete cross-validation, which means i) to train SVMs on different training points for a given set of learning parameters, ii) to validate the trained SVMs on different test points and iii) to compute the quality measure. Based on these values, the optimizer selects new promising search directions in the parameter space and checks for convergence. Good load balancing is achieved by using an asynchronous scheme. Currently we exploit only the parallelism provided by *APPSPACK*; the training of single SVMs and the validation routine have not yet been parallelized.

Mapping of SVM Learning onto the *APPSPACK* Environment. The *APPSPACK* software package is freely available. A *configure* script automatically locates the commands and system files that are required for compiling and installing the package, and automatically generates appropriate *makefiles*. Due to some IBM-specific settings these *makefiles* could not be used directly for building the libraries and executables for our machine, but a few additional steps had to be done. More details on the JUMP supercomputer will be given in Chapter 5. Once we had successfully configured *APPSPACK* we built the libraries and executables by using the *makefiles*. All in all the installation of the software is easy and the *APPSPACK* developer team gives instructions if requested.

The second step consisted of integrating SVM learning into the APPSPACK framework. This required only minor changes of the SVM code because the executable just has to read a file containing values for all parameters, to evaluate the objective function E_β, and to provide an output file which should contain either a single numeric entry that is the function value or an error string. *APPSPACK* is able to generate the input files and to read the output without additional instructions. Please note that the optimizer examines the function values exclusively, whereas the underlying simulation is not of any interest. Thus its usage is easy to realize for support vector learning and any other supervised machine learning algorithm. The users' final task is to provide an *apps*-file containing the relevant solver information like

- the number of parameters,
- lower and upper bounds for them (infinite bounds are allowed), and the
- executable name.

Optionally one can set

- the initial parameter vector for a hot start,
- the maximum number of evaluations,

and many more. Some examples are provided, too. For the parallel version the number of workers ω is deduced from the submission of the MPI job via $\omega = \text{proc} - 1$, where proc is the number of processors. A single CPU, the master, is used to assign work, i.e., trial points, to the workers. *APPSPACK* is robust due to the toleration of error strings. Even if a single function evaluation fails, the optimization won't stop. For our quality measure (5) such a situation may occur in the case pr = se = 0, when E_β is not defined.

In the following section we will also compare results of the serial and parallel version to show drawbacks and advantages of both methods. To our knowledge this is the first presentation of work on parallel numerical optimization of non-standard SVM parameters.

5   Results and Discussion

The numerical experiments were performed with the so-called thyroid data set available from [16]. There are 7200 instances with 15 binary and 6 continuous attributes. The task is to determine whether a patient is hypothyroid. Therefore one class, representing 93% of the data, has the characteristic *"not hypothyroid"*. The remaining instances are considered to belong to a single class *"hypothyroid"*, even if a closer inspection would allow to classify them further as either *"hyperfunction"* or *"subnormal functioning."* This merging is usually proposed, and sometimes the dataset is even distributed in this form with the task of finding hypothyroid persons. Note that the merging of classes is somewhat critical as we do not know the level of similarity between them. However we try to design a sensitive binary classifier that is able to find as many hypothyroid points as possible. In addition to grossly unequal class sizes, the data set is unbalanced with high cost for false negative results. In [16] the data is already partitioned into a training set of 3772 points and a test set of 3428 points. Since the percentage of positive and negative instances in the proposed training set is compatible with the overall distribution we didn't change this partitioning.

The thyroid data set was used in [17] for performance analysis of multilayer neural networks. The best net reached a classification performance of 95%. It was also stated that due to the imbalance of the data a learning method must perform better than 93%. This is true only for scenarios where errors have always the same weight and are not considered separately. Unfortunately, [17] does not give data concerning the distribution of the errors. In [18] the performance of SVMs for the same data set is given. Standard SVMs achieved between 93% and 95% accuracy on the test set. There the SVM results were compared with results on fuzzy SVM learning. The latter approach led to classification rates between 95% and 97%. Again, the distribution of the errors was not specified.

Our numerical experiments were performed on the Juelich Multi Processor (JUMP) at Research Centre Juelich [19]. JUMP is a distributed shared memory parallel computer consisting of 41 frames (nodes). Each node contains 32 IBM Power4+ processors running at 1.7 GHz and 128 GB shared main memory. All in all the 1312 processors have an aggregate peak performance of 8.9 TFlop/s. Since we used a single node of JUMP for our tests, the *APPSPACK* manager process could assign jobs to 31 workers.

Throughout the tests, some control parameters were kept fixed. For the decomposition method in the SVM training we chose a working set size of 100, and the stopping criterion was defined according to [6] with $\epsilon = 0.001$. The quality measure E_β was computed via a simple twofold cross validation. We did not specify a starting point or a maximum number of evaluations for *APPSPACK*. In contrast to some published results, we kept training data and test data strictly separated. The former were used only for validating and training the SVMs, and the latter were used only for assessing the quality of the final optimized SVM.

The Influence of the Quality Measure. One of the most challenging problems for unbalanced data sets is to find a reasonable trade-off between high sensitivity and high precision. In our quality measure E_β the relative weight of these two important goals is controlled by the parameter β. Since increasing β gives more weight to sensitivity, we expect a reduction of the false negative points, at the cost of a potential growth of false positive results. Indeed the data obtained with a single-σ model confirm this expectation; see Table 1.

Since in our example the cost for false negative classifications is significantly higher than for false positive, we are primarily interested in sensitivity, so a higher value of β should be used. Small values for β led to good overall results with increasing numbers of false negative points. Note that the 98% test per-

Table 1. Test results for different quality measures

β	0.5	0.75	1.0	1.5	2.5
trial points	125	79	75	78	62
function evaluations	101	62	58	60	46
E_β	0.092	0.102	0.108	0.099	**0.072**
training errors	**48**	51	88	120	170
σ	91.15	52.05	28.84	25.75	64.42
C^+	100000	100000	10280	39670	100000
C^-	21790	19560	1000	1000	1000
ratio C^+/C^-	4.6	5.1	10.3	39.7	100
false negative test points	7	5	4	3	**1**
false positive test points	**63**	68	99	134	196
test sensitivity	97%	98%	98%	99%	**100%**
overall test errors	**70**	73	103	137	197
test performance	**98%**	98%	97%	96%	94%

formance compares favorably with the results obtained with neural networks or fuzzy SVMs.

Even if sensitivity is important, at some point the attempt to reduce the false negatives further leads to so many additional false positives that the overall cost increases again. This reflects the fact that for large and very unbalanced data sets it is dangerous to optimize only sensitivity because this can lead to weak classifiers. In certain situations, however, it can be important to be able to design very sensitive classifiers, e.g., when false positive points can be located by experiments after classification. Possibly the overall performance might also be improved further with another quality measure or with very small values β, but these issues have not yet been investigated.

The Optimal Ratio of C^+ and C^-. While unequal evaluation of slack variables during training seems to be accepted universally in the field of support vector learning [20], detailed descriptions of results or of the effects of this model generalization are not available. Thus tuning of these parameters is not trivial, if done by hand. Since the ratio of positive and negative points is about 7% the natural weighting choice [9] would be $C^+/C^- \approx 14$. However the data in Table 1 indicate that this ratio is not adequate for minimizing either the total number of errors or the sensitivity. The C^+/C^- ratios given in the table were delivered automatically by the numerical optimizer. One can see that the ratio increases for larger values of β, which is exactly what one would expect.

Computational Cost. Simple tuning methods like grid search are very popular due to the predictable number of training stages. In Table 1 we show the number of steps for automatic parameter tuning. Not all trial points generated by *APPSPACK* led to a new function evaluation (cross validation) because sometimes points were regularly pruned or function values in the cache could be reused. For optimizing 3 parameters we had to do between 46 and 101 cross validations, which is at least one order of magnitude less than any reasonable grid search would need.

Generalized vs. Standard Kernel. In Section 3 we showed how the standard Gaussian kernel can be extended by using different standard deviations for (groups of) the features. The thyroid data have 21 features, of which 15 are binary and 6 are continuous. Therefore we used two kernel parameters σ_{bin} and σ_{cont}. In Table 2 we compare the results to those for the standard kernel with a single parameter σ. For both runs we used $\beta = 1.5$. The number of function evaluations for optimizing the generalized model is more than twice as large as for the 3-parameter model. On the other hand, the cost-sensitive quality measure E_β could be reduced. This improvement could be seen in the final test, too. *Both* the false negative and false positive classifications could be lowered.

Table 2. Comparison of the standard and the generalized kernel

model	standard	generalized
function evaluations	60	140
E_β	0.108	0.098
σ	25.75	—
σ_{bin}	—	72.16
σ_{cont}	—	31.38
C^+	39670	13380
C^-	1000	1000
ratio C^+/C^-	39.7	13.4
false negative test points	3	2
false positive test points	134	110
overall errors	137	112

From $\sigma_{cont} < \sigma_{bin}$ we conclude that the significance of the binary features is high in comparison to the continuous values. It is interesting to see that $\sigma = 25.75$ is not between the two new σ values.

Serial vs. Parallel Optimization. The training time for a single SVM can vary significantly depending on the values of the learning parameters. For example, it is known that larger values of C lead to longer training times. Asynchronous parallel pattern search (APPS) is a parallel optimization approach that is well suited to such situations since it does not synchronize the system at the end of every single iteration. The cost for the good load balancing is some additional function evaluations in the parallel mode.

Results in [21] indicate a small number of additional function evaluations for multi-processor APPS. By contrast, our results in Table 3 for $\beta = 0.75$ show that the number of function evaluations in parallel mode can be significantly larger than in serial mode so that the efficiency is reduced. Usage of 8 processors led to 80 function evaluations, which is 60% more than with the serial version. However, usage of 8, 16 or 32 CPUs decreases overall running time of SVM parameter tuning.

During the tests we observed an increasing number of workers without a job, i.e., more and more processors did no longer receive trial points for function

Table 3. Overhead for parallel optimization

mode	serial	7 workers	15 workers	31 workers
function evaluations	49	80	62	62
E_β	0.102	0.101	0.102	0.102
training errors	50	50	51	51
σ	62.88	72.16	52.05	52.05
C^+	69060	100000	100000	100000
C^-	13380	19560	19560	19560
ratio C^+/C^-	5.2	5.1	5.1	5.1
false negative test points	4	4	5	5
false positive test points	74	74	68	68

evaluations. This is due to caching effects and the decreasing number of new trial points during the final steps. Thus the asynchronous scheme cannot sustain a large degree of parallelism when the system is near convergence.

The optimization results in terms of accuracy, however, depend only slightly on the number of processors. The number of training errors is nearly the same for all tests. Misclassifications in the test set differ only a little bit and the values of our quality measure are nearly equal for all tests. Note that the parallel mode yields larger values for C^+ and C^-, whereas their ratio remains almost constant. This is a very interesting detail and gives evidence to the assumption that the ratio C^+/C^- should always be considered, too. The most significant differences between serial and parallel optimization with *APPSPACK* can be seen in the σ values. They differ in both directions up to 20%.

Please note that SVM training for a fixed set of parameter values can be formulated as a global optimization problem with a single optimum, but the task of parameter tuning might lead to a large number of local minima. Since we are interested in robust methods for SVM parameter tuning it might be interesting to analyze in future tests *APPSPACK*'s ability to avoid local minima.

6 Conclusions and Future Directions

We have introduced an automated parameter optimization scheme for support vector learning. Our scheme is to a wide degree portable and can be adapted very easily. The *APPSPACK* software is freely available and runs on different platforms in serial and parallel mode. While we have used our own implementation of support vector learning, using publicly available SVM software is also possible; [14] might be a good choice. We have shown results for different quality measures, different models with varying numbers of parameters, and serial and parallel computing mode.

In the future we plan integrating different kernels into a single SVM model and a hierarchical parallelization combining parallel SVM training with parallel parameter optimization to speed up the model selection even more.

Acknowledgements

We would like to thank Tamara Kolda for continuous help with *APPSPACK*-related questions. We are grateful to Wolfgang Frings, Inge Gutheil, Ruth Zimmermann and the ZAM team at Juelich for technical support, several remarks and careful reading. We also would like to thank the unknown referees for their valuable comments.

References

1. Vapnik, V.N.: Statistical learning theory. Wiley & Sons, New York (1998)
2. Platt, J.: Fast training of support vector machines using sequential minimal optimization. In Schölkopf, B., Burges, C.J.C., Smola, A.J., eds.: Advances in Kernel Methods — Support Vector Learning, Cambridge, MA, MIT Press (1999) 185–208
3. Poulet, F.: Multi-way distributed SVM algorithms. In: Proc. of ECML/PKDD 2003 Int. Workshop on Parallel and Distributed Algorithms for Data Mining. (2003)
4. Cristianini, N., Shawe-Taylor, J.: An Introduction to Support Vector Machines and Other Kernel-Based Learning Methods. Cambridge University Press, Cambridge, UK (2000)
5. Schölkopf, B., Smola, A.J.: Learning With Kernels. MIT Press, Cambridge, MA (2002)
6. Hsu, C.W., Lin, C.J.: A simple decomposition method for support vector machines. Machine Learning **46** (2002) 291–314
7. Serafini, T., Zanghirati, G., Zanni, L.: Gradient projection methods for quadratic programs and applications in training support vector machines. Optimization Methods and Software **20** (2005) 353–378
8. Pardalos, P.M., Kovoor, N.: An algorithm for a singly constrained class of quadratic programs subject to upper and lower bounds. Mathematical Programming **46** (1990) 321–328
9. Eitrich, T., Lang, B.: Efficient optimization of support vector machine learning parameters for unbalanced datasets. Preprint BUW-SC 2005/2, University of Wuppertal (2005)
10. Zanghirati, G., Zanni, L.: A parallel solver for large quadratic programs in training support vector machines. Parallel Computing **29** (2003) 535–551
11. Collobert, R., Bengio, S., Bengio, Y.: A parallel mixture of SVMs for very large scale problems. Neural Computation **14** (2002) 1105–1114
12. Selikoff, S.: The SVM-tree algorithm (2003) http://scott.selikoff.net/papers/CS678_-_Final_Report.pdf.
13. Celis, S., Musicant, D.R.: Weka-parallel: machine learning in parallel. Computer Science Technical Report 2002b, Carleton College (2002)
14. Chang, C.C., Lin, C.J.: LIBSVM: a library for support vector machines. (2001) Software available at http://www.csie.ntu.edu.tw/~cjlin/libsvm.
15. Gray, G.A., Kolda, T.G.: APPSPACK 4.0: asynchronous parallel pattern search for derivative-free optimization. Sandia Report SAND2004-6391, Sandia National Laboratories, Livermore, CA (2004)
16. Hettich, S., Blake, C.L., Merz, C.J.: UCI Repository of machine learning databases. (1998) http://www.ics.uci.edu/~mlearn/MLRepository.html.
17. Schiffmann, W., Joost, M., Werner, R.: Synthesis and performance analysis of multilayer neural network architectures. Technical Report 16/1992, University of Koblenz (1992)

18. Inoue, T., Abe, S.: Fuzzy support vector machines for pattern classification. In: Proc. Intl. Joint Conf. Neural Networks (IJCNN'01). (2001) 1449–1454
19. Detert, U.: Introduction to the JUMP architecture. (2004) http://jumpdoc.fz-juelich.de.
20. Markowetz, F.: Support vector machines in bioinformatics. Master's thesis, University of Heidelberg (2001)
21. Hough, P.D., Kolda, T.G., Torczon, V.J.: Asynchronous parallel pattern search for nonlinear optimization. SIAM Journal on Scientific Computing **23** (2001) 134–156

The Architecture of a Proteomic Network in the Yeast*

Emad Ramadan[1], Christopher Osgood[2], and Alex Pothen[1]

[1] Computer Science Department, Old Dominion University,
Norfolk, VA 23529, USA
{eramadan, pothen}@cs.odu.edu
[2] Biological Sciences Department, Old Dominion University,
Norfolk, VA 23529, USA
cosgood@odu.edu

Abstract. We describe an approach to clustering the yeast protein-protein interaction network in order to identify functional modules, groups of proteins forming multi-protein complexes accomplishing various functions in the cell. We have
developed a clustering method that accounts for the small-world nature of the network. The algorithm makes use of the concept of k-cores in a graph, and employs
recursive spectral clustering to compute the functional modules. The computed
clusters are annotated using their protein memberships into known multi-protein
complexes in the yeast. We also dissect the protein interaction network into a
global subnetwork of hub proteins (connected to several clusters), and a local
network consisting of cluster proteins.

1 Introduction

Systems biology involves the study of complex biological structures and processes
by identifying their molecular components and the interactions among them. Looking
across the evolutionary landscape, biological subsystems performing discrete functions
are capable of being linked together in different ways without lethality to an organism, and often with positive gains in complexity and adaptation. Among the properties
that are now recognized in multiple biological systems are: modularity (sets of semi-
autonomous molecules that perform specific functions); robustness (the ability of biological systems to tolerate perturbations and noise); and emergence (new properties that
emerge from the interaction of functional modules) [14].

One of the challenges in computational systems biology is to create tools that enable
biologists to identify functional modules and the interactions among them from large-
scale genomic and proteomic data. We report the results of a study on an organism-scale
protein-protein interaction network in the yeast with the goal of identifying proteins that
form functional modules, (i.e., multiple proteins involved that have identical or related
biological function), by clustering techniques. Furthermore, we propose a hierarchical
organization of the proteomic network.

* Research supported by NSF grant CCR0306334, by subcontract B542604 from the Lawrence
Livermore National Laboratory, and by a grant from the Office of Research at Old Dominion
University.

M.R. Berthold et al. (Eds.): CompLife 2005, LNBI 3695, pp. 265–276, 2005.
© Springer-Verlag Berlin Heidelberg 2005

Methods for clustering proteomic networks have to cope with several features specific to protein interaction data. High-throughput experiments such as the yeast 2-hybrid system and the tagged affinity purification (TAP) [1,15,16], have high error rates, nearing 50% in some instances. Proteomic networks are *modified power-law* networks and *small-world* networks [6]. That is, the distribution of the fraction of vertices with a given degree follows a modified power-law; and the average path length between vertices is of the order of $\ln n$ (or smaller), where n is the number of vertices in the network. Hence there is a large number of low degree proteins, and a significant number of high degree proteins. The latter make it harder to discover clusters in the data, while the former increase the computational requirements. Cluster validation is hampered by the fact that there is often little overlap between different experimental studies due to the limited coverage of the interactome [13]. Finally, the predicted clusters must be biologically significant: e.g., functionally homogeneous.

In spite of these difficulties, we believe that we have successfully clustered a yeast proteomic network, with the predicted clusters overlapping well with multi-protein complexes and organelles. Our approach is based on identifying *hub proteins*, proteins that connect to a large number of clusters, and low-shell proteins (defined in the next section), and clustering the residual network. Low-shell proteins can be added to the cluster network at a later stage. We validate the clusters by comparing the clusters against experimental data on multi-protein complexes.

The hub proteins carry interesting information about the architecture of proteomic network, and are organized into a subnetwork of their own. Thus we propose a two level architecture for the yeast proteomic network, consisting of a global subnetwork of hubs, and a local subnetwork of clusters and low-shell proteins. A schematic of this architecture is shown in Fig. 1, where the top level corresponds to the global hub network, and the lower level corresponds to the local cluster network.

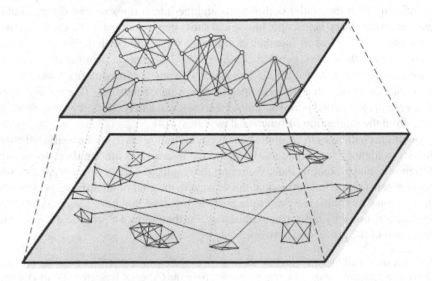

Fig. 1. A schematic representation of the yeast proteomic network as a hub-cluster interaction network. The top level corresponds to a global network of hub proteins, and the bottom level to a local network of cluster proteins.

2   Materials and Methods

2.1 k-Cores and k-Shells in Graphs

We begin by describing the concepts of a k-core and a k-shell in a graph, since our clustering method makes use of these.

Given a natural number k, the *k-core* of a graph G is the maximal subgraph of G in which every vertex has degree at least k in the subgraph (provided it is not the empty graph). The k-cores in a graph are nested: the $(k+1)$-core is contained in the k-core, for $k = 0, 1, \ldots, K - 1$, where K is the value of the maximum core in the graph. The k-core of a graph need not be a connected subgraph even if the original graph is connected. Note that if a graph contains a k-vertex connected component or a clique on $k + 1$ vertices, then it is contained in a k-core; however, the k-core need not contain a k-connected subgraph or a clique on $k + 1$ vertices.

The *k-shell* of a graph is the set of vertices that belong to the k-core, but not to the $(k+1)$-core. The k-shell includes vertices with degree k from the k-core, but also other vertices whose degree in the residual graph becomes less than $(k + 1)$ when low degree vertices are removed.

There is a well-known linear-time algorithm, in the number of edges, for computing the k-core (indeed, for finding all k-cores, for $k = 0$ to the maximum core value) of a graph. The idea is to repeatedly remove vertices v of degree less than k from the graph and all edges incident on v, updating the degrees of the neighbors of v in the residual graph as edges are deleted. The algorithm repeats this step until all vertices that remain have degree k or higher in the residual subgraph.

We have extended the concept of a k-core to a hypergraph in earlier work [19]. k-cores have been used earlier for clustering proteomic networks as a way of identifying highly connected subnetworks and for removing proteins belonging to low shell values [4].

We claim that clustering the k-core of a network removes noise in the data, in the same spirit as computing a shared nearest neighbor similarity (SNN) network. In an SNN network, two vertices are joined by an edge with weight equal to the number of their common neighbors at a distance less than or equal to d, where d is a natural number parameter. The SNN network includes only those edges that have weight higher than a threshold, and clustering algorithms have been designed to work with this network [17]. Unfortunately for large networks of small average path lengths, the computation of the SNN network can be prohibitively expensive. We suggest that the k-core is an efficient way to compute a network that approximates an SNN network. Every vertex in a k-core is adjacent to at least k other vertices in the subgraph, each of which is adjacent to k vertices with high core values.

2.2   Clustering Algorithms

Three major clustering approaches have been employed to identify functional modules in proteomic networks. The first approach searches for subgraphs with specified connectivities, called network motifs, and characterizes these as functional modules or parts of them. A complete subgraph (clique) is one such candidate, but other network motifs on small numbers of vertices have been identified through exhaustive searching or statistical methods [21]. This approach is not scalable for finding larger clusters in large-scale networks. The second approach, recently proposed in this context by Bader and Hogue [4], computes a weight for each vertex (depending on the density of a maximum core in the neighborhood of the vertex); it then grows a cluster around a seed vertex, a vertex with the largest weight in the currently unclustered graph. A vertex in the neighborhood of a cluster is added to it as long as its weight is close (within a threshold) to the weight of the seed vertex. Once a cluster has been identified, the procedure is repeated with a vertex of largest weight that currently does not belong to a cluster as the seed vertex. However, our experience comparing this approach with the spectral algorithms that we describe next shows that this method is less stable than the latter (i.e., the clusters depend on the seed vertices chosen).

We now discuss a spectral algorithm for clustering.

Let $G = (V, E, W)$ denote a weighted graph with vertex set V, edge set E, and weights on the edges W. Consider the problem of partitioning V into two sets $V_1 \cup V_2$. We consider the weights

$$W_{il} \equiv W(V_i, V_l) = \sum_{j \in V_i, k \in V_l, (j,k) \in E} w_{jk},$$

where $i, l = 1, 2$. Minimizing the objective function

$$J(V_1, V_2) = \frac{W_{12}}{W_{11}} + \frac{W_{12}}{W_{22}}$$

minimizes the sum of weights of the edges between distinct clusters, while simultaneously maximizing the sum of the weights of the edges within each cluster. This objective function for clustering has been called the MinMaxCut [10], and it measures a ratio related to the *separability* of a cluster to its *cohesion*. We prefer this function to related objective functions that have been proposed such as Normalized Cut.

Let Q denote the Laplacian matrix of a graph with weights w_{ij} on its edges (i, j); thus $q_{ij} = -w_{ij}$ for $i \neq j$, and each diagonal element q_{ii} is the sum of the weights of the edges incident on the vertex i. Let D be a diagonal matrix with its i-th component $d_{ii} = \sum_{(i,j) \in E} w_{ij}$; $d_1 = \sum_{i \in V_1} d_{ii}$, and $d_2 = \sum_{i \in V_2} d_{ii}$. Let p be a 'generalized partition vector' with $p_i = \sqrt{d_2/d_1}$ for $i \in V_1$; and $p_i = -\sqrt{d_1/d_2}$ for $i \in V_2$; let e be the n-vector of all ones. Then we have $p^T D e = 0$, and $p^T D p = d_1 + d_2$. Ding et al. [9] have shown that

$$\min_{V_1, V_2} J(V_1, V_2)$$

is equivalent to

$$\min_p p^T Q p / p^T D p, \quad \text{subject to} \quad p^T D e = 0.$$

This minimization problem is NP-hard since the generalized partition vector p is restricted to have elements from one of two values. However, we can relax this constraint and let p take values from $[-1, +1]$ to obtain an approximate solution. This problem is solved by the eigenvector x corresponding to the smallest positive eigenvalue of the generalized eigenproblem $Qx = \lambda Dx$.

The partition is obtained by choosing the vertices in one part to consist of vertices with eigenvector components smaller than a threshold value, while the other part has the remaining vertices. The threshold value could be chosen so as to locally minimize the MinMaxCut objective function. For details, see [8,10].

A clustering method is obtained by recursively applying the spectral partitioning method, by splitting each current cluster into two subclusters. The MinMaxCut objective function can be used to determine if a given cluster should be split further.

2.3   Algorithm

The yeast protein interaction network under study has 2610 proteins and 6236 interactions; we work with its largest connected component, which has 2406 proteins and 6117 interactions.

In the first step, we separate the high degree proteins, which are candidates for *hub proteins*. A hub is a protein that connects several different clusters in the network together, and these form a subset of the high degree proteins. After some experimentation, we chose candidate hub proteins to be those with degree 15 or higher in the network we study. The residual network has 2241 proteins and 3057 interactions, and consists of 397 connected components. The largest connected component of the residual graph has 1773 proteins and 2974 interactions (and hence most of the other components have few or no edges). We chose the largest component for further analysis.

In the second step, we compute the 3-core of the residual graph in order to remove the low- shell proteins (the 0-, 1-, and 2-shells) from the network. As discussed earlier, we believe that this step removes some of the noise from the experimental protein interaction data. We have found that this step has two advantages. First, the clustering algorithms generate better clusters of the residual network; the low shell proteins can be assigned to a cluster after it has been identified. Second, this step reduces the graph size substantially since this is a modified power law network with a large number of low degree proteins.

In the third step, we have applied the spectral clustering recursively to cluster the subgraph and identify the clusters, employing the MinMaxCut objective function. Once the clusters are identified, then the high-degree proteins which were removed as candidate hub proteins can be confirmed as hub proteins if they connect multiple clusters, or can be included among the cluster proteins.

Our spectral clustering code is currently written in Matlab for quick prototyping. The current code takes 65 seconds on a PC with a 1.3 MHz Intel processor and 768 MB memory. The hub and k-core computations are faster. Here we have greatly reduced the run times needed by removing the low-shell and hub proteins before clustering.

We have been concerned in this paper with identifying a methodology that can successfully deliver biologically significant clusters in proteomic networks. Distributed computations will be needed when we consider larger proteomic networks such as the human, and networks consisting of heterogeneous data.

We are also concerned with scalable clustering algorithms. The proposed approach requires $O(|E| \log |V|)$ time, where $|E|$ is number of edges in the network, and $|V|$ is the number of vertices. The k-core computation and the eigenvector computation at each clustering step can be performed in time $O(|E|)$; and there are $\log |V|$ partitioning steps needed to cluster. The spectral clustering could be replaced with a multi-level clustering approach that can also be implemented in time $O(|E|)$.

3   Results

3.1   Data Source and Analysis

Among the protein interactions produced by high-throughput methods such as the yeast 2-hybrid experiment or tagged affinity purification (TAP) [1,15,16], there are many false positives due to experimental limitations as well as biological factors (proteins that are not expressed at the same time or in the same cellular locale) [13]. In order to reduce the interference by false positives, we focused on the protein interaction network from the Database of Interacting Proteins (DIP), circa. April 2004 (URL: dip.doe-mbi.ucla.edu/dip/), consisting of the reliable dataset, which includes only data determined by a small-scale experiment, confirmed by independent high-throughput experiments, or scored highly by a probabilistic method that estimates the reliability of an interaction. This dataset has 2610 proteins that involve 6236 interactions considered to be reliable with high confidence.

3.2   The Cluster and Hub Networks

The local network computed by the clustering algorithm on the yeast protein interaction network, from which high degree proteins (hubs) and low-shell proteins have been removed, is shown in Fig. 2. Colors are used to distinguish the proteins belonging to a cluster, although some colors are reused to color proteins belonging to clusters that are drawn sufficiently far from each other. Thirty-eight clusters are displayed; for clearer presentation, we have omitted the edges joining two clusters when fewer than three edges join a cluster to another. All edges joining proteins within each cluster are shown.

The sum of the numbers of within-cluster edges is 984, while the sum of the between-cluster edges is 239, and the largest number of edges joining one cluster to another is 9. These measures are related to the concepts of *cohesion* and *separation* of the clustering [22], and thus we believe that our method has been able to cluster the residual network well. Each of the clusters is assigned to multi-protein complexes using the Munich Information Center for Protein Sequences (MIPS) database (URL: mips.gsf.de), as described in the next subsection. Each low-shell protein can now be easily assigned to a cluster with whose proteins it has the most number of interactions.

From a topological point of view, our approach to clustering helps to uncover the hidden topological structure of a proteomic network. We found that there are two major subnetworks within the protein-protein interaction network. In addition to the *cluster network*, we also construct a *hub network*, the subnetwork formed by the hub proteins in the protein interaction network; a subnetwork formed by the 5-core of the hub network is shown in Fig. 3. Four 'super-clusters' are clearly evident in the hub interaction network: from top to bottom, these correspond to the spliceosome, proteins involved in mRNA export and the nuclear pore complex, the regulatory subunit of the proteasome, and proteins that are transcription factors.

We now consider various subnetworks of the yeast protein interaction network to illustrate the differences between the 'global' hub network, and the local 'cluster' network. Table 1 lists the sizes of these networks, the average path lengths, the diameters, and the cluster coefficients. (The cluster coefficient measures how likely two neighbors of a vertex are to be adjacent to each other in the network, on the average.) The row 'C + S' denotes the 'cluster and shell' subnetwork obtained by removing the hub proteins from the whole network. Note that this subnetwork has the highest diameter and average path length, due to the presence of the large number of low-shell proteins. Once they are removed, the cluster network exhibits the highest clustering coefficient, supporting our premise that this is a local network. The hub network has the lowest diameter and average path length due to the edges joining the hub proteins to each other (cf. Fig. 3). The tight clustering seen in the hub network was surprising to us, but it is clear that hub proteins preferentially interact with cluster proteins and with each other, rather than the low shell proteins. We discuss the hub subnetwork and clusters in it in more detail in the next subsection.

The average path lengths in these networks are compared against $\ln n$, where n is the number of vertices in each subnetwork. Random power-law networks with exponent

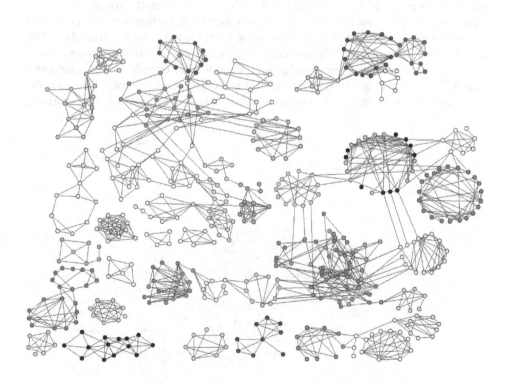

Fig. 2. Clusters in the yeast proteomic network from which hub and low-shell proteins have been removed. When fewer than three edges join a pair of clusters, such edges have not been drawn in this figure, for clarity in presentation.

Table 1. Properties of various subnetworks of the yeast protein interaction network

Subnetwork	No. of vertices	edges	Average Path Length (ln n)	Diameter	Cluster coefficient
Hub	165	507	3.5 (5.1)	7	0.37
Cluster	495	1223	6.5 (6.2)	16	0.43
C +S	1773	2974	7.6 (7.5)	19	0.15
Whole	2406	6117	5.1 (7.8)	13	0.21

β satisfying $2 < \beta < 3$ have expected average path lengths of order $\ln \ln n$, while if the exponent $\beta > 3$, it is $\ln n$ [7]. We see that $\ln n$ is a good approximation for the average path length of the cluster and cluster-shell networks; but any network that includes the hub proteins has an even lower average path length.

3.3 Functional Annotation of Clusters

One way to validate the clusters we discovered is to check how homogeneous the proteins in each cluster are with respect to function or the biological process that they are involved in. Each cluster should consist of one or more multi-protein complexes, which are molecular machines responsible for various cellular functions. We compared 38 clusters that we found with multi-protein complexes listed in the MIPS database. We found that in thirteen of the MIPS protein complexes, every protein in the complex was also identified in a cluster corresponding to it; for nine more complexes, we found more than half the proteins involved in the complex in a corresponding cluster. These results are despite the facts that hub and low-shell proteins are not included in this comparison,

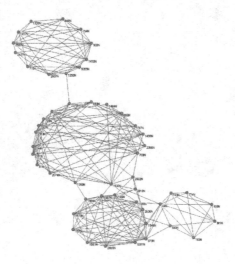

Fig. 3. The 5-core of the global hub network. The four clusters evident in this figure correspond to the spliceosome, mRNA export, the proteasome, and various transcription factors.

and that many proteins in the MIPS database are not included in the DIP protein network under study here. When the hubs and low-shell proteins are included, the coverage will increase further. A table containing the number of each cluster, a corresponding MIPS complex name and its MIPS ID, the number of proteins the cluster and the complex have in common, and the names of these proteins, is included in file `table1.xls`, see the Supplementary Materials at
`www.cs.odu.edu/~pothen/Papers/Cluster/DIP/`.

We should note that, in general, the clusters that we have discovered contain more proteins than those reported to belong to a corresponding MIPS complex. This suggests possible biological roles and functional assignments for such proteins, many of which are not currently functionally annotated.

The protein interaction graph of each cluster and a biological process annotation for it, using a directed acyclic subgraph (DAG) derived from the Gene Ontology (URL: `www.geneontology.org`), are also included in the Supplementary Materials. The cluster subgraphs are included in the `jpg` files, while GO DAGs are listed in the `png` files. While some of the cluster graphs are near-cliques or subgraphs with high edge connectivity, many of them are not. We believe that this validates our approach of finding complexes by a general clustering approach rather than searching for specific subgraph motifs.

3.4   Interactions Between the Hub and the Local Networks

We now consider the hub protein subnetwork and its interaction with the local network in more detail.

One of the complexes in cluster 8, the U4/U6 x U5 tri-snRNP complex, (listed in the file `table1.xls` in the Supplementary Materials), is comprised of a group of proteins involved in spliceosome processing of mRNA. This is the top-most cluster represented in Fig. 3. The spliceosome is required for the ordered and accurate removal of intronic sequences from pre-mRNA and thus plays a key role in alternative splicing, a process of great importance in higher eukaryotes whereby a single gene can generate multiple transcripts (alternatively spliced mRNAs) and thus multiple proteins [12]. The PRP (pre-mRNA processing) and Sm family proteins make up most of the proteins found in cluster 8. Some of the key proteins involved in mRNA processing, including those belonging to the LSM family, are not found in that cluster, but among the hub proteins that interact with multiple clusters.

One of the complexes in cluster 24, the first mRNA cleavage factor complex (represented in the file `cl24.jpg` in the Supplementary Materials), includes five proteins involved in mRNA cleavage in preparation for the addition of the eukaryotic signature poly-A tail. Thus, proteins including CLP1 (involved in cleavage of the 3' end of the mRNA prior to tailing), and RNA 14 and 15 (two proteins that participate with CLP1 in formation of the 3' end of mRNA), are collectively implicated in alternative selection of the poly-A addition site [18].

Now we focus our attention on the single protein-protein edge which joins the topmost hub cluster in the figure, corresponding to *mRNA splicing*, to a second hub cluster involving mRNA export and nuclear pore formation proteins, corresponding to *mRNA export*, in Fig. 3. The two hub proteins that form the bridge between these clusters are

PRP6, a component of the mRNA splicing machinery, and PAB1, the poly-A binding protein involved in the final step in mRNA processing. We note that PRP6 is involved in the later stages of mRNA splicing and is in that sense the penultimate step prior to poly-A tailing. Thus, the overall logic of joining these two complexes by these particular hub proteins is compelling.

We now examine the connections formed by these two hub proteins with the local clusters that we picture as lying below them in the hierarchy of global (hub) and local (cluster) networks (see Fig. 1). PRP6 interacts with a single cluster (cluster 8) through the protein SMD1. SMD1 further interacts with splicing proteins PRP3 and SMD3 in the hub complex that includes PRP6. SMD1 is involved in the early stages of mRNA splicing and is highly conserved, showing greater than 40% amino acid identity between yeast and human [20]. PRP6 interacts with PAB1 in the second hub complex. PAB1 in turn interacts with three clusters, 22, 24 and 34. As noted above, cluster 24 includes RNA14 and RNA15, both involved in mRNA cleavage, and it is these proteins that interact with PAB1. PAB1 also forms connections with cluster 22 (via its interaction with TIF4632 = eIF4F, file c122.jpg), and with cluster 34 (via PKC1, file c134.jpg). These latter interactions (eIF4F and PKC1) are at first glance puzzling, but in fact they are entirely consistent with emerging evidence of interactions and regulatory loops that exist between distinct components in the gene expression machinery. The poly-A terminus of mRNA, and the associated PAB1, not only interacts with the 5' end of the mRNA, ensuring structural integrity of the transcript prior to its participation in protein translation [5] , but the PAB1 terminus also interacts with the translation machinery itself, and specifically with eIF4 initiation factors. Finally, PKC1, a protein kinase crucial to cell signaling pathways, is also implicated in functional interactions with PAB1 and eIFs [3], suggesting global level regulation of protein synthesis from metabolites through mRNA processing.

At the global level of our network model, we find key proteins involved in rate-limiting steps of gene expression, linked in logical order; these are connected to the local network consisting of clusters of proteins involved in execution level functions. Whether this overall pattern is typical of the proteome organizational structure we have identified here, remains to be further investigated.

3.5   Incorporation of Protein Domain Data

Proteins interact with each other through regions that have a specific sequence and fold, called domains. Here we further validate the protein complexes predicted from our clustering approach using information on the domain structure of proteins.

The study of proteins involved in processing eukaryotic mRNAs indicate that virtually all steps involved in gene expression are coordinated and integrated via protein-protein interactions. The LSM proteins provide an informative example of the integration of cellular protein machinery to couple synthesis and quality control in gene expression [2,11]. LSM proteins form heptameric complexes that bind to RNA molecules; one such complex is found primarily in the nucleus where it coordinates splicing of mRNAs, while a second, related, complex of LSM proteins assembles in the cytoplasm to monitor mRNA quality control. LSM proteins have been extensively characterized and include two highly conserved protein interaction domains, SM1 and SM2.

It is proposed that these conserved domains permit each LSM protein to interact with two other LSM proteins in forming the heptameric, doughnut shaped ring structure that is implicated in mRNA splicing. LSM proteins comprise a gene family in which successive rounds of gene duplication have increased LSM copy numbers. LSM proteins have also been shown to form stable interactions with other protein types, including the PRP proteins discussed below.

The PRP proteins similarly carry a protein-protein interaction motif, the tetratrico peptide repeat (TPR) [23]. PRP proteins typically contain multiple copies of the 34 amino acid repeat; Prp1 for example contains 19 repeats of the TPR (ibid). Some PRP proteins contain a second conserved site at the C-terminus of the protein that facilitates interactions between them and LSM proteins, thus coupling two complexes with key roles in mRNA splicing. Our hub network predicts the formation of a complex containing LSM proteins 1-5, 7 and 8, as well as PRP proteins 4, 6, 8, and 31. The specific interactions implicated in our sub-network are, to our knowledge, the first explicit assignments of interactions between these two families of proteins.

We were surprised to find that the hub proteins form a highly interconnected subnetwork. Biological evidence indicates that LSM proteins do indeed form multi-protein complexes in the course of performing their key cellular functions. The fact that each LSM protein has at least two protein-protein interaction domains helps us understand how the complexes are formed. Whether similar binding interactions can account for other closely knit hub networks is under investigation.

4 Conclusions

We have proposed a two-level architecture for a yeast protein-protein interaction network. We place a small set of hub proteins, each with at least fifteen interaction partners and involved in gene expression, mRNA export, the proteasome, and transcription factors, into a global subnetwork. A local subnetwork of proteins is organized into clusters that correspond well with multi-protein complexes in the MIPS database. We used the computed clustering to examine the biological significance of some of the interactions observed between the hub and local subnetworks. If the proposed two-level architecture exists in other proteomic networks, then it would be interesting to discover properties that distinguish hub proteins from the proteins in the local network.

In future work, we will consider the computation of an overlapping clustering rather than the exclusive clustering approach considered in this paper, so that a protein could be included in more than one cluster in the local network. We will also investigate additional clustering approaches and biological networks involving heterogeneous data.

References

1. A-C. Gavine, M. Bosche, and R. Krause et al. Functional organization of the yeast proteome by systematic analysis of protein complexes. *Nature*, 415:141–147, 2002.
2. T. Achsel et al. The Sm domain is an ancient RNA-binding motif with oligo(U) specificity. *Procs. Natl. Acad. Sci.*, 98:3685–3689, 2001.
3. F. Angenstein et al. A receptor for activated C kinase is part of messenger ribonucleoprotein complexes associated with polyA-mRNA in neurons. *J. Neurosci.*, 22:8827–37, 2002.

4. G. D. Bader and C. W. Hogue. An automated method for finding molecular complexes in large protein interaction networks. *BMC Bioinformatics*, 4(2):27 pp., 2003.
5. X. Bi and D. Goss. Wheat germ poly (A)-binding protein increases ATPase and the RNA helicase activity of translation initiation factors eIF4A, eIF4B and eIF-iso-4F. *J. Biol. Chem.*, 275:17740–6, 2000.
6. S. Bornholdt and H. G. Schuster, editors. *Handbook of Graphs and Networks*. Wiley VCH, 2003.
7. F. Chung and L. Lu. The average distances in random graphs with given expected degrees. *Procs. Natl. Acad. Sci.*, 99(25):15879–15882, 2002.
8. I. S. Dhillon. Co-clustering documents and words using bipartite spectral graph partitioning. In *Procs. ACM Internatl. Conf. Knowledge Discovery in Data Mining (KDD)*, 2001.
9. C. Ding, X. He, R. F. Meraz, and S. R. Holbrook. A unified representation of multi-protein complex data for modeling interaction networks. *Proteins: Structure, Function, and Genetics*, 2004. To appear.
10. C. Ding, X. He, H. Zha, M. Gu, and H. Simon. A MinMaxCut spectral method for data clustering and graph partitioning. In *Procs. IEEE Internatl. Conf. Data Mining (ICDM)*, pages 107–114, 2001.
11. M. Fromont-Racine et al. Genome-wide protein interaction screens reveal functional networks involving Sm-like proteins. *Yeast*, 17:95–110, 2000.
12. F. Galisson and P. Legrain. The biochemical defects of PRP4-1 and PRP6-1 yeast splicing mutants reveal that the PRP6 protein is required for the accumulation of the [U4/U6.U5] tri-snRNP. *Nucl. Acids Res.*, 21:1555–62, 1993.
13. J. D. J. Han, D. Dupuy, N. Bertin, et al. Effect of sampling on topology predictions of protein-protein interaction networks. *Nature Biotechnology*, 23:839–844, 2005.
14. L. H. Hartwell, J. J. Hopfeld, S. Leibler, and A. W. Murray. From molecular to modular cell biology. *Nature*, 402:C47–C52, 1999.
15. Y. Ho, A. Gruhler, and A. Heilbut et al. Systematic identification of protein complexes in *Saccharomyces cerevisiae* by mass spectrometry. *Nature*, 415:180–193, 2002.
16. T. Ito, T. Chiba, and R. Ozawa et al. A comprehensive two hybrid analysis to explore the yeast protein interactome. *Procs. Natl. Acad. Sci.*, 98:4569–4574, 2001.
17. A. A. Jarvis and E. A. Patrick. Clustering based on a similarity measure based on shared nearest neighbors. *IEEE Trans. Computers*, C-22:1025–1034, 1973.
18. C. Noble et al. Rna14-Rna15 assembly mediates the RNA-binding capability of *Saccharomyces cerevisiae* cleavage factor IA. *Nucl. Acids Res.*, 32:3364–75, 2004.
19. E. Ramadan, A. Tarafdar, and A. Pothen. A hypergraph model for the yeast protein complex network. In *Procs. Workshop High Performance Computational Biology (HICOMB)*. IEEE / ACM, 2004. 8 pp., (CDROM).
20. B. C. Rymond. Convergent transcripts of the yeast PRP38-SMD1 locus encode two essential splicing factors, including the D1 core polypeptide of small nuclear ribonucleoprotein particles. *Procs. Natl. Acad. Sci.*, 90:848–52, 1993.
21. V. Spirin and L. A. Mirny. Protein complexes and functional modules in molecular networks. *Procs. Natl. Acad. Sci.*, 100:12123–12128, 2003.
22. P. Tan, M. Steinbach, and V. Kumar. *Introduction to Datamining*. Addison Wesley, 2005.
23. S. Urushiyama et al. The prp1+ gene required for pre-mRNA splicing in *Schizosaccharomyces pombe* encodes a protein that contains TPR motifs and is similar to Prp6p of budding yeast. *Genetics*, 147:101–115, 1997.

Author Index

Lecture Notes in Bioinformatics